Computational Approaches: Drug Discovery and Design in Medicinal Chemistry and Bioinformatics, 3rd Edition

Computational Approaches: Drug Discovery and Design in Medicinal Chemistry and Bioinformatics, 3rd Edition

Guest Editors

Anna Maria Almerico
Marco Tutone

MDPI

Basel • Beijing • Wuhan • Barcelona • Belgrade • Novi Sad • Cluj • Manchester

Guest Editors
Anna Maria Almerico
Dipartimento di Scienze e Tecnologie Biologiche Chimiche e Farmaceutiche (STEBICEF)
Università di Palermo
Palermo
Italy

Marco Tutone
Dipartimento di Scienze e Tecnologie Biologiche Chimiche e Farmaceutiche (STEBICEF)
Università degli Studi di Palermo
Palermo
Italy

Editorial Office
MDPI AG
Grosspeteranlage 5
4052 Basel, Switzerland

This is a reprint of the Special Issue, published open access by the journal *Molecules* (ISSN 1420-3049), freely accessible at: https://www.mdpi.com/journal/molecules/special_issues/9I9ZP9D26K.

For citation purposes, cite each article independently as indicated on the article page online and as indicated below:

Lastname, A.A.; Lastname, B.B. Article Title. *Journal Name* **Year**, *Volume Number*, Page Range.

ISBN 978-3-7258-6898-8 (Hbk)
ISBN 978-3-7258-6899-5 (PDF)
https://doi.org/10.3390/books978-3-7258-6899-5

Contents

About the Editors

Anna Maria Almerico

Anna Maria Almerico graduated in Chemistry with full marks and honors in 1977 at the University of Palermo. During her academic career, she has been a Visiting Professor at the School of Chemical Sciences, University of East Anglia (UK, 1983–1984), at the College of Pharmacy, Ohio State University (USA, 1986–1987), and at Masaryk University, Brno (Czech Republic, 1993).

During her nearly fifty-year tenure at the University of Palermo, she was a Full Professor of Medicinal Chemistry at the Faculty of Pharmacy. Her scientific interests focused primarily on the design and synthesis of novel nitrogen heterocycles of biological interest. In particular, she discovered new compounds related to well-known anticancer/antiviral/antiparasitic drugs, evaluating their medicinal properties also through *in silico* analyses. She published more than 150 research papers, as well as several book chapters.

Marco Tutone

Marco Tutone graduated *summa cum laude* in Medicinal Chemistry and Pharmaceutical Technology from the University of Palermo, Italy, in 2001. He obtained his Ph.D. in Pharmaceutical Sciences from the same university in 2006. From 2006 to 2008, he carried out postdoctoral research at the University of Palermo as a research fellow funded by the Italian Ministry of University and Research (MIUR).

In 2008, he was appointed Assistant Professor of Medicinal Chemistry at the University of Palermo, where he taught Advanced Methodologies in Medicinal Chemistry. He is currently a lecturer in Advanced Medicinal Chemistry and Drug Design. In October 2019, he was promoted to Associate Professor of Medicinal Chemistry at the University of Palermo. Marco Tutone is the inventor of one international patent: US20210002238B2, entitled *"Oxadiazole derivatives for the treatment of genetic diseases due to nonsense mutations"*.

His research activity is mainly focused on the application of molecular modeling techniques in Medicinal Chemistry, including molecular dynamics, pharmacophore modeling, molecular docking, machine learning approaches, and high-throughput virtual screening (HTVS) for the identification of bioactive compounds.

In recent years, his research interests have also expanded to the study of natural compounds. He is actively involved in research on translational readthrough-inducing drugs (TRIDs).

Preface

Modern drug discovery has received a significant boost thanks to the use of computational strategies. In addition to well-established computer-assisted drug discovery, *in silico* strategies have been further enhanced by the application of machine learning and artificial intelligence tools. This Special Issue brings together recent discoveries in various fields of medicinal chemistry, addressing classic diseases that have so far been poorly treated with traditional approaches or proposing new natural compounds for novel medical applications.

Machine learning tools have been used for the discovery of new selective serotonin and serotonin–norepinephrine reuptake inhibitors, providing a strong support for the treatment of depression, a serious mood disorder, affecting about 5% of the population. In the Quantitative Structure–Activity Relationship (QSAR) study, models for serotonin (SERT) and norepinephrine (NET) transporters were developed to predict the affinity and inhibition potential of new molecules. Models were developed using the Automated Machine Learning tool Mljar. The resulting QSAR models for serotonin and norepinephrine transporters have been made available in a new module of the SerotoninAI application to enhance usability for scientists.

Additionally, Leishmaniasis, a neglected tropical disease that constitute a serious public health challenges due to limited treatment options, toxicity, high costs, and drug resistance, can be approached using *in silico* methods. In fact several antileishmanial agents, such as malvidin and echioidinin have been examined against different types of parasite species, in comparison with a standard drug, amphotericin B (AmpB). Malvidin exhibited lower cytotoxicity than AmpB and a favorable selectivity index. These *in silico* studies revealed strong binding between malvidin and Leishmania arginase, with the residues HIS139 and PRO258 playing key roles. Moreover gene expression analysis indicated that malvidin is implicated in the modulation of oxidative stress and DNA repair pathways, in which GLO1 and APEX1 genes are involved. The combined *in vitro, in silico,* and gene expression results highlight malvidin's potential as a low-toxicity antileishmanial candidate with specific ARG enzyme interactions.

The interactions of small organic molecules with highly active alkaline phosphatase (ALP) proteins are currently being investigated through molecular docking in order to predict their mechanism of action, binding efficiency, and inactivation potency. Among them, the natural 5-azaindoles, marine sponge guitarrin, extracted from the Northwest Pacific marine sponge Guitarra fimbriata, were found to be highly active against the alkaline phosphatase CmAP from the marine bacterium C. amphilecti KMM 296. The superimposition of CmAP complexes with p-nitrophenyl phosphate (pNPP), a commonly used chromogenic aryl substrate for ALP, and the inhibitory guitarrins C, D, and the non-inhibitory guitarrins A, B, and E revealed that the presence of a carboxyl group at C6 together with a hydroxyl group at C8 is a prerequisite for the inhibitory effect of these series of 5-azaindoles. However, the inhibition of CmAP and calf intestinal ALP (CIAP) by guitarrin C was observed to occur *via* a non-competitive mode of action, whereas, in contrast, the kinetic model with guitarrin D, with an additional OH group at C7, reflected a mixed type of inhibition.

Computational approaches are also involved in the discovery of neuromuscular blocking agents (NMBAs) as an adjunct to general anesthesia. In fact NMBAs are routinely used during anesthesia to relax skeletal muscles. Nicotinic acetylcholine receptors (nAChRs) are ligand-gated ion channels; NMBAs can induce muscle paralysis by preventing the neurotransmitter acetylcholine (ACh) from binding to nAChRs situated on the postsynaptic membranes. Despite widespread efforts, it is still a great challenge to find new NMBAs since the introduction of cisatracurium. An effective ensemble-based virtual screening method, including molecular property filters, 3D pharmacophore model, and molecular docking, was applied to discover potential NMBAs from the ZINC15 database. The results showed that screened hit compounds had better docking scores than the reference compound d-tubocurarine. Molecular dynamics simulations revealed that ZINC257459695 can stably bind to nAChRs' active sites and interact with the key residue Asp165. The binding free energies were also calculated for the obtained hits using the MM/GBSA method, whereas *in silico* ADMET calculations allowed for the pharmacokinetic properties of hit compounds in the human body to be accessed. Overall, the identified ZINC257459695 may be a promising lead compound for developing new NMBAs as an adjunct to general anesthesia, necessitating further investigations.

In the search for non-invasive detection of biomarkers in vitro and in vivo with high sensitivity, fluorescence-based imaging was proposed for integrated diagnosis and treatment application, in the case of compounds with both biological activities and fluorescent properties. Thus, Alectinib and Rilpivirine, two drugs clinically approved in therapy for ALK-rearranged NSCLC, presenting suitable optical properties, were deeply explored, through the simulation of molecular structures, electrostatic potential, OPA/TPA, and emission spectral properties and experiments on UV-vis spectra, fluorescence, and cell imaging. A deep relationship between molecular structures and optical properties for Alectinib and Rilpivirine was evidenced on the basis of molecular modeling. The results suggested that the drugs Alectinib and Rilpivirine with planar parent skeleton, stronger π-conjugation effect in S1 and electron acceptors connecting conjugated structures could be excited by UV-vis spectra, and subsequently emitted fluorescence. Moreover, it was found that Alectinib and Rilpivirine displayed green fluorescence in HeLa cells, suggesting the potential ability for biological imaging and for applications in developing integrated diagnosis and treatment.

Combined classical *in silico* approaches to drug design (ligand-based and structure-based pharmacophore models, molecular docking screening, ADMET, and kinetic simulations) were used to identify novel compounds as potential inhibitors of mPGES-1, a key enzyme, which, when activated by inflammatory factors, can cause prostaglandin E synthesis, thus causing gastrointestinal reactions and coagulation disorders. High-throughput virtual screening (HTVS) of the traditional Chinese medicine database allowed us to select the best ten compounds, which were chosen by comparing their score against the reference ligand 4U9. At the end of the analyses, a compound in all aspects superior to the reference ligand was identified.

The classical partition coefficient (logP), a physicochemical parameter used in various fields such as drug and pharmaceutical product design and, substance toxicology, was evaluated either in three distinct micellar systems such as hexadecyltrimethylammonium bromide (HTAB), sodium cholate (SC), and lithium perfluorooctanesulfonate (LPFOS), and theoretically derived. In particular Quantum Mechanics (QM) and machine learning approaches were proposed for the prediction of the partition coefficients for a set of 63 molecules in different solvent mixtures. The combined data from the experimental logP in the three micellar formulations showed that the 1-propanol/water combination demonstrated the best agreement with the experimental partition coefficients for the SC and HTAB micelles. Moreover, the SVM approach and k-means clustering based on the generation of the chemical descriptor space were used. The analysis revealed distinct partitioning patterns associated with specific characteristic features within each identified class. Overall the results indicated that the combined techniques allowed us to obtain an efficient and quicker model for predicting partition coefficients in diverse micelles.

Throughout this Reprint, all the recent aspects of the computational approaches applied to several research fields are reported. We express our deep gratitude to all the contributors to this Special Issue for their commitment, hard work, and outstanding papers. We also thank all the reviewers involved in the manuscript revisions for their unpaid contributions to improve any aspects of the submitted works. Last but not least, we deeply thank Ms. Jessica Wang for her assistance during the period in which we served as Guest Editors.

Anna Maria Almerico and Marco Tutone
Guest Editors

molecules

MDPI

Editorial

Computational Strategies Reshaping Modern Drug Discovery

Marco Tutone * and Anna Maria Almerico

Dipartimento di Scienze e Tecnologie Biologiche Chimiche e Farmaceutiche (STEBICEF), Università degli Studi di Palermo, Via Archirafi 28, 90123 Palermo, Italy
* Correspondence: marco.tutone@unipa.it

Conventional drug development remains a protracted, costly, and high-risk endeavor, typically spanning 10–15 years from target identification to post-marketing surveillance. This multistage process encompasses target discovery and validation, hit identification and optimization, preclinical testing, clinical trials, regulatory approval, and ongoing safety monitoring. In this way, only 10–12% of candidates entering clinical trials reach regulatory approval. Despite being well established, this linear model of drug development is increasingly recognized as unsustainable. Each approved drug is estimated to cost over USD 2.5 billion, nearly 90% of candidates fail during clinical development, and meaningful breakthroughs, particularly for orphan diseases and targeted therapies, remain limited [1]. These shortcomings have catalyzed growing interest in computational approaches as transformative tools capable of accelerating discovery, reducing costs, and improving success rates in the search for viable drug candidates [2].

In traditional methods, once lead compounds are identified through high-throughput screening, structural optimization aims to enhance binding affinity, selectivity, bioavailability, and safety. This process is typically guided by structure–activity relationships (SAR) analysis [3]. Preclinical evaluation then assesses toxicity and efficacy in animal models [4], with only the most promising compounds advancing to human trials. Collectively, these stages underscore the inefficiencies of the conventional approach and highlight the urgent need for innovation to streamline drug development timelines [5]. Computational methods have emerged as a powerful catalyst in modern drug discovery, offering novel strategies to accelerate pharmaceutical innovation [6]. Traditional compound optimization methods are often labor-intensive and experimentally driven. In contrast, computational methods enable researchers to navigate vast chemical spaces efficiently, generating novel molecules with high affinity for specific biological targets in a fraction of the time [7].

AI-driven target identification and validation leverage large-scale biological datasets—including genomics, proteomics, and transcriptomics—to uncover disease-associated pathways, particularly in complex, polygenic disorders such as cancer and neurodegenerative diseases [8]. Deep learning models can predict biological activity and validate targets computationally, reducing reliance on costly and time-consuming laboratory assays [9]. One of the most transformative applications of computational methods lies in de novo compound generation. Deep learning architectures can design novel drug-like molecules that extend beyond existing chemical libraries while adhering to principles of chemical stability, drug-likeness, and bioactivity [10]. This capability significantly compresses early discovery timelines.

Computational methods also enhance drug optimization by predicting pharmacokinetic and pharmacodynamic properties, thereby improving candidate selection and reducing late-stage failures. By analyzing large datasets, Computational models can identify off-target effects and toxicity risks, enabling safer and more effective compound refinement [3]. Moreover, machine learning techniques such as random forests, support vector machines, and neural networks help prioritize the most promising candidates for downstream testing [11].

Beyond discovery, computational models are increasingly applied to preclinical and clinical development. Predictive models can assess toxicity and side effects without extensive animal or human exposure, while AI-assisted trial design improves patient recruitment, monitoring, and overall trial efficiency [12,13]. Generative AI is also a cornerstone of precision medicine, enabling therapies tailored to individual genetic and molecular profiles. By predicting patient-specific drug responses, AI supports personalized treatment strategies—particularly in oncology, where tumor heterogeneity demands highly individualized interventions [14–16]. Additionally, computer-driven virtual screening and drug repurposing allow rapid identification of new therapeutic uses for existing compounds. By mining large clinical and preclinical datasets, computer-assisted drug design can uncover previously unrecognized drug–target relationships, accelerating the availability of treatments for unmet medical needs [17–19].

In this context, this Special Issue consolidates a broad array of computational strategies—from structure-based methods like docking, molecular dynamics (MD) and free-energy profiling, to ligand-based techniques like pharmacophore modeling, similarity analysis, and QSAR (quantitative structure–activity relationships). This content curation sends a clear message: computational chemistry and bioinformatics are no longer niche—they are a central, mainstream pillar of modern drug discovery. As in earlier editions, the ambition is to blur the boundary between "in silico only" and "wet-lab + in silico" workflows, encouraging hybrid studies that combine computational predictions with experimental validation. The papers collected in this issue illustrate how computational tools are applied across a wide spectrum of real-world drug-discovery problems.

Conflicts of Interest: The authors declare no conflict of interest.

References

1. Dowden, H.; Munro, J. Trends in clinical success rates and therapeutic focus. *Nat. Rev. Drug Discov.* **2019**, *18*, 495–496. [CrossRef]
2. Ru, J.; Li, P.; Wang, J.; Zhou, W.; Li, B.; Huang, C.; Li, P.; Guo, Z.; Tao, W.; Yang, Y.; et al. TCMSP: A database of systems pharmacology for drug discovery. *J. Cheminform.* **2014**, *6*, 13. [CrossRef] [PubMed]
3. Jorgensen, W.L. Efficient drug lead discovery and optimization. *Acc. Chem. Res.* **2009**, *42*, 724–733. [CrossRef]
4. Teicher, B.A.; Andrews, P.A. Anticancer drug development guide. *Cancer Chemother. Pharmacol.* **2004**, *54*, S1–S8.
5. Johnson, J.R.; Williams, G.; Smith, R. Accelerating drug discovery using computational approaches. *Drug Discov. Today* **2001**, *6*, 239–246.
6. Roomi, M.S.; Culletta, G.; Longo, L.; Filgueira de Azevedo, W., Jr.; Perricone, U.; Tutone, M. Docking in the Dark: Insights into Protein–Protein and Protein–Ligand Blind Docking. *Pharmaceuticals* **2025**, *18*, 1777. [CrossRef]
7. Sellwood, M.A.; Ahmed, M.; Segler, M.H.S.; Brown, N. Artificial intelligence in drug discovery. *Future Med. Chem.* **2018**, *10*, 2025–2028. [CrossRef]
8. Kadurin, A.; Aliper, A.; Kazennov, A.; Mamoshina, P.; Vanhaelen, Q.; Khrabrov, K.; Zhavoronkov, A. The cornucopia of meaningful leads: Applying deep adversarial autoencoders for new molecule development in oncology. *Oncotarget* **2017**, *8*, 10883–10890. [CrossRef] [PubMed]
9. Murphy, K.P. *Machine Learning: A Probabilistic Perspective*; MIT Press: Cambridge, MA, USA, 2011.
10. Bleicher, K.H.; Böhm, H.J.; Müller, K.; Alanine, A.I. Hit and lead generation: Beyond high-throughput screening. *Nat. Rev. Drug Discov.* **2003**, *2*, 369–378. [CrossRef]
11. Isarankura-Na-Ayudhya, C.; Na-Bangchang, K.; Prachayasittikul, V. Machine learning techniques for drug discovery. *EXCLI J.* **2009**, *8*, 74–88.
12. Harrer, S.; Shah, P.; Antony, B.; Hu, J. Artificial Intelligence for Clinical Trial Design. *Trends Pharmacol. Sci.* **2019**, *40*, 577–591. [CrossRef]

 https://doi.org/10.3390/molecules31020200

13. Banik, S. Artificial intelligence in clinical trials: Opportunities and challenges. *Perspect. Clin. Res.* **2015**, *6*, 215–217.
14. Lentini, L.; Perriera, R.; Corrao, F.; Melfi, R.; Tutone, M.; Carollo, P.S.; Fiduccia, I.; Pace, A.; Ricci, D.; Genovese, F.; et al. A precision medicine approach to primary immunodeficiency disease: Ataluren strikes nonsense mutations once again. *Mol. Ther.* **2025**, *33*, 3231–3241. [CrossRef] [PubMed]
15. Ashley, E.A. Towards precision medicine. *Nat. Rev. Genet.* **2016**, *17*, 507–522. [CrossRef] [PubMed]
16. Ginsburg, G.S.; McCarthy, J.J. Personalized medicine: Revolutionizing drug discovery and patient care. *Trends Biotechnol.* **2001**, *19*, 491–496. [CrossRef] [PubMed]
17. Culletta, G.; Almerico, A.M.; Tutone, M. Comparing molecular dynamics-derived pharmacophore models with docking: A study on CDK-2 inhibitors. *Chem. Data Collect.* **2020**, *28*, 100485. [CrossRef]
18. Brindha, G.; Vinodhini, R.; Sathya, R. Drug repurposing: A review. *J. Pharm. Sci. Res.* **2016**, *8*, 1293–1297.
19. Ma, D.L.; Chan, D.S.; Leung, C.H. Molecular docking for virtual screening of natural product databases. *Chem. Sci.* **2013**, *4*, 3366–3383. [CrossRef]

 https://doi.org/10.3390/molecules31020200

molecules

MDPI

Article

Machine Learning Tool for New Selective Serotonin and Serotonin–Norepinephrine Reuptake Inhibitors

Natalia Łapińska [1], Jakub Szlęk [1,2,*], Adam Pacławski [1] and Aleksander Mendyk [1]

[1] Department of Pharmaceutical Technology and Biopharmaceutics, Jagiellonian University Medical College, 30-688 Kraków, Poland; natalia.czub@uj.edu.pl (N.Ł.); adam.paclawski@uj.edu.pl (A.P.); aleksander.mendyk@uj.edu.pl (A.M.)

[2] Bioinformatics and In Silico Analysis Laboratory, Center for the Development of Therapies for Civilization and Age-Related Diseases (CDT-CARD), 8 Skawińska St., 31-066 Kraków, Poland

* Correspondence: j.szlek@uj.edu.pl

Abstract: Depression, a serious mood disorder, affects about 5% of the population. Currently, there are two groups of antidepressants that are the first-line treatment for depressive disorder: selective serotonin reuptake inhibitors and serotonin–norepinephrine reuptake inhibitors. The aim of the study was to develop Quantitative Structure–Activity Relationship (QSAR) models for serotonin (SERT) and norepinephrine (NET) transporters to predict the affinity and inhibition potential of new molecules. Models were developed using the Automated Machine Learning tool Mljar based on 80% of the dataset according to 10-fold cross-validation and externally validated on the remaining 20% of data. The molecular representation featured two-dimensional Mordred descriptors. For each model, Shapley additive explanations analysis was performed to clarify the influence of the descriptors on the models' predictions. Based on the final QSAR models, the following results were obtained: NET and pIC50 value $RMSE_{test}$ = 0.678, R^2_{test} = 0.640; NET and pKi $RMSE_{test}$ = 0.590, R^2_{test} = 0.709; SERT and pIC50 $RMSE_{test}$ = 0.645, R^2_{test} = 0.678; SERT and pKi value $RMSE_{test}$ = 0.540, R^2_{test} = 0.828. QSAR models for serotonin and norepinephrine transporters have been made available in a new module of the SerotoninAI application to enhance usability for scientists.

Keywords: depression; QSAR model; SERT; NET; artificial Intelligence; SerotoninAI

1. Introduction

Mental health is a key aspect of well-being that can affect daily life. Among various mental disorders, one of the most common is depression, which is also known as depressive disorder. According to the World Health Organization, around 280 million people suffer from it with women affected more frequently than men [1]. Depression is characterized as low mood, loss of pleasure and interest in activities for long period of time, which affects personal and professional life [2]. Untreated depression can tragically lead to suicide, emphasizing the critical need for effective treatment of the disease. In recent years, treatment of depression is based on drugs with different mechanisms of action. However, a group of drugs used as first-line treatment is selective serotonin reuptake inhibitors (SSRIs), e.g., Fluoxetine, Paroxetine, and Sertraline. By inhibiting the serotonin transporter, they increase the concentration of serotonin in the synaptic cleft. In case of a low treatment efficacy of SSRIs, Duloxetine, Venlafaxine or Milnacipran, from the serotonin and norepinephrine reuptake inhibitor (SNRI) group is prescribed for a patient. SNRI increases the concentration of these two neurotransmitters, acting to improve the patient's mood [3,4]. Due to the importance of serotonin (SERT) and norepinephrine (NET) transporters in the mechanism of action of the most commonly used antidepressants, they form the foundation of the conducted studies presented in this manuscript.

The significant development of computational methods, especially machine learning, allows the search for advanced relationships between the structure of ligands and activity toward the selected protein. Such dependencies, called QSAR (Quantitative Structure–Activity Relationship) models, enable the discovery of new molecules. QSARs are used to find new molecules that interact with a chosen biological target with high activity [5,6].

The serotonin transporter in terms of QSAR model analysis or development has been a foundation of research over the past 20 years. In the first paper that appeared in 2004, Kulkarni et al. described N-substituted 3-alpha-(bis[4-fluorophenyl]methoxy)tropane derivatives for 76 molecules [7]. In subsequent years, studies including N- arylmethylpiperidinamine [8], phenylpiperidine [9] and phenethylamine [10] derivatives were published. The size of the databases ranged from 7 to 213 serotonin transporter ligands [7–22]. On the other hand, QSAR models for the norepinephrine transporter were rarely the centerpiece of published studies. Usually, these papers were part of the serotonin transporter-related work already presented above [7,8,10,13,16,20]. The article that focused exclusively on the norepinephrine transporter is a paper by Olasupo et al. presenting a study for a set of 50 ligands using statistical analysis by genetic function approximation (GFA) to develop a QSAR model [23]. In addition, the authors of the above studies were concerned with various forms of measuring the activity of ligands against the selected transporter; these were an inhibition constant Ki [8,13,17,22] and its negative logarithm—pKi (affinity value) [14,16,20], a half maximal effective concentration (EC50) [15] or a half maximal inhibitory concentration (IC50) [21] and the negative logarithm of its concentration—pIC50 [9,18].

A review of the literature reveals a lack of general, cutting-edge models using advanced machine learning techniques that can predict the activity of previously unseen molecules with low error. In a previous paper presenting the application of SerotoninAI, we presented a model predicting the affinity of ligands for the serotonin transporter (pKi value predictions) [24]. In the present study, we wanted to extend this database for serotonin transporter and pKi values with additional data sources. Focusing on the mechanism of action of selective serotonin inhibitors and serotonin–norepinephrine inhibitors, we also created an affinity model for the norepinephrine transporter. The next two QSAR models focused on the level of inhibition of the two transporters. The final result of the study was the development of four QSAR models for the serotonin and norepinephrine transporter and their relevant properties.

In the subsequent sections of the manuscript, we present the obtained results, data acquisition and selection process, a description of the methodology for developing QSAR models along with an analysis of features' effect on prediction using SHAP analysis.

2. Results

2.1. Databases

The first step was to design four databases. Figures 1 and 2 below show the overview method of data curation, and Table 1 shows the ligand abundance. The data sources used in this research contained molecules that were found also in other databases.

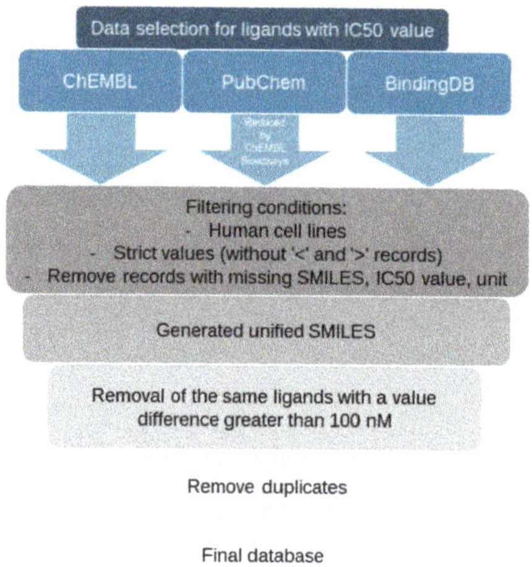

Figure 1. Overview of data selection process for IC50 value for serotonin and norepinephrine transporters.

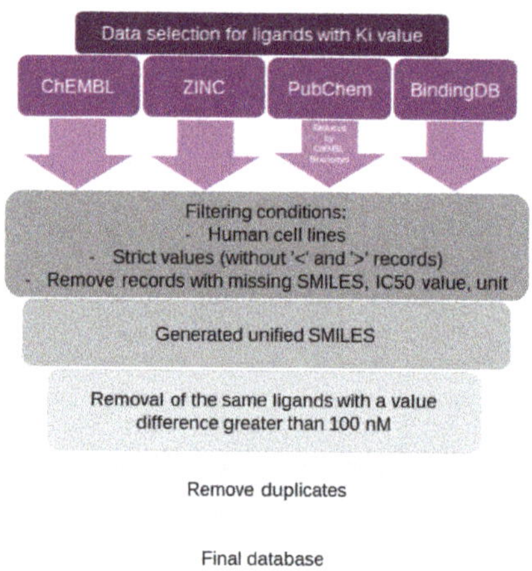

Figure 2. Overview of data selection process for Ki value for serotonin and norepinephrine transporters.

The study used four open databases as a source of molecule data: BindingDB [25], ChEMBL [26], PubChem [27] and ZINC [28]. However, the ZINC database contains molecule affinity values not inhibitory concentrations. Moreover, all data sources contain overlapping molecules. The BindingDB database contained a high percentage of molecules with the database identifier ChEMBL and ZINC. The PubChem database, in nearly all cases, references ChEMBL as its primary data source. Since the ChEMBL database was accessed separately for this study, duplicating data retrieval from PubChem was deemed unnecessary (accessed on 31 October 2024).

The next database cleaning step included the removal of records with ">" and "<" tags, missing activity values or units, missing SMILES (Simplified Molecular Input Line Entry System), data from studies using non-human cell lines, a set of the same molecules between which differences in Ki or IC50 values exceeded 100 nM, and duplicate molecules based on a unified SMILES generation system. For the serotonin transporter and pKi values, we used our previously curated database [24], which contained 8091 unique molecules, as the primary database. As an additional value, we provided new molecules from the BindingDB and PubChem databases. This yielded an additional 781 molecules. Final databases are available in Supplementary Materials S1–S4.

Table 1. Summary of databases for serotonin (SERT) and norepinephrine (NET) transporters.

Target	Property	ChEMBL	BindingDB	ZINC	PubChem (Without ChEMBL)	Molecules in Curated Database
NET	pIC50	1631	2136 (1486-ZINC; 1869-ChEMBL)	-	165	1879
NET	pKi	1598	2077 (1469-ZINC; 1390-ChEMBL)	3553	302	4304
SERT	pIC50	2827	3420 (2307-ZINC; 2782-ChEMBL)	-	533	2952
SERT	pKi	3059	3068 (2071-ZINC; 2263-ChEMBL)	7312	615	8872

Divided into training and test sets in a ratio of 80:20, the datasets had a similar distribution of the dependent variable, as seen in the histograms shown in Figure 3. The obtained databases are characterized by diverse molecules. Figure 4 shows the results of a t-SNE analysis performed using ChemPlot [29], which is a Python package designed to visualize molecular diversity based on SMILES structures. The t-SNE algorithm is a dimensionality reduction technique that projects multidimensional data into a two-dimensional space. The results presented here show the high diversity of molecules in the databases.

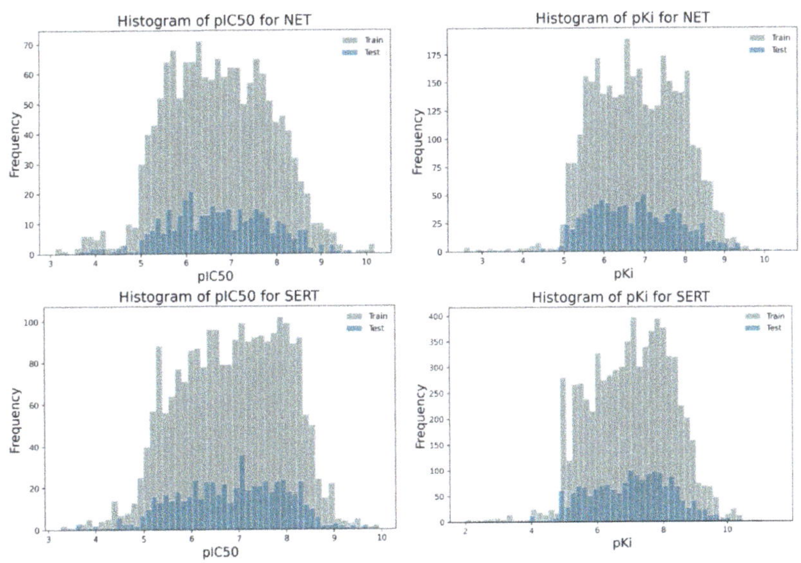

Figure 3. Histograms for training and test sets for biological targets and their corresponding properties.

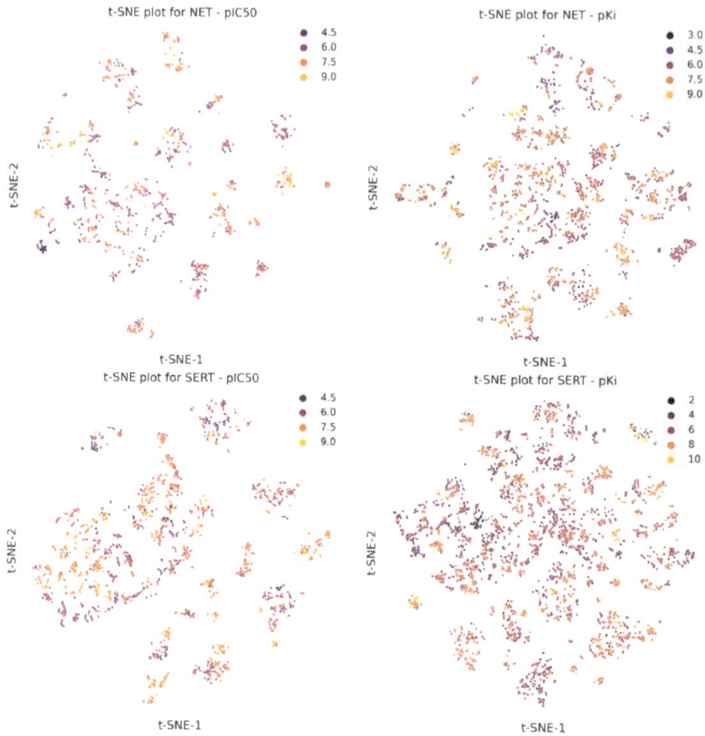

Figure 4. Scatter plots of t-SNE analysis for four databases.

2.2. QSAR Models

The best-performing models and their metric values are summarized in Table 2. The recap includes the root mean square error (RMSE) and coefficient of determination R^2 values for both the training set according to 10-fold cross validation and the test set.

Table 2. Summary of results for QSAR models of serotonin and norepinephrine transporters.

Target	Property	RMSE 10-CV	R^2 10-CV	RMSE Test	R^2 Test	Ensemble Model Structure
NET	pIC50	0.621	0.708	0.678	0.640	2 × Xgboost, 5 × LightGBM, Neural Network, 3 × CatBoost
NET	pKi	0.522	0.764	0.590	0.709	3 × Xgboost, 2 × LightGBM, 3 × Neural Network, CatBoost
SERT	pIC50	0.597	0.715	0.645	0.678	Xgboost, 4 × LightGBM, 4 × Neural Network
SERT	pKi	0.474	0.863	0.540	0.828	3 × Xgboost, 7 × LightGBM, 6 × Neural Network, 2 × CatBoost

2.2.1. Norepinephrine Transporter—pIC50

For the database built up with 1879 unique molecules, a QSAR model was developed with RMSE values equal to 0.621 and 0.678 for training and test sets, respectively, where the coefficient of determinations were 0.708 (train set) and 0.640 (test set). Model performance is presented as a scatter plot of predicted versus true values for both the training and test sets (Figure 5). For the obtained ensemble model (2 × Xgboost, 5 × LightGBM, Neural Network, 3 × CatBoost), SHAP analysis was performed based on 20% of the training dataset. Table 3 presents the values for top ten molecular descriptors for the model. Moreover, Figure 6 presents the impact of descriptor values on the derivative of half maximal inhibitory concentration (pIC50).

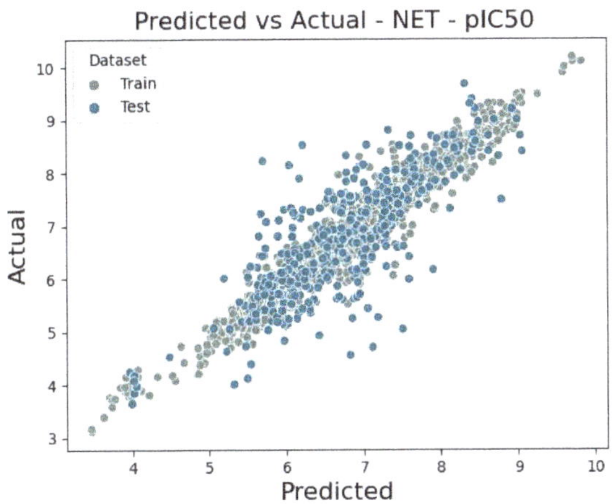

Figure 5. Scatter plot of a comparison of actual vs. predicted values for noradrenaline transporter of pIC50 value.

Table 3. Shapley values for ten most important descriptors for the NET-pIC50 QSAR model.

Descriptor	Description [30]	Shapley Value
AATS8d	Averaged Moreau–Broto autocorrelation of lag 8 weighted by sigma electrons	0.08196
Xp-1dv	1-ordered Chi path weighted by valence electrons	0.06080
nBase	Basic group count	0.05944
nAHRing	Aromatic hetero ring count	0.05231
SRW09	Walk count (leg-9, only self-returning walk)	0.04940
CIC5	5-ordered complementary information content	0.04560
SssNH	Sum of nitrogen atoms bonded by two single bonds	0.04211
MDEC-33	Molecular distance edge between tertiary C and tertiary C	0.04023
SlogP_VSA1	MOE logP VSA Descriptor 1 (-inf < x < −0.40)	0.03818
AATSC1Z	Averaged and centered Moreau–Broto autocorrelation of lag 1 weighted by atomic number	0.03776

The most relevant descriptor for predicting pIC50 values for the norepinephrine transporter is the AATS8d descriptor, which, according to Mordred's documentation [30], is described as the averaged Moreau–Broto autocorrelation of lag 8 weighted by sigma electrons. This two-dimensional feature of ligands at low values has a positive effect on pIC50 values, while at high values, it has a decreasing effect on pIC50. Similar behavior can be observed for the Xp-1dv descriptor (1-ordered Chi path weighted by valence electrons), which describes molecular branching [31]. A high number of base groups (nBase) positively increases the inhibitory activity of ligands, while lower pIC50 values are observed at low ones. This principle can be seen in the case of Milnacipran, where the structure of the drug includes both a primary amine and a tertiary amine (N, N-diethyl), contributing to its interaction at the binding site and its inhibitory activity [32]. For a low number of aromatic rings with a heteroatom (nAHRing), a negative effect on the predicted value is observed, but it cannot be said that a molecule with a number of such rings increases the pIC50 value. The fifth descriptor relevant to this analysis is descriptor SRW09 (walk count (leg-9, only self-returning walk)). It was described in 1993 [33] and for the inhibitory properties of the norepinephrine transporter; high values of it positively affect pIC50 values, and correspondingly low values decrease this dependent variable. The CIC5 descriptor (five-ordered complementary information content) at low values has a positive effect on the pIC50 value. On the other hand, high values have no significant effect, which could be described as neutral. This neutral effect is also observed at low values of the descriptor SssNH (descriptor describing the sum of secondary amines) [34]. On the other hand, high values of the sum of the nitrogen atom have an increasing effect on the value of the derived inhibitory concentration. The molecular distance between tertiary carbon atoms (MDEC-33) is difficult to interpret unambiguously. In contrast, the descriptor SlogP_VSA1 positively influences pIC50 values at high descriptor values. The descriptor SlogP_VSA relates to the molecular property of hydrophobicity and its spatial distribution across a ligand's surface area. It could link to how the norepinephrine transporter binding site interacts with ligands. The last descriptor, AATSC1Z, mostly positively affects pIC50 values at low values and negatively at high values.

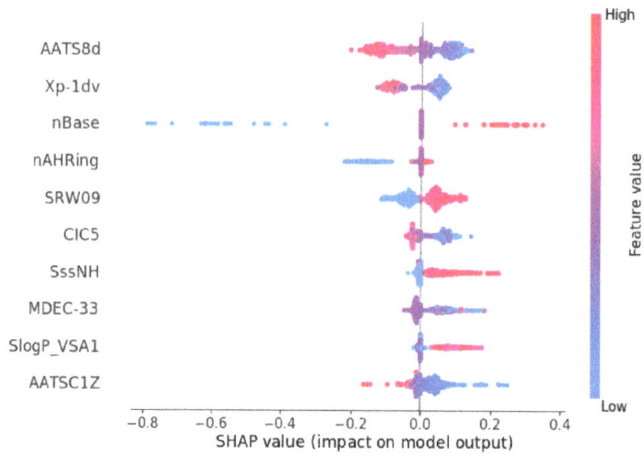

Figure 6. Summary plot of SHAP analysis for ten of the most important descriptors for NET pIC50 QSAR model.

2.2.2. Norepinephrine Transporter—pKi

The number of norepinephrine transporter ligands based on pKi values is 4304. The final ensemble model (3 × Xgboost, 2 × LightGBM, 3 × Neural Network, CatBoost) results of metrics are for the training set RMSE = 0.522 and R^2 = 0.764. Interestingly, for the test set, the error value is lower, RMSE = 0.590 and R^2 = 0.709. The performance of the model is visualized as a scatter plot of predicted and actual values for both the training and test sets (Figure 7). Similar to the model predicting pIC50 values, SHAP analysis was also performed. The results are shown in Table 4 and Figure 8.

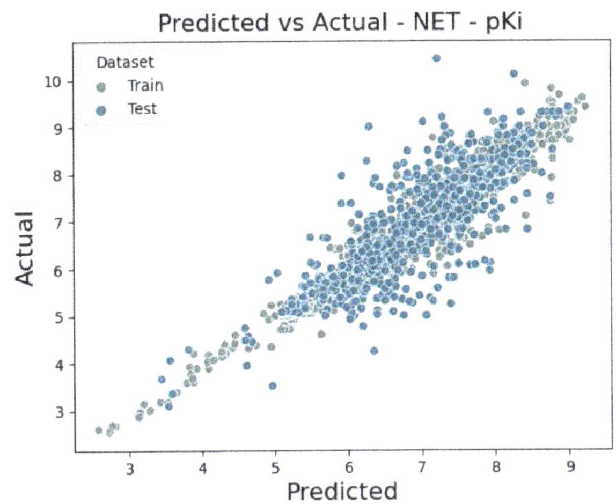

Figure 7. Scatter plot of a comparison of actual vs. predicted values for noradrenaline transporter of pKi value.

Table 4. Shapley values for ten most important descriptors for NET-pKi QSAR model.

Descriptor	Description [30]	Shapley Value
ATS5d	Moreau–Broto autocorrelation of lag 5 weighted by sigma electrons	0.05284
GGI10	10-ordered raw topological charge	0.04617
nBase	Basic group count	0.04613
AATS8d	Averaged Moreau–Broto autocorrelation of lag 8 weighted by sigma electrons	0.04608
Xch-5d	5-ordered Chi chain weighted by sigma electrons	0.04140
SMR_VSA6	MOE-type descriptors using Wildman–Crippen MR and surface area contribution	0.04028
AATS6p	Averaged Moreau–Broto autocorrelation of lag 6 weighted by polarizability	0.03630
SssNH	Atom type e-state descriptor—sum of nitrogen atoms bonded by two single bonds	0.03408
AATS5v	Averaged Moreau–Broto autocorrelation of lag 5 weighted by vdw volume	0.02857
GATS6i	Geary coefficient of lag 6 weighted by ionization potential	0.02807

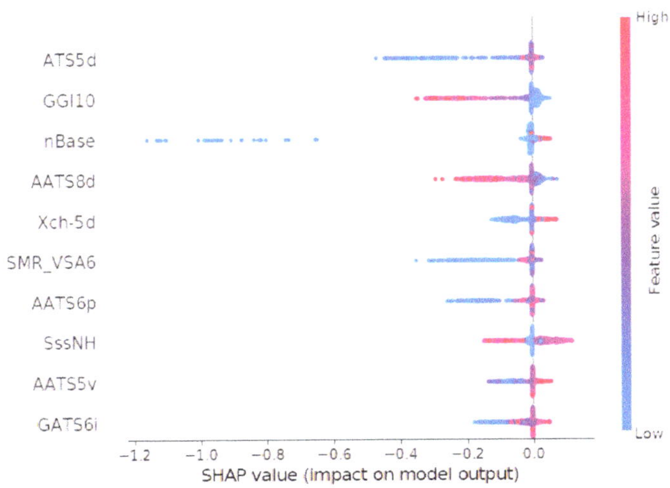

Figure 8. Summary plot of SHAP analysis for ten of the most important descriptors for NET pKi QSAR model.

The SHAP analysis revealed significant relationships between specific descriptors and the affinity values. The ATS5d descriptor, representing the Moreau–Broto autocorrelation of lag 5 weighted by sigma electrons, was identified as the most influential variable in the QSAR model. Lower values of ATS5d were associated with a reduction in pKi, while higher values did not exhibit a clear positive or enhancing effect on affinity. The second descriptor, GGI10, corresponds to the 10-ordered raw topological charge and demonstrated a contrasting pattern. High values of this descriptor were found to negatively influence pKi, whereas low values exhibited a neutral or slightly positive effect. This behavior aligns with findings reported by Olasupo et al. [23], where this descriptor also contributed negatively to the calculated pKi values for NET. For the nBase descriptor, which reflects the count of basic groups in the molecule, low values showed both neutral and negative effects on pKi, while higher values were associated with a minimal increase in affinity. Similarly, the AATS8d descriptor, representing the averaged Moreau–Broto autocorrelation of lag 8 weighted by sigma electrons, was observed to negatively impact pKi at high values, while low values were associated with neutral or marginally positive effects. In the case of Xch-5d, a five-ordered Chi chain weighted by sigma electrons, low values correlated with reduced pKi, whereas higher values positively influenced ligand affinity. The SMR_VSA6 descriptor, a MOE-type variable based on Wildman–Crippen molar refractivity and surface area contributions, exhibited a complex pattern. Low values of SMR_VSA6 had a negative impact on pKi, while higher values were largely neutral with some indication of a slight increase in affinity. The AATS6p descriptor, an averaged Moreau–Broto autocorrelation of lag 6 weighted by polarizability, showed a similar trend to SMR_VSA6. Low values were associated with a reduction in pKi, whereas high values resulted in neutral or slightly positive effects. The SssNH descriptor, an atom-type e-state representing the sum of ssNH, presented an ambiguous relationship with pKi. While high values exhibited both negative and positive influences, low values predominantly showed a neutral effect. It is worth noting that the ssNH atom type refers to nitrogen atoms bonded by two single bonds. For AATS5v, an averaged Moreau–Broto autocorrelation of lag 5 weighted by van der Waals volume, low values were associated with a negative effect on pKi. High values, in contrast, were neutral or contributed to a slight increase in affinity. Finally, the GATS6i descriptor, a Geary coefficient of lag 6 weighted by ionization potential, demonstrated no consistent relationship with pKi. Low values were primarily linked to a negative effect, with occasional neutrality, while high values ranged from slightly negative to marginally positive influences on affinity. Interestingly, for the model predicting pIC50 values for NET, the SHAP analysis similarly identified nBase and AATS8d as significant descriptors, displaying a comparable influence to that observed in the pKi model. However, the SssNH descriptor did not exhibit the same pattern in its effect on pIC50 values, indicating variability in its contribution across different predictive models [30,34].

2.2.3. Serotonin Transporter—pIC50

The number of serotonin transporter unique ligands with pIC50 values was 2952. The model developed was also an ensemble model built with three types of machine learning algorithms (Xgboost, 4 $\times$ LightGBM, 4 $\times$ Neural Network). The RMSE values were 0.597 and 0.645 for the training set and test set, respectively. The coefficient of determination, in turn, had values of 0.715—train set and 0.678—test set. The scatter plot in Figure 9 depicts the model's performance by showing predicted versus true values for the training and test sets.

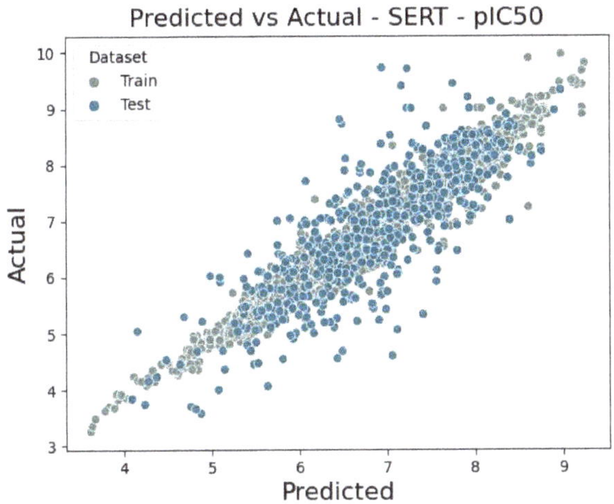

Figure 9. Scatter plot of a comparison of actual vs. predicted values for serotonin transporter of pIC50 value.

The second stage of the QSAR model analysis was to analyze the significance of the descriptors affecting the pIC50 value. In Table 5 and Figure 10, we presented a summary of the ten most important descriptors. The most significant descriptor is the number of aromatic carbon atoms of NaaCH. At high values of this descriptor, a decreasing effect on predicted pIC50 values is observed, but at low values, the effect is ambiguous. The same situation is observed for the second descriptor SlogP_VSA6 (logP value and the influence of surface area). The third descriptor GATS2p at low values positively affects the value of pIC50, and high values lower the value of the inhibition derivative. The number of carbon atoms in the aromatic ring additionally forming a single bond (NaasC) at high values positively affects the pIC50 value. On the other hand, a decreasing or neutral effect is seen at low values of NaasC. Neighborhood information content descriptor 2 (IC2) at low values has a negative effect, and at high values, it has a positive effect. SMR_VSA3, described as Wildman–Crippen molar refractivity and surface area, at high values has a positive effect on the pIC50 value but no clear effect at low values. Low values of Chi descriptor (Xch-7d) and AATS5i (average Autocorrelation of Topological Structure descriptor) have an effect on increasing the derived inhibition. The last two descriptors, NaaN (aromatic nitrogen atom) and SsssCH (sum of tertiary carbon atoms), do not represent unambiguous influences. In the case of the aromatic nitrogen atom, a characteristic negative effect of high descriptor values on the reduction in pIC50 values can be observed [30,34].

Table 5. Shapley values for ten most important descriptors for SERT-pIC50 QSAR model.

Descriptor	Description [30]	Shapley Value
NaaCH	Number of aromatic carbon atoms	0.06192
SlogP_VSA6	MOE logP VSA Descriptor 6 ($0.15 \leq x < 0.20$)	0.04926
GATS2p	Geary coefficient of lag 2 weighted by polarizability	0.04684
NaasC	Number of aromatic carbon atoms with single bond	0.04545
IC2	2-ordered neighborhood information content	0.04496
SMR_VSA3	MOE Molar Refractivity VSA Descriptor 3 ($1.82 \leq x < 2.24$)	0.04269
Xch-7d	7-ordered Chi chain weighted by sigma electrons	0.04099
AATS5i	Averaged Moreau–Broto autocorrelation of lag 5 weighted by ionization potential	0.03823
NaaN	Number of nitrogen aromatic atoms	0.03417
SsssCH	Sum of tertiary carbon atoms	0.03109

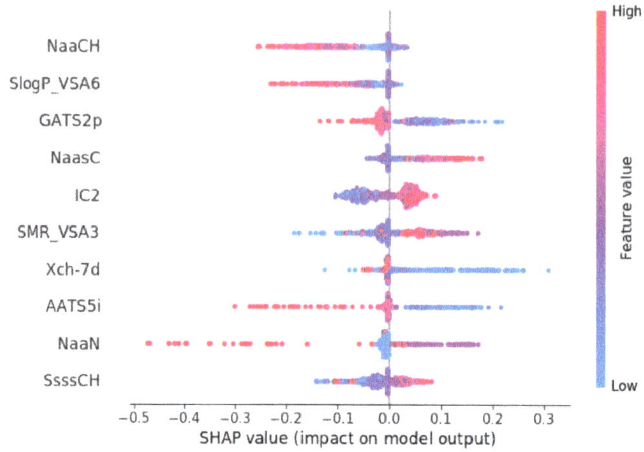

Figure 10. Summary plot of SHAP analysis for ten of the most important descriptors for SERT pIC50 QSAR model.

2.2.4. Serotonin Transporter—pKi

The database containing the serotonin transporter ligands and affinities is represented by 8872 unique molecules. The error values of the ensemble model created with Mljar (3 × Xgboost, 7 × LightGBM, 6 × Neural Network, 2 × CatBoost) are 0.474 and 0.540, and the coefficient of determination is 0.863 and 0.828 for the training and test set, respectively. Model performance is illustrated below using a scatter plot comparing predicted and true values for both the training and test sets (Figure 11).

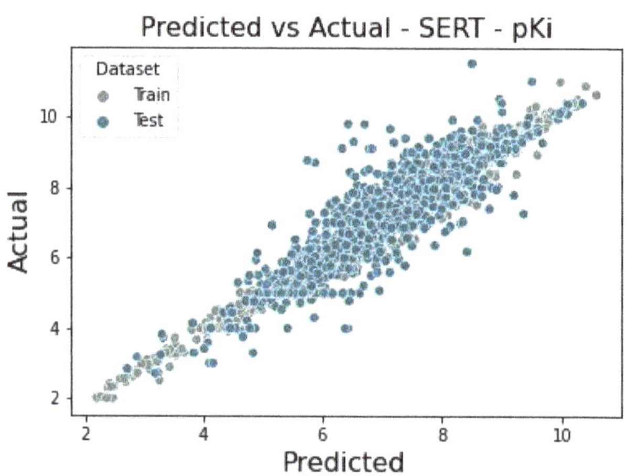

Figure 11. Scatter plot of a comparison of actual vs. predicted values for serotonin transporter of pKi value.

SHAP analysis identified a set of most important descriptors; the ten most important are shown in Table 6 and Figure 12 below. The most important descriptor, SssNH, representing the sum of nitrogen atoms bonded by two single bonds, was found to positively influence pKi at high values. Low values, in contrast, showed a slightly negative or neutral effect. Similarly, the nBase descriptor, which quantifies the number of basic groups in a molecule, exhibited a dual influence. Low values corresponded to a negative effect on pKi, while high values were associated with a positive impact. The GATS5c descriptor, a Geary coefficient of lag 5 weighted by Gasteiger charge, showed a nuanced pattern. Low values had a positive or neutral effect, whereas high values were generally neutral with a slight negative influence on pKi. The SLogP descriptor, representing the Wildman–Crippen LogP, indicated that low values negatively impacted pKi, though neutrality was also observed. High SLogP values, however, were predominantly associated with neutral effects. For the ATS1se descriptor, which is the Moreau–Broto autocorrelation of lag 1 weighted by Sanderson electronegativity, high values exhibited a neutral effect on pKi, while low values were associated with a positive influence, enhancing the predicted affinity. The NaaCH descriptor, which denotes the number of carbon atoms bonded by two aromatic bonds (typically within aromatic rings), displayed a more complex behavior. Low values of NaaCH positively influenced pKi, whereas high values had a negative effect. Nonetheless, both low and high values also demonstrated neutral contributions to pKi under certain conditions. The GGI7 descriptor, a seven-ordered raw topological charge, showed an ambiguous influence on pKi. Both low and high values were linked to negative or neutral effects without a clear pattern emerging from the data. The BIC2 descriptor, a two-ordered bonding information content, revealed a clearer trend. Low values were predominantly associated with a reduction in pKi, while high values exhibited a slightly positive effect, indicating a potential role in enhancing ligand affinity at higher descriptor values. For the JGI4 descriptor, a four-ordered mean topological charge, low values had a negative effect on pKi, while high values positively influenced the predicted affinity. Lastly, the SsssCH descriptor, representing the sum of carbon atoms bonded by three single bonds, exhibited a mixed pattern. Low values negatively influenced pKi, while high values showed a positive effect [30,34].

Table 6. Shapley values for ten most important descriptors for SERT-pKi QSAR model.

Descriptor	Description [30]	Shapley Value
SssNH	Atom type e-state descriptor—sum of nitrogen atoms with two single bonds	0.09159
nBase	Basic group count	0.08458
GATS5c	Geary coefficient of lag 5 weighted by Gasteiger charge	0.05913
SLogP	Wildman–Crippen LogP	0.05444
ATS1se	Autocorrelation of Topological Structure descriptor—Moreau–Broto autocorrelation of lag 1 weighted by Sanderson EN	0.05037
NaaCH	Atom type e-state descriptor—number of aaCH	0.04245
GGI7	7-ordered raw topological charge descriptor	0.04082
BIC2	2-ordered bonding information content	0.04035
JGI4	4-ordered mean topological charge	0.04015
SsssCH	Atom type e-state descriptor—sum of tertiary carbon atoms	0.03947

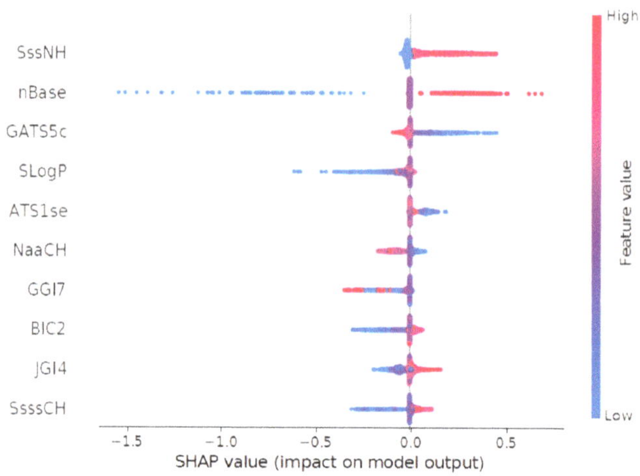

Figure 12. Summary plot of SHAP analysis for ten of the most important descriptors for SERT pKi QSAR model.

The results of the SHAP analysis summarized, given the ranges of the most important descriptors for the training set, formed the foundation for establishing the applicability domain. If at least seven ranges of the 10 descriptors are met, the molecule under test falls within the applicability domain, and then the prediction is defined as reliable.

2.2.5. Case Study

To illustrate the mechanism of action of an antidepressant and validate model predictions for the activity of SSRIs and SNRIs, we chose Duloxetine as an example. Duloxetine is a registered drug belonging to the serotonin and norepinephrine reuptake inhibitor class. To obtain predictions, users need the SMILES representation of the molecule of interest. For Duloxetine, the SMILES string is shown below:

"CNCC[C@@H](C1=CC=CS1)OC2=CC=CC3=CC=CC=C32".

The user simply needs to input the SMILES string into the designated cell in the SerotoninAI application. Within seconds, the results are displayed, as illustrated in Figure 13 below. The QSAR models predicted the following affinity values: pKi for SERT = 8.777 and pKi for NET = 7.757, which are very close to the experimental values reported in the literature (pKi for SERT = 9.1 and pKi for NET = 8.27) [35]. For the predicted pIC50 values, the application provided pIC50 for SERT = 8.126 and pIC50 for NET = 7.542. These predictions are consistent with the experimental values reported by Van Orden et al., where pIC50 for SERT = 8.2 and pIC50 for NET = 9.5 [36].

Additionally, the application provides information about the domain of applicability (see below). Duloxetine meets all predefined criteria, as its top 10 most important descriptors fall within the ranges defined by the training dataset. Finally, the application generates a comment regarding the potential use of the molecule within the appropriate mechanism, which is based on Table 7. For this case, the output states: "The molecule is a potential SSRI and SNRI drug", which aligns with the known pharmacological properties of Duloxetine. Figure 13 shows the predictions and applicability domains for Duloxetin, which is a registered SNRI drug. The view shown is the result of the SerotoninAI application.

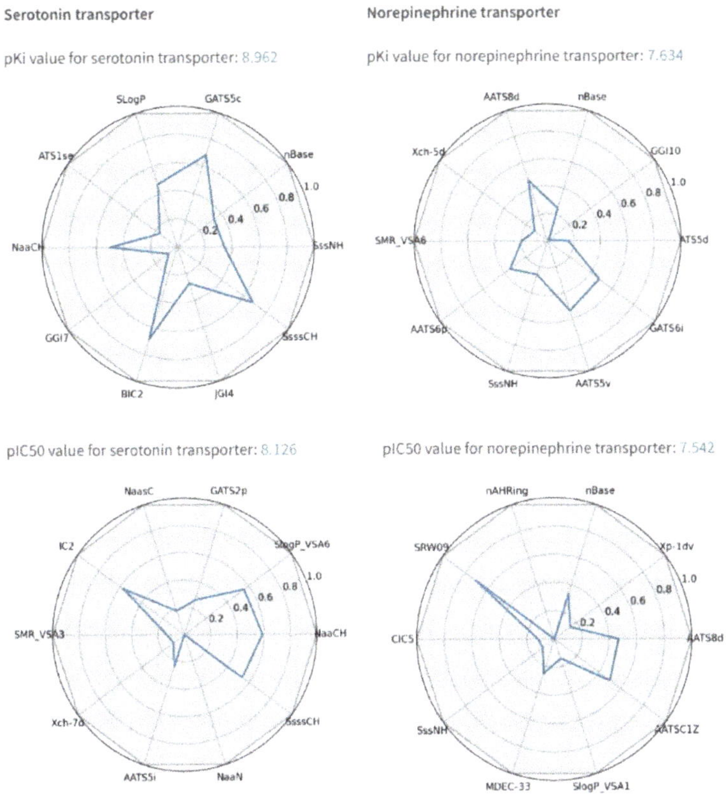

Figure 13. SerotoninAI predictions with applicability domain for Duloxetin.

Table 7. Predictions values of pKi and pIC50 for serotonin and norepinephrine transporters and corresponding comment from SerotoninAI application.

NET pKi	NET pIC50	SERT pKi	SERT pIC50	Comment
≥7	≥7	≥7	≥7	Molecule is a potential SSRI and SNRI drug
		≥7	≥7	Molecule is a potential SSRI drug
≥6 & <7	≥6 & <7	≥6 & <7	≥6 & <7	Molecule might have SSRI and SNRI mechanism of action
		≥6 & <7	≥6 & <7	Molecule might have SSRI mechanism of action
<6	<6	<6	<6	The molecule under study does not exhibit antidepressant activity by the mechanism of SSRIs and SNRIs

3. Discussion

In the presented work, we developed the general QSAR models to enable the search for new antidepressants that act like SSRIs or SNRIs. From the literature review presented in the introduction of the manuscript, to date, there are no models that consider the serotonin and norepinephrine transporter for both pKi and pIC50 values. The largest databases are a set of 213 ligands for the serotonin transporter [18] and 76 ligands for the norepinephrine transporter [7]. None of the models presented in the research articles is a simple user-accessible model.

Moreover, another aspect that underscores the relevance of our research is the lack of structural limitations of the obtained databases. They do not represent a set of molecules that are derivatives of one selected chemical structure, which stands in contrast to already published studies [7–10].

We also would like to mention the importance of developing two separate models for a given biological target for pKi and pIC50 values. An article from 2013 [37], in which the authors conduct an analysis of the ChEMBL database, suggests that the databases for pIC50 and pKi values should be merged to increase the size of the collection without losing data quality. The paper's authors point out that there is a significant difference between the same molecules and their pIC50 values. A successive comparison of pIC50 and pKi values, after appropriate mathematical transformations, leads to values with similar error to pIC50 values only. In our opinion, consolidating the dataset in a single computational experiment would lead to unnecessary error. For this reason, we decided to limit the database to ligands whose differences in activity values do not exceed 100 nM and not to combine the databases of pIC50 and pKi values. In our research, we wanted to achieve the highest possible degree of consistency in the data, where experiments measuring inhibition and affinity values were carried out exclusively using human cell lines.

The models presented in this study have been implemented in a new module of the SerotoninAI application. The antidepressant activity module will allow the user to obtain a prediction and receive feedback on whether the molecule may show the potential of a future SSRI or SNRI drug in an accessible and quick way. In addition, the applicability domain is presented for each model.

4. Materials and Methods

4.1. Databases

The object of this manuscript is to develop QSAR models for the serotonin and norepinephrine transporters. The first step was to construct selected databases separately for these transporters and for two values of the inhibition constant (Ki) and the half-maximal inhibitory concentration (IC50). We used the ChEMBL [26], BindingDB [25], PubChem [27] and ZINC [28] platforms to curate as many compounds as possible. It was important to obtain ligands of these transporters with measured activity values based on human cell lines. Based on activity value conversions, the dependent variables in the study were pKi and pIC50.

4.2. Molecular Representation

The method of representing transporter ligands for machine learning methods to develop QSAR models was 2D Mordred descriptors [38]. Based on the experience gained with QSAR models for various serotonergic system databases, the chosen representation is the best form to train the model while accurately describing each descriptor [30]. The cleaning phase of the Mordred descriptors representing the ligands of biological targets consisted of removing columns with fixed values, and for columns that contained errors in the calculation of descriptors, the value was filled in with the column's mean.

4.3. Machine Learning

Machine learning methods, particularly supervised machine learning, focus on training a model on a dataset so that it is both sufficiently fit to the training set and general enough to be applied to previously unseen data. In this way, they want to achieve a variance–bias trade-off. The experience gained [39] shows that the simplest methods of model development are insufficient to handle the complex nature of the ligand–protein relationship. We have developed QSAR models using the automated machine learning tool Mljar [40]. This system enables the development of complex ensemble models. The models were trained on 80% of the dataset according to 10-fold cross-validation (10-CV) and tested externally on the remaining set of ligands. Cross-validation is a method of evaluating and verifying model performance that is the gold standard for developing predictive models.

The method involves dividing the data into several equal parts to repeatedly train and test the model on different subsets of the data.

4.4. Evaluation Metrics

We used the root mean square error (RMSE) and coefficient of determination (R^2) to evaluate the obtained regression models. The calculations of these metrics are shown below in Equations (1) and (2). The best model was selected based on the lowest RMSE value.

$$\mathrm{RMSE} = \sqrt{\sum_{i=1}^{n} \frac{(\mathrm{pred_i} - \mathrm{obs_i})^2}{N}} \quad (1)$$

where RMSE is root mean square error, and N is the number of observations.

$$R^2 = 1 - \frac{SS_{res}}{SS_{tot}} = 1 - \frac{\sum_{i=1}^{n} (\mathrm{pred_i} - \overline{\mathrm{obs}})^2}{\sum_{i=1}^{n} (\mathrm{obs_i} - \overline{\mathrm{obs}})^2} \quad (2)$$

where R^2 = the coefficient of determination, SS_{res} = the sum of squares of the residual errors, SS_{tot} = the total sum of the errors, obs_i and $pred_i$ = observed and predicted values, and $\overline{obs}$ = arithmetical mean of observed values.

4.5. Model Interpretation

According to principle five of the Organization for Economic Cooperation and Development's Guidelines for regulatory purposes on valid QSAR models, if possible, QSAR models should have a mechanistic interpretation [41]. A keyword that appears in relation to advanced machine learning models is Explainable Artificial Intelligence [42]. When using automated machine learning methods to develop QSAR models that have a high degree of advancement, the aspect of explainability of the entire model becomes difficult. However, a method that partially allows the interpretation of the results obtained and the influence of individual features on the prediction obtained is SHAP (Shapley additive explanations) analysis. The method is named after game theory inventor and Nobel Memorial Prize winner Lloyd Shapley. His considerations tested the influence of an individual participant in a team game on the final outcome [43,44].

For the top four QSAR models, the analysis was conducted based on a randomly selected 20% of the training set, using an adapted tool of SHAP analysis [45]. The obtained results formed the basis for establishing an applicability domain (AD), which informs the reliability of the prediction. When the molecule under test is at least within seven ranges of the training set values for the 10 most important descriptors for a given model, then such a prediction is reliable.

4.6. Software

Antidepressant activity is another module of the SerotoninAI application (https://serotoninai.streamlit.app/, accessed on 17 January 2025) [46]. The application was developed through the Streamlit platform and allows the user to easily obtain predictions of affinity and inhibitory concentration values for the serotonin and norepinephrine transporter. An additional aspect is the accompanying commentary for the user on whether the molecule could be a future drug in the SSRI or SNRI mechanism of action. Table 7 shows the ranges of pKi and pIC50 values and the concluding commentary. We would like to emphasize that the commentary is conditional, due to the need to limit the numerical values. The SerotoninAI source code and QSAR models are available on the GitHub platform (https://github.com/nczub/SerotoninAI_streamlit, accessed on 17 January 2025).

5. Conclusions

This study presents QSAR models for the serotonin and norepinephrine transporters, specifically predicting affinity (pKi) and inhibitory concentration (pIC50) values for experimentally measured unique sets of ligands. The developed models demonstrate low prediction errors and high coefficients of determination, indicating their robustness and reliability. These characteristics make the models valuable tools for application in the drug discovery process, particularly for identifying novel compounds with mechanisms of action as selective serotonin reuptake inhibitors (SSRIs) or serotonin–norepinephrine reuptake inhibitors (SNRIs).

Supplementary Materials: The following supporting information can be downloaded at https://www.mdpi.com/article/10.3390/molecules30030637/s1. In the Supplementary Materials, the authors share four databases of pKi and pIC50 values for serotonin and norepinephrine transporters S1–S4.

Author Contributions: Conceptualization, N.Ł., A.P., J.S. and A.M.; methodology, N.Ł.; software, N.Ł.; validation, N.Ł.; formal analysis, N.Ł.; investigation, N.Ł.; resources, J.S. and A.M.; data curation, N.Ł.; writing—original draft preparation, N.Ł.; writing—review and editing, N.Ł., A.P., J.S. and A.M.; visualization, N.Ł.; supervision, A.M.; project administration, A.M.; funding acquisition, N.Ł., J.S. and A.M. All authors have read and agreed to the published version of the manuscript.

Funding: This research was carried out with the use of research infrastructure co-financed by the Smart Growth Operational Programme POIR 4.2, project no. POIR.04.02.00-00-D023/20.

Institutional Review Board Statement: Not applicable.

Informed Consent Statement: Not applicable.

Data Availability Statement: The data presented in this study are available on request from the corresponding author.

Conflicts of Interest: The authors declare no conflicts of interest.

References

1. Depressive Disorder (Depression), World Health Organization. Available online: https://www.who.int/news-room/fact-sheets/detail/depression (accessed on 11 July 2024).
2. McCarron, R.M.; Shapiro, B.; Rawles, J.; Luo, J. Depression. *Ann. Intern. Med.* **2021**, *174*, ITC65–ITC80. [CrossRef] [PubMed]
3. Dionisie, V.; Filip, G.A.; Manea, M.C.; Manea, M.; Riga, S. The anti-inflammatory role of SSRI and SNRI in the treatment of depression: A review of human and rodent research studies. *Inflammopharmacology* **2021**, *29*, 75–90. [CrossRef] [PubMed]
4. Shelton, R.C. Serotonin and Norepinephrine Reuptake Inhibitors. In *Handbook of Experimental Pharmacology*; Springer: Cham, Switzerland, 2019; Volume 250, pp. 145–180. [CrossRef]
5. Gupta, R.; Srivastava, D.; Sahu, M.; Tiwari, S.; Ambasta, R.K.; Kumar, P. Artificial intelligence to deep learning: Machine intelligence approach for drug discovery. *Mol. Divers.* **2021**, *25*, 1315–1360. [CrossRef] [PubMed]
6. de Moura, É.P.; Fernandes, N.D.; Monteiro, A.F.M.; de Medeiros, H.I.R.; Scotti, M.T.; Scotti, L. Machine Learning, Molecular Modeling, and QSAR Studies on Natural Products Against Alzheimer's Disease. *Curr. Med. Chem.* **2021**, *28*, 7808–7829. [CrossRef]
7. Kulkarni, S.S.; Grundt, P.; Kopajtic, T.; Katz, J.L.; Newman, A.H. Structure-activity relationships at monoamine transporters for a series of N-substituted 3alpha-(bis [4-fluorophenyl]methoxy)tropanes: Comparative molecular field analysis, synthesis, and pharmacological evaluation. *J. Med. Chem.* **2004**, *47*, 3388–3398. [CrossRef]
8. Dessalew, N. QSAR study on dual SET and NET reuptake inhibitors: An insight into the structural requirement for antidepressant activity. *J. Enzym. Inhib. Med. Chem.* **2009**, *24*, 262–271. [CrossRef]
9. Zare, S.; Fereidoonnezhad, M.; Afshar, D.; Ramezani, Z. A comparative QSAR analysis and molecular docking studies of phenyl piperidine derivatives as potent dual NK1R antagonists/serotonin transporter (SERT) inhibitors. *Comput. Biol. Chem.* **2017**, *67*, 22–37. [CrossRef]

10. Karabulut, S.; Kaur, H.; Gauld, J.W. Uncovering Structure-Activity Relationships of Phenethylamines: Paving the Way for Innovative Mental Health Treatments. *ACS Chem. Neurosci.* **2024**, *15*, 972–982. [CrossRef]
11. Pratuangdejkul, J.; Schneider, B.; Jaudon, P.; Rosilio, V.; Baudoin, E.; Loric, S.; Conti, M.; Launay, J.M.; Manivet, P. Definition of an uptake pharmacophore of the serotonin transporter through 3D-QSAR analysis. *Curr. Med. Chem.* **2005**, *12*, 2393–2410. [CrossRef]
12. Mente, S.; Gallaschun, R.; Schmidt, A.; Lebel, L.; Vanase-Frawley, M.; Fliri, A. Quantitative structure-activity relationship of phenoxyphenyl-methanamine compounds with 5HT2A, SERT, and hERG activities. *Bioorganic Med. Chem. Lett.* **2008**, *18*, 6088–6092. [CrossRef]
13. Kharkar, P.S.; Reith, M.E.; Dutta, A.K. Three-dimensional quantitative structure-activity relationship (3D QSAR) and pharmacophore elucidation of tetrahydropyran derivatives as serotonin and norepinephrine transporter inhibitors. *J. Comput. Aided Mol. Des.* **2008**, *22*, 1–17. [CrossRef] [PubMed]
14. Avram, S.; Buiu, C.; Duda-Seiman, D.M.; Duda-Seiman, C.; Mihailescu, D. 3D-QSAR design of new escitalopram derivatives for the treatment of major depressive disorders. *Sci. Pharm.* **2010**, *78*, 233–248. [CrossRef] [PubMed]
15. Sakloth, F.; Kolanos, R.; Mosier, P.D.; Bonano, J.S.; Banks, M.L.; Partilla, J.S.; Baumann, M.H.; Negus, S.S.; Glennon, R.A. Steric parameters, molecular modeling and hydropathic interaction analysis of the pharmacology of para-substituted methcathinone analogues. *Br. J. Pharmacol.* **2015**, *172*, 2210–2218. [CrossRef] [PubMed]
16. Ortore, G.; Orlandini, E.; Betti, L.; Giannaccini, G.; Mazzoni, M.R.; Camodeca, C.; Nencetti, S. Focus on Human Monoamine Transporter Selectivity. New Human DAT and NET Models, Experimental Validation, and SERT Affinity Exploration. *ACS Chem. Neurosci.* **2020**, *11*, 3214–3232. [CrossRef] [PubMed]
17. Cerda-Cavieres, C.; Quiroz, G.; Iturriaga-Vásquez, P.; Rodríguez-Lavado, J.; Alarcón-Espósito, J.; Saitz, C.; Pessoa-Mahana, C.D.; Chung, H.; Araya-Maturana, R.; Mella-Raipán, J.; et al. Synthesis, Docking, 3-D-Qsar, and Biological Assays of Novel Indole Derivatives Targeting Serotonin Transporter, Dopamine D2 Receptor, and Mao-A Enzyme: In the Pursuit for Potential Multitarget Directed Ligands. *Molecules* **2020**, *25*, 4614. [CrossRef]
18. Ivanova, L.; Karelson, M.; Dobchev, D.A. Multitarget Approach to Drug Candidates against Alzheimer's Disease Related to AChE, SERT, BACE1 and GSK3β Protein Targets. *Molecules* **2020**, *25*, 1846. [CrossRef]
19. Avram, S.; Stan, M.S.; Udrea, A.M.; Buiu, C.; Boboc, A.A.; Mernea, M. 3D-ALMOND-QSAR Models to Predict the Antidepressant Effect of Some Natural Compounds. *Pharmaceutics* **2021**, *13*, 1449. [CrossRef]
20. Botha, M.J.; Kirton, S.B. In Silico Investigations into the Selectivity of Psychoactive and New Psychoactive Substances in Monoamine Transporters. *ACS Omega* **2022**, *7*, 38311–38321. [CrossRef]
21. Petković, M.; Đorđević, V.; Rančić, S.; Stevanović, M.; Uroš, G.; Veselinovic, A.M. QSAR Modelling, Molecular Docking Studies and ADMET Predictions of Serotonin Transporter Inhibitors. SSRN. Available online: https://papers.ssrn.com/sol3/papers.cfm?abstract_id=4270253 (accessed on 1 September 2024).
22. Jaramillo, D.N.; Millán, D.; Guevara-Pulido, J. Design, synthesis and cytotoxic evaluation of a selective serotonin reuptake inhibitor (SSRI) by virtual screening. *Eur. J. Pharm. Sci. Off. J. Eur. Fed. Pharm. Sci.* **2023**, *183*, 106403. [CrossRef]
23. Olasupo, S.B.; Uzairu, A.; Shallangwa, G.; Uba, S. QSAR analysis and molecular docking simulation of norepinephrine transporter (NET) inhibitors as anti-psychotic therapeutic agents. *Heliyon* **2019**, *5*, e02640. [CrossRef]
24. Łapińska, N.; Pacławski, A.; Szlęk, J.; Mendyk, A. SerotoninAI: Serotonergic System Focused, Artificial Intelligence-Based Application for Drug Discovery. *J. Chem. Inf. Model.* **2024**, *64*, 2150–2157. [CrossRef] [PubMed]
25. BindingDB. Available online: https://www.bindingdb.org/rwd/bind/index.jsp (accessed on 1 September 2024).
26. Mendez, D.; Gaulton, A.; Bento, A.P.; Chambers, J.; de Veij, M.; Félix, E.; Magariños, M.P.; Mosquera, J.F.; Mutowo, P.; Nowotka, M.; et al. ChEMBL: Towards direct deposition of bioassay data. *Nucleic Acids Res.* **2018**, *47*, D930–D940. [CrossRef] [PubMed]
27. PubChem. Available online: https://pubchem.ncbi.nlm.nih.gov/ (accessed on 1 September 2024).
28. Irwin, J.; Shoichet, B.K. ZINC—A Free Database of Commercially Available Compounds for Virtual Screening. *J. Chem. Inf. Model.* **2004**, *45*, 177–182. [CrossRef] [PubMed]
29. Cihan Sorkun, M.; Mullaj, D.; Koelman, J.; Er, S. ChemPlot, a Python Library for Chemical Space Visualization. *Chem. Methods* **2022**, *2*, e202200005. [CrossRef]
30. Descriptor List. Available online: https://mordred-descriptor.github.io/documentation/master/descriptors.html (accessed on 1 October 2024).
31. Hu, Q.N.; Liang, Y.Z.; Yin, H.; Peng, X.L.; Fang, K.T. Structural interpretation of the topological index. 2. The molecular connectivity index, the Kappa index, and the atom-type E-State index. *J. Chem. Inf. Comput. Sci.* **2004**, *44*, 1193–1201. [CrossRef]
32. Pidathala, S.; Mallela, A.K.; Joseph, D.; Penmatsa, A. Structural basis of norepinephrine recognition and transport inhibition in neurotransmitter transporters. *Nat. Commun.* **2021**, *12*, 2199. [CrossRef]
33. Ruecker, G.; Ruecker, C. Counts of all walks as atomic and molecular descriptors. *J. Chem. Inf. Comput. Sci.* **1993**, *33*, 683–695. [CrossRef]
34. Kier, L.B.; Hall, L.H. An electrotopological-state index for atoms in molecules. *Pharm. Res.* **1990**, *7*, 801–807. [CrossRef]

35. Fan, Q.Y.; Xue, R.; Li, Y.; Zhang, T.T.; He, X.H.; Fan, S.Y.; Li, Y.F.; Zhong, B.H.; Zhang, Y.Z.; Li, J. Antidepressant-like Effects of ZBH2012001, a Novel Potent Serotonin and Norepinephrine Reuptake Inhibitor. *CNS Neurosci. Ther.* **2016**, *22*, 700–706. [CrossRef]
36. Van Orden, L.J.; Van Dyke, P.M.; Saito, D.R.; Church, T.J.; Chang, R.; Smith, J.A.; Martin, W.J.; Jaw-Tsai, S.; Stangeland, E.L. A novel class of 3-(phenoxy-phenyl-methyl)-pyrrolidines as potent and balanced norepinephrine and serotonin reuptake inhibitors: Synthesis and structure-activity relationships. *Bioorganic Med. Chem. Lett.* **2013**, *23*, 1456–1461. [CrossRef]
37. Kalliokoski, T.; Kramer, C.; Vulpetti, A.; Gedeck, P. Comparability of mixed IC_{50} data—A statistical analysis. *PLoS ONE* **2013**, *8*, e61007. [CrossRef] [PubMed]
38. Moriwaki, H.; Tian, Y.-S.; Kawashita, N.; Takagi, T. Mordred: A molecular descriptor calculator. *J. Cheminform.* **2018**, *10*, 4. [CrossRef] [PubMed]
39. Czub, N.; Pacławski, A.; Szlęk, J.; Mendyk, A. Curated Database and Preliminary AutoML QSAR Model for 5-HT1A Receptor. *Pharmaceutics* **2021**, *13*, 1711. [CrossRef]
40. Plonska, A.; Plonski, P. MLJAR: State-of-the-Art Automated Machine Learning Framework for Tabular Data, Version 0.10.3. Available online: https://github.com/mljar/mljar-supervised (accessed on 1 September 2024).
41. OECD Principles for the Validation, for Regulatory Purposes, of (Quantitative) Structure-Activity Relationship Models. Available online: https://www.oecd.org/en/publications/guidance-document-on-the-validation-of-quantitative-structure-activity-relationship-q-sar-models_9789264085442-en.html (accessed on 8 October 2024).
42. James, T.; Hristozov, D. Deep Learning and Computational Chemistry. *Methods Mol. Biol.* **2022**, *2390*, 125–151. [CrossRef] [PubMed]
43. Shapley, L.S. *A Value for N-Person Games*; RAND Corporation: Santa Monica, CA, USA, 1952. Available online: https://www.rand.org/pubs/papers/P295.html (accessed on 1 September 2024).
44. Lundberg, S.; Lee, S.I. A Unified Approach to Interpreting Model Predictions. *arXiv* **2017**, arXiv:1705.07874.
45. Szlęk, J. Model Interpretation. Available online: https://github.com/jszlek/MODEL_INTERPRETATION (accessed on 15 September 2023).
46. SerotoninAI. Available online: https://serotoninai.streamlit.app/ (accessed on 19 November 2024).

molecules

MDPI

Article

Exploring the Potential of Malvidin and Echiodinin as Probable Antileishmanial Agents Through In Silico Analysis and In Vitro Efficacy

Luis Daniel Goyzueta-Mamani [1], Daniela Pagliara Lage [2], Haruna Luz Barazorda-Ccahuana [1], Margot Paco-Chipana [1], Mayron Antonio Candia-Puma [1,3], Gonzalo Davila-Del-Carpio [3], Alexsandro Sobreira Galdino [4,5], Ricardo Andrez Machado-de-Avila [6], Rodolfo Cordeiro Giunchetti [7,8], Edward L. D'Antonio [9], Eduardo Antonio Ferraz Coelho [2] and Miguel Angel Chávez-Fumagalli [1,*]

[1] Computational Biology and Chemistry Research Group, Vicerrectorado de Investigación, Universidad Católica de Santa María, Arequipa 04000, Peru; lgoyzueta@ucsm.edu.pe (L.D.G.-M.); hbarazorda@ucsm.edu.pe (H.L.B.-C.); 74252022@ucsm.edu.pe (M.P.-C.); mcandia@ucsm.edu.pe (M.A.C.-P.)
[2] Programa de Pós-Graduação em Ciências da Saúde: Infectologia e Medicina Tropical, Faculdade de Medicina, Universidade Federal de Minas Gerais, Belo Horizonte 31270-901, Brazil; dpagliarara@icb.ufmg.br (D.P.L.); eduardoferrazcoelho@yahoo.com.br (E.A.F.C.)
[3] Facultad de Ciencias Farmacéuticas, Bioquímicas y Biotecnológicas, Universidad Católica de Santa María, Arequipa 04000, Peru; gdavilad@ucsm.edu.pe
[4] Laboratório de Biotecnologia de Microrganismos, Universidade Federal São João Del-Rei, Divinópolis 35501-296, Brazil; asgaldino@ufsj.edu.br
[5] Instituto Nacional de Ciência e Tecnologia em Biotecnologia Industrial (INCT-BI), Distrito Federal, Brasilia 70070-010, Brazil
[6] Programa de Pós-Graduação em Ciências da Saúde, Universidade do Extremo Sul Catarinense, Criciúma 88806-000, Brazil; r_andrez@unesc.net
[7] Laboratório de Biologia das Interações Celulares, Instituto de Ciências Biológicas, Universidade Federal de Minas Gerais, Belo Horizonte 31270-901, Brazil; giunchetti@icb.ufmg.br
[8] Instituto Nacional de Ciência e Tecnologia de Doenças Tropicais (INCT-DT), Salvador 40110-160, Brazil
[9] Department of Natural Sciences, University of South Carolina Beaufort, 1 University Boulevard, Bluffton, SC 29909, USA; edantonio@uscb.edu
* Correspondence: mchavezf@ucsm.edu.pe

Abstract: Leishmaniasis, a neglected tropical disease caused by *Leishmania* species, presents serious public health challenges due to limited treatment options, toxicity, high costs, and drug resistance. In this study, the in vitro potential of malvidin and echioidinin is examined as antileishmanial agents against *L. amazonensis*, *L. braziliensis*, and *L. infantum*, comparing their effects to amphotericin B (AmpB), a standard drug. Malvidin demonstrated greater potency than echioidinin across all parasite stages and species. Against *L. amazonensis*, malvidin's IC_{50} values were 197.71 ± 17.20 µM (stationary amastigotes) and 258.07 ± 17 µM (axenic amastigotes), compared to echioidinin's 272.99 ± 29.90 µM and 335.96 ± 19.35 µM. AmpB was more potent, with IC_{50} values of 0.06 ± 0.01 µM and 0.10 ± 0.03 µM. Malvidin exhibited lower cytotoxicity (CC_{50}: 2920.31 ± 80.29 µM) than AmpB (1.06 ± 0.12 µM) and a favorable selectivity index. It reduced infection rates by 35.75% in *L. amazonensis*-infected macrophages. The in silico analysis revealed strong binding between malvidin and *Leishmania* arginase, with the residues HIS139 and PRO258 playing key roles. Gene expression analysis indicated malvidin's modulation of oxidative stress and DNA repair pathways, involving genes like GLO1 and APEX1. These findings suggest malvidin's potential as a safe, natural antileishmanial compound, warranting further in vivo studies to confirm its therapeutic efficacy and pharmacokinetics in animal models.

Keywords: leishmaniasis; antileishmanial activity; malvidin; echioidinin; cytotoxicity; arginase inhibition

1. Introduction

Leishmaniasis, a vector-borne disease caused by the parasitic protozoa of the *Leishmania* genus, affects millions of people globally, particularly in tropical and subtropical regions. Spread by sandflies, it manifests in three main clinical forms [1]. The most severe, visceral leishmaniasis (VL), is characterized by fever, weight loss, and organ enlargement. The second clinical category, tegumentary leishmaniasis (TL), encompasses both cutaneous (CL) and mucosal leishmaniasis (ML) presentations. ML, the most destructive form, causes significant injuries [2]. Several *Leishmania* species are implicated in the disease. *L. infantum*, *L. amazonensis*, and *L. braziliensis* are some of the most prevalent in South America [3]. *L. infantum* has a broad geographic distribution, causing VL in regions like the Mediterranean, East Africa, Asia, and Latin America. This parasite primarily infects dogs in regions where the disease is a zoonosis, and can infect humans where it is an anthroponosis [4]. *L. amazonensis* demonstrates a wider range of clinical manifestations being responsible for both CL and ML [5] and it is also suggested as the cause of VL [6]. Finally, *L. braziliensis* is associated with a particularly aggressive form of CL. The lesions induced by *L. braziliensis* can be large, chronic, and lead to substantial scarring [7]. According to the World Health Organization, an estimated 0.7 to 1 million new leishmaniasis cases occur globally annually [8]. This neglected disease significantly impacts public health, causing disability, disfigurement, and even death if left untreated. Notably, 85% of global VL cases are reported from just the following seven countries: Brazil, Ethiopia, India, Kenya, Somalia, South Sudan, and Sudan [6,9]. The transmission process begins with a sandfly bite. Sandflies deposit *Leishmania* parasites in their promastigote form (extracellular, with a flagellum) into the mammalian host's skin. Macrophages then phagocytose these parasites, the body's first line of defense [10]. Inside a macrophage compartment for degrading engulfed particles (phagolysosome), the parasites transform into amastigotes (non-flagellated, replicative) [11]. *Leishmania* parasites employ a complex interplay of strategies to establish infection. They can suppress macrophage activation, hindering the production of pro-inflammatory cytokines (signaling molecules) that typically recruit and activate other immune cells to eliminate the parasite. This creates a permissive environment for parasite survival [12]. *Leishmania*'s repertoire of immune evasion strategies extends to molecular mimicry. Certain parasite species express molecules that closely resemble host components. This ingenious disguise allows for evading recognition and elimination by the immune system [13].

The phagolysosome presents a unique challenge for *Leishmania* due to its acidic environment, typically ranging from a pH of 4.5 to 5.5. However, amastigotes, the replicative form within the mammalian host, demonstrate a higher tolerance for this acidic niche. Amastigotes possess various mechanisms to adapt and thrive in harsh conditions [14]. Some studies have proposed that *Leishmania* might be able to manipulate the phagolysosome's pH to create a more favorable environment for its growth [15–17]. This manipulation could involve the parasite actively expelling protons from the phagolysosome, reducing its acidity [18]. The precise mechanisms underlying *Leishmania*'s tolerance of the acidic environment and its potential manipulation of phagolysosomal pH remain under investigation. However, elucidating this intricate interplay between parasites and the environment holds immense potential for developing novel treatment strategies. By targeting the parasite's ability to survive in the acidic environment or disrupting its pH manipulation mechanisms, researchers aim to disrupt its lifecycle and ultimately prevent successful infection.

Leishmaniasis treatment presents a complex challenge despite some recent improvements. While effective, the standard medications, such as pentavalent antimonials and amphotericin B (AmpB), have limitations that restrict their optimal use. These medications are associated with significant side effects [19]. Pentavalent antimonials cause severe reactions like nausea, vomiting, muscle pain, and irregular heartbeats. AmpB poses a challenging problem due to its nephrotoxicity and requires hospitalization for administration because of its toxicity [20]. These side effects can significantly impact a patient's ability to adhere to the treatment regimen and complete the entire course, potentially leading to treatment failure. The efficacy of leishmaniasis treatment is additionally hindered by the variability in medication response among different *Leishmania* species and the coexistence of overlapping species [21]. Certain species have a differing susceptibility to the same drug, which poses a challenge in selecting the appropriate treatment. The emergence of drug-resistant *Leishmania* strains exacerbates this problem, presenting a substantial challenge to existing treatments [22]. These problems highlight the importance of developing new treatment approaches focusing on specific pathways and targets crucial for the parasite's survival and growth inside the host. It is essential to consider the variation in species and the development of drug resistance when working to develop improved and safer treatments for leishmaniasis.

One promising approach involves targeting specific enzymes crucial for the parasite's metabolism. *Leishmania* parasites rely on the enzyme arginase (ARG) for polyamine biosynthesis, a vital process for their growth and survival [23]. ARG has recently obtained considerable attention since new studies have highlighted it as a potential therapeutic target in leishmaniasis [24–26]. ARG is the first enzyme of the polyamine pathway. It catalyzes the conversion of L-arginine to L-ornithine and urea, and L-ornithine continues as the next metabolite in polyamine biosynthesis to form putrescine and spermidine [27]. Studies have shown that inhibiting ARG activity with specific compounds can effectively lead *Leishmania* parasites to cell death in laboratory studies (in vitro) and animal models. The inhibition results in a lack of protection against reactive oxygen species (ROS), which damages *Leishmania*'s genetic material. These examples highlight a few exciting new research directions in leishmaniasis treatment [28]. By targeting specific vulnerabilities in the parasite's biology, researchers hope to develop safer, more effective therapies with minimal side effects and a reduced risk of resistance. The search for alternatives to traditional leishmaniasis medications has led researchers to explore natural products derived from plants and other organisms. These natural products often have a lower toxicity profile and might offer new avenues for combating the parasite. Our group's recent research has been particularly encouraging, shedding light on the potential of specific natural compounds like malvidin (PubChem CID: 159287) and echioidinin (PubChem CID: 15559079) (Figure 1) [29].

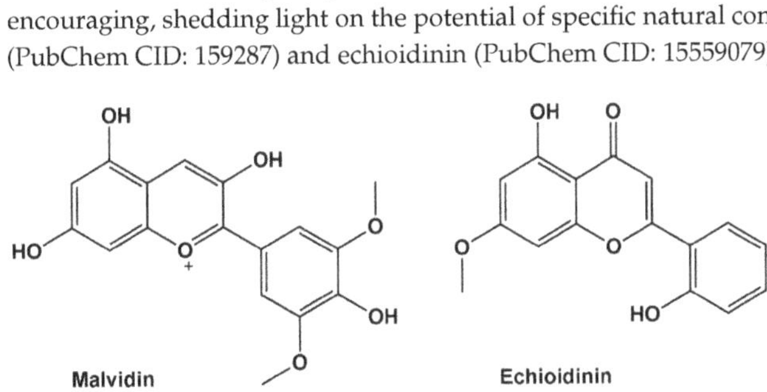

Figure 1. Chemical structures of malvidin (**left**) and echioidinin (**right**).

Our previous study investigated malvidin using computational modeling predicting interactions with ARG which is vital for *Leishmania* survival and may disrupt parasite growth [29,30]. However, no in vitro or in vivo studies have yet confirmed its efficacy against *Leishmania*. Another promising natural product, echioidinin derived from *Andrographis echioides* and *Andrographis alata*, has demonstrated a cytotoxic effect on leukemic cell lines while showing low toxicity to mammalian cells, suggesting it could be a therapeutic alternative against leishmaniasis [30–32]. Echioidinin also exhibited strong antifungal activity against several fungi, potentially disrupting DNA synthesis and mitochondrial function, though the exact mechanisms remain under investigation [33,34]. A crucial aspect of exploring natural products, like malvidin and echioidinin, is ensuring their safety and computer-based methods play a vital role. These methods allow for rapid and cost-effective screening of various natural compounds [35]. Researchers can prioritize the most promising candidates for further evaluation by analyzing potential anti-leishmanial activity and predicted toxicity. In silico methods can also model the interactions between natural compounds and human biomolecules at an atomic level. This helps predict potential adverse effects by identifying how a compound might interact with essential proteins, enzymes, or DNA, potentially causing disruptions or damage [36].

Furthermore, in silico tools are invaluable for elucidating compounds' potential mechanisms of action against *Leishmania* parasites. Researchers can design and optimize targeted therapies more effectively by understanding how natural products disrupt parasitic processes. In this study, the aim was to evaluate the in vitro inhibitory effects of malvidin and echioidinin on the ARG of three Leishmania species. Subsequently, in silico docking studies were conducted to analyze the binding interactions of the best candidate with the ARG at a pH of 5 and 7 by molecular dynamics (MD) simulations, assessing its behavior and stability. Also, network pharmacology and single-cell RNA sequencing data mining were used to analyze the potential molecular targets and pathways of NPs on different human organ datasets to predict potential tissue-specific toxicity. In this research, the promise of innovative enzyme inhibitors in expanding the field of drug discovery for leishmaniasis is highlighted and it provides critical molecular-level insights. The findings suggest that natural products could significantly advance the development of new therapeutic agents against leishmaniasis.

2. Results and Discussion

2.1. Antileishmanial Activity

In this study, the in vitro antileishmanial activity was evaluated of malvidin and echioidinin, two natural compounds, against three *Leishmania* species of *L. amazonensis*, *L. braziliensis*, and *L. infantum*. The experiments assessed the effectiveness against promastigote and axenic amastigotes forms [37]. AmpB, a well-established antileishmanial alternative, was used as positive control (Table 1).

Malvidin demonstrated consistently greater potency than echioidinin across both parasite stages and all three *Leishmania* species tested. The effectiveness of the compounds was measured using IC_{50} data, the concentration required to inhibit 50% of parasite growth, with lower IC_{50} values indicating higher potency. Malvidin exhibited lower IC_{50} values compared to echioidinin in all cases. The IC_{50} values for malvidin were as follows: 197.71 ± 17 μM for stationary promastigotes and 258.07 ± 17 μM for axenic amastigotes in *L. amazonensis*; 164.5 ± 19 μM for stationary promastigotes and 227.89 ± 17 μM for axenic amastigotes in *L. braziliensis*; and 141.26 ± 26 μM for stationary promastigotes and 173.86 ± 19 μM for axenic amastigotes in *L. infantum*. The IC_{50} values for *L. infantum* were slightly lower than the other species.

The IC_{50} values for echioidinin were as follows: 272.99 ± 29 μM for stationary promastigotes and 335.96 ± 19 μM for axenic amastigotes in *L. amazonensis*; 251.18 ± 22 μM for stationary promastigotes and 293.04 ± 20.76 μM for axenic amastigotes in *L. braziliensis*;

and 212.83 ± 16 μM for stationary promastigotes and 261.73 ± 17μM for axenic amastigotes in *L. infantum*.

Table 1. In vitro antileishmanial activity IC_{50}, selectivity index (SI), cytotoxicity (CC_{50}), and hemolytic activity (RBC_{50}) of malvidin, echioidinin, and pure amphotericin B (AmpB). The inhibitory concentrations at 50% observed for *Leishmania* spp. (IC_{50}) and macrophages (CC_{50}) were calculated using a sigmoidal regression to the dose–response curve. The results are presented as the mean ± standard deviation.

Compounds	IC_{50} (μM) Stationary Promastigotes			Selectivity Index (SI) Stationary Promastigotes			CC_{50} (μM)	RBC_{50} (μM)
	L. amazonensis	*L. braziliensis*	*L. infantum*	*L. amazonensis*	*L. braziliensis*	*L. infantum*		
Malvidin	197.71 ± 17	164.5 ± 19	141.26 ± 26	14.70	17.60	20.70	2920.31 ± 80	3120.13 ± 107
Echioidinin	272.99 ± 29	251.18 ± 22	212.83 ± 16	10.80	12.10	14.00	3010.62 ± 114	3400.76 ± 78
AmpB	0.08 ± 0.02	0.09 ± 0.03	0.06 ± 0.01	12.60	11.00	14.70	1.06 ± 0.12	14.28 ± 1
	Axenic amastigotes			Axenic amastigotes				
	L. amazonensis	*L. braziliensis*	*L. infantum*	*L. amazonensis*	*L. braziliensis*	*L. infantum*		
Malvidin	258.07 ± 17	227.89 ± 1	173.86 ± 19	11.30	12.70	16.70		
Echioidinin	335.96 ± 19	293.04 ± 20	261.73 ± 17	8.90	10.30	11.60		
AmpB	0.09 ± 0.02	0.10 ± 0.03	0.08 ± 0.02	11.00	9.80	12.60		

IC_{50}, the concentration needed to inhibit 50% of the parasites' viability; hSI, selectivity index, calculated by the ratio between CC_{50} and IC_{50}; fCC_{50}, the concentration required to inhibit 50% of the macrophages' viability; $gRBC_{50}$, the concentration needed to lysis 50% of the O+ human red blood cells.

The IC_{50} values for AmpB were 0.08 ± 0.02 μM/0.09 ± 0.02 μM, 0.09 ± 0.03 μM/0.10 ± 0.03 μM, and 0.06 ± 0.01 μM/0.08 ± 0.02 μM against *L. amazonensis*, *L. braziliensis*, and *L. infantum* for stationary promastigotes/axenic amastigotes, respectively. Malvidin exhibited superior in vitro efficacy against stationary promastigotes compared to axenic amastigotes. Also, natural anthocyanins like malvidin are known for their diverse biological properties, including antiparasitic activity. Studies have shown that some flavonoids extracted from *Byrsonima coccolobifolia*, such as isoquercetin, catechin, and epicatechin, act as non-competitive inhibitors of ARG [38]. These flavonoids bind to sites other than the active site, disrupting the production of polyamines, which are crucial for parasite survival. Additionally, they may interact with parasite membranes, disrupt essential cellular processes, or impair *Leishmania*'s ability to survive within host cells [39,40].

2.2. Cytotoxicity Assay

The CC_{50} values were 2920.31 ± 80, 3010.62 ± 114, and 1.06 ± 0.1 μM for malvidin, echioidinin, and AmpB, respectively. The CC_{50} value, which indicates the concentration required to kill 50% of the cells, is an important measure of cytotoxicity. A higher CC_{50} value, as seen with malvidin, suggests lower toxicity to healthy cells, whereas a lower CC_{50} value, as seen with AmpB, indicates higher toxicity.

Using these data, the selectivity index (SI) was calculated by the ratio between the CC_{50} and IC_{50} values. For malvidin, the SI values were 14.70/12.60, 17.60/11.00, and 20.70/14.70 against *L. amazonensis*, *L. braziliensis*, and *L. infantum* for stationary promastigotes/axenic amastigotes, respectively. For echioidinin, the SI values were 10.80/8.90, 12.10/10.30, and 14.00/11.60 against *L. amazonensis*, *L. braziliensis*, and *L. infantum* for stationary promastigotes/axenic amastigotes, respectively. For AmpB, the SI values were 11.30/11.00, 12.70/9.80, and 16.70/12.60 against *L. amazonensis*, *L. braziliensis*, and *L. infantum* against the same species in the same stages. An SI value of 10 indicates high selectivity [41], implying that the compound is more toxic to the parasites than normal cells, making it a promising candidate for drug development. Consequently, these results show that malvidin has significant antileishmanial activity with lower mammalian cell toxicity than AmpB, as evidenced by its higher SI values. Among the *Leishmania* species tested, *L. infantum* showed the highest susceptibility to malvidin, reflected by the highest SI values.

These results are comparable to, and in some cases exceed, the selectivity indices reported by other authors for various natural compounds derived from different plant parts and essential oils. Compounds such as oleanic and maslinic acids [42], as well as those from *Connarus suberosus* [43], *Pseudelephantopus spiralis* [44], and certain species from Saudi Arabia [45], Mexico [46], and Brazil [47], have shown SI values ranging from 1 to 10. This suggests that malvidin demonstrates competitive or enhanced selectivity. None of the tested natural products significantly damaged the O^+ human red blood cells (Table 1). The RBC_{50} values were determined to be 14.28 ± 1 μM for AmpB and 3120.13 ± 107 μM for malvidin. Consequently, malvidin exhibited significantly lower hemolytic activity compared to pure AmpB.

2.3. *Treatment of Infected Macrophages*

The efficacy of malvidin and echioidinin against *Leishmania* spp. was evaluated by measuring infection rates, the number of amastigotes per macrophage, and infectiveness reduction (Table 2). These assessments are crucial for determining how effectively these compounds inhibit the parasite's lifecycle and may reveal underlying mechanisms of action. Additionally, this model allows comparative analysis with existing treatments and other experimental compounds. Strong efficacy results could justify further clinical investigations.

Table 2. Treatment of infected macrophages and infection inhibition. The analysis determined the percentage of infection and the number of recovered amastigotes per infected cell after counting 200 macrophages in triplicate.

Compounds	Concentration (μM)	Infection After Treatment (%)	Number of Amastigotes per Macrophage	Infectiveness Reduction (%)
L. amazonensis				
Malvidin	120.74	59.30	3.80	35.75
	60.37	61.10	2.80	33.80
	30.18	75.60	3.30	18.09
	Control	92.30	4.40	0.00
Echioidinin	140.72	63.40	6.30	31.31
	70.36	71.80	3.70	22.21
	35.18	80.20	5.40	13.11
	Control	92.30	4.40	-
AmpB	5.41	24.20	2.80	73.78
	2.71	34.30	2.60	62.84
	Control	92.30	4.40	-
L. braziliensis				
Malvidin	120.74	39.80	3.30	42.07
	60.37	48.70	5.40	29.11
	30.18	52.10	4.40	24.16
	Control	68.70	5.50	-
Echioidinin	140.72	43.20	3.70	37.12
	70.36	50.80	3.70	26.06
	35.18	56.50	6.40	17.76
	Control	68.70	5.50	-
AmpB	5.41	16.50	2.40	75.98
	2.71	24.40	2.60	64.48
	Control	68.70	5.50	-
L. infantum				
Malvidin	120.74	49.80	2.70	33.15
	60.37	55.60	3.70	25.37
	30.18	61.20	4.70	17.85
	Control	74.50	3.70	-
Echioidinin	140.72	50.70	4.20	31.95
	70.36	58.70	2.80	21.21
	35.18	65.50	4.80	12.08
	Control	74.50	3.70	-
AmpB	5.41	19.80	2.50	73.42
	2.71	28.70	3.00	61.48

For *L. amazonensis*, malvidin exhibited a concentration-dependent reduction in infection rates. At the highest concentration (120.74 μM), the infection rate decreased to 59.30%, with the number of amastigotes per macrophage reduced to 3.80, resulting in a 35.75% reduction in infectiveness. Lower concentrations showed diminished efficacy. Echioidinin also demonstrated concentration-dependent effects, with the highest concentration (140.72 μM) resulting in a 63.40% infection rate.

For *L. braziliensis*, malvidin at 120.74 μM reduced infection rates to 39.80%, with 3.30 amastigotes per macrophage, resulting in a 42.07% reduction in infectiveness. Lower concentrations had less impact. Echioidinin at 140.72 μM led to a 43.20% infection rate, with 3.70 amastigotes per macrophage and a 37.12% reduction in infectiveness. Lower concentrations were less effective.

Regarding *L. infantum*, malvidin at 120.74 μM reduced infection rates to 49.80%, with 2.70 amastigotes per macrophage, achieving a 33.15% reduction in infectiveness. Lower concentrations were less effective. Echioidinin at 140.72 μM achieved a 50.70% infection rate and 4.20 amastigotes per macrophage, leading to a 31.95% reduction in infectiveness. AmpB was used as a control, showing a reduction in infectiveness of 73.78%, 75.98%, and 73.41% for *L. amazonensis*, *L. braziliensis*, and *L. infantum*, respectively.

Moreover, the treatment of infected macrophages was assessed to investigate the potential in vitro therapeutic efficacy of this compound against *Leishmania* spp. in mammalian cells. The results demonstrated that malvidin also exhibited a satisfactory selective index compared to the data obtained from using AmpB.

Screening of antileishmanial candidates is usually performed on parasite-stationary promastigote cultures, primarily due to the ease of cultivating the parasites and the resulting yield. However, evaluations using axenic amastigotes were also conducted since this parasite stage directly interacts with the host immune system and is responsible for the active disease [19]. Based on these in vitro results, malvidin emerged as the optimal candidate due to its superior SI and overall performance in key parameters, outperforming echioidinin in terms of selectivity and efficacy. Given this, we prioritized malvidin for further in silico analysis, aiming to explore its molecular interactions in more detail and better understand its potential as an antileishmanial agent.

2.4. In Silico Analysis

2.4.1. Electrostatic Potential Surface and Frontier Orbitals of Malvidin and ABH Compounds

To fully understand the potential of malvidin (Figure 2A) for ARG inhibition, it is essential to explore its electronic properties and binding interactions comprehensively. In this study, the computational methods were leveraged to analyze malvidin's electrostatic potential surface (EPS) and frontier molecular orbitals (FMOs). The EPS map will pinpoint regions on malvidin that could engage with the ARG active site. The FMO analysis, which focuses on the HOMO–LUMO molecular orbitals, will provide insight into its electron-donating and electron-accepting capabilities during binding.

In drug discovery, these computational approaches are important for characterizing the molecular interactions and binding affinities of potential drug candidates. The EPS map identifies regions of a molecule likely to participate in electrostatic interactions with target sites, such as ARG [48]. By highlighting areas with positive or negative charges, it identifies prospective sites for hydrogen bonding or ionic interactions with specific residues in the target's active site. Simultaneously, analyzing the HOMO and LUMO of malvidin helps clarify its electronic behavior during binding, offering critical insights into its molecular interactions with the enzyme's active site [49,50]. These analyses can guide structural modifications to enhance binding efficiency, stability, and selectivity.

We used ABH as a control molecule to assess the specificity of these interactions and gain insights into crucial functionalities for inhibition (Figure 2D). ABH has a well-defined structure with a boronic acid moiety known to participate in reversible covalent interactions with enzymes [51]. However, it lacks the extensive network of conjugated double bonds and aromatic rings present in malvidin. By comparing the electronic properties and docking results of malvidin and ABH, we can differentiate between general electrostatic interactions and specific interactions crucial for ARG inhibition by malvidin. This comparative approach will provide valuable insights for the targeted design of novel ARG inhibitors with improved potency and selectivity.

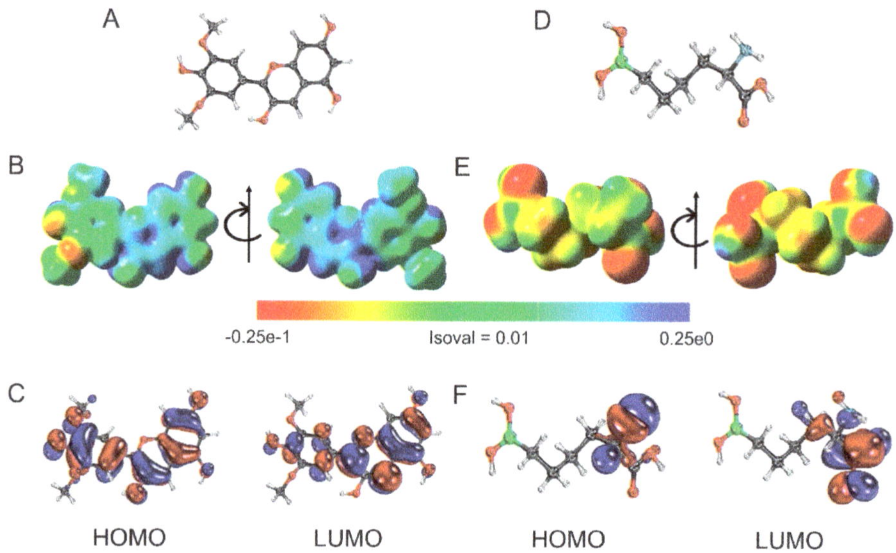

Figure 2. Analysis of ESP and frontier orbitals of malvidin and 2(S)-amino-6-boronohexanoic acid (ABH) compounds. (**A**) Optimized structure of malvidin. (**B**) ESP of malvidin. The range of colors corresponds to that described above. (**C**) HOMO and LUMO of malvidin. (**D**) Optimized structure of ABH. (**E**) ESP of ABH. The range of colors corresponds to that described above. (**F**) HOMO and LUMO of ABH.

The ESP of malvidin (see Figure 2B) reveals a heterogeneous electron density distribution, with high-density regions (red) predominantly located on the lower methoxy group and some hydroxyl groups. The hydroxyl groups on the two conjoined rings show medium electron density (yellow/green), while low-density regions (blue) are near the hydrogen atoms. This distribution indicates that malvidin can form specific protein interactions through hydrogen bonds and electrostatic interactions. High electron density areas can interact with positively charged amino acid residues, while low-density regions can interact with negatively charged residues [52].

The HOMO of malvidin is distributed over aromatic rings and hydroxyl groups, suggesting that these sites are prone to donating electrons and acting as nucleophiles. The LUMO, also localized on the aromatic rings and functional groups, identifies electrophilic regions that can accept electrons (see Figure 2C). These properties suggest that malvidin can stabilize through specific interactions with amino acid residues in proteins, enhancing its affinity for active sites and its role in biological processes.

The ESP of ABH (see Figure 2E) shows a high electron density (red) on the two hydroxyl groups of the boronic acid moiety, in addition to the α-carboxylic acid group. The high electron density regions of the compound are either negatively charged or have permanent dipoles that produce a partial negative charge character (δ^-). These electrostatic effects can enable ABH to engage in hydrogen bonding as well as electrostatic interactions with positively charged residue side chains. The medium electron density region (yellow/green) is spread along the aliphatic chain, indicating a more neutral charge distribution that can interact with hydrophobic regions of proteins. As expected, the low electron density (positive electrical field) (blue) is found in a smaller area of the molecule, such as the α-amino group. This observation suggests this region is more involved in participating in strong electrostatic interactions with negatively charged residue side chains of the enzyme's active site. The HOMO of ABH is primarily localized around the amino group and adjacent carbon atoms, indicating that interacting residue side chains of the enzyme would act as nucleophilic sites capable of donating electrons to the amino group. On the other hand, the LUMO is mainly situated in the carboxyl group, indicating that interacting residue side chains would act as electrophilic sites and be capable of accepting electrons (see Figure 2F). This combination of nucleophilic and electrophilic characteristics allows ABH to interact with various amino acid residues in the active site of arginase, specifically a region well known as the L-α-amino acid recognition site and thus allowing for specific L-amino acid stereochemistry [53,54].

2.4.2. Chemical Reactivity Properties of Malvidin and ABH Compounds

Malvidin and ABH exhibit distinct profiles in global chemical reactivity (Table 3). These key reactivity descriptors suggest distinct behaviors and potential biological impacts [55,56]. With its smaller energy gap and higher chemical potential, malvidin exhibits increased chemical reactivity and a pronounced tendency to attract electrons, facilitating robust interactions with the electron-rich and polar regions of proteins. The greater global softness of this compound indicates a high degree of electron cloud deformability [57], allowing for flexible binding to various protein sites, including those that are sterically hindered. The high electrophilicity index of malvidin suggests that it can readily accept electrons from nucleophilic amino acid residues, forming stable complexes that could significantly affect the activity and function of the protein. Additionally, the substantial dipole moment of malvidin enhances its interactions with polar regions of the protein, potentially leading to strong dipole–dipole interactions that further stabilize binding and influence protein conformation and dynamics [58].

Table 3. Global reactivity of malvidin and ABH.

Compound	HOMO (eV)	LUMO (eV)	ΔE Gap	μ	η	S	χ	ω	$\vec{\mu}$ *
Malvidin	−6.32	−3.47	2.84	−4.89	1.42	0.35	4.89	8.41	5.57
ABH	−7.00	−0.41	6.59	−3.71	3.30	0.15	3.71	2.08	1.69

* $\vec{\mu}$ corresponds to dipole moment (Debye).

Conversely, ABH, characterized by a larger energy gap and lower chemical potential for the mid-section of this chemical structure, demonstrates lower reactivity and a reduced tendency to transfer electrons, resulting in fewer and weaker interactions with proteins. ABH's higher global hardness and lower global softness indicate a less flexible electronic structure, which limits its ability to adapt to various binding sites on the protein, thus reducing the likelihood of strong or specific interactions. The lower electrophilicity index of ABH suggests it is less prone to forming stable complexes with nucleophilic residues, diminishing its potential to alter protein function. Furthermore, the lower dipole moment of ABH indicates reduced polarity, which may limit its interactions with polar and charged regions of the protein, leading to minimal binding affinity and specificity.

Based on our computational analysis, malvidin appears to have high potential for engaging in intermolecular interactions in an enzyme's active site based on its charge character, flexibility, and polarity. These interactions could have biological implications, such as modulating enzyme activity, altering signaling pathways, or affecting protein stability [59–62]. On the other hand, from a computational basis considering electrostatics alone, ABH appears to have higher stability and lower reactivity. These features may suggest that ABH is more modest in forming a protein–ligand complex, thus further suggesting a more inert behavior with minimal biological impact. However, this is indeed not the case with ABH and ARG. The boron chemistry involved with ABH allows for an interesting structure–analog inhibition case in the active site. The boron atom (of boronic acid) has a deficient octet by two electrons, and when ABH binds in the L-α-amino acid recognition site in ARG, the Y-shaped boronic acid moiety (showcasing sp^2 hybridization) is immediately subject to nucleophilic attack by bridging hydroxide stemming off a $Mn^{2+}{}_2$ cluster. A tetrahedral boronate anion (sp^3 hybridization) is produced that resembles the unstable sp^3 hybridized transition state that L-arginine undergoes during catalysis. In the case of ABH, the boronate anion is highly stable and serves as one of the strongest arginase inhibitors known and may not be able to be predicted by a computational assessment.

If we consider a comparative analysis on primary electrostatics between ABH and malvidin, there appears to be more potential for malvidin to serve as a more effective modulator in biochemical applications, while ABH may be more suited for its more specific role requiring chemical transition state chemistry involving the boron atom. From a quantum analysis comparison of ABH and malvidin, malvidin reveals high reactivity and a pronounced tendency to attract electrons, facilitating robust interactions with proteins through hydrogen bonds and electrostatic interactions. Its high electron density areas allow it to interact with positively charged residue side chains, while its electronic flexibility allows greater adaptability to various protein binding sites. In contrast, ABH is characterized by a larger energy gap and lower chemical potential that demonstrates reduced reactivity and a less flexible electronic structure, limiting its specific interactions.

2.4.3. Homology Modeling, Docking, and Molecular Dynamics of Leishmania ARG

The interaction between malvidin and ARG enzymes that catalyze the hydrolysis of arginine was studied in our previous work to understand the potential therapeutic effects of malvidin [29]. MD simulations analyzed the stability of the ARG–malvidin complex during an equilibrium simulation time of 100 ns. Table 4 shows the average root-mean squared deviation (RMSD) values, where an average value of less than 0.3 nm is observed, indicating that the enzyme–ligand complex is in equilibrium. The compactness of the ARGs was analyzed by calculating the radius of gyration (RG) and the solvent-accessible surface area (SASA). The average SASA value of the ARGs remained compact throughout the trajectory, meaning that their secondary and tertiary structures remained stable throughout the simulation without losing protein folding. This information can be observed especially through the values of the protein surface area, which did not show significant variations and remained constant around 333, 345, and 336 nm^2 for *L. amazonensis*, *L. braziliensis*, and *L. infantum*, respectively. The RG values of the enzyme models did not vary in any of the simulations. This analysis allowed us to evaluate the structural dimension of the complexes where these measures remained constant throughout the trajectory in all simulations performed, confirming that the complexes remained compact and protein folding remained stable.

We employed the MM/GBSA method to determine the binding free energy from the trajectories obtained during the last 10 ns of MD simulations. The results, summarized in Table 5, highlight the interaction between ARG and malvidin at different pH levels. At a pH of 5, the total binding free energies (ΔG_{TOTAL}) of malvidin were −9.73, −8.41, and −12.83 kcal/mol for *L. amazonensis*, *L. braziliensis*, and *L. infantum*, respectively. At a pH of 7, the ΔG_{TOTAL} values were −12.82, −12.05, and −10.85 kcal/mol for *L. amazonensis*, *L. braziliensis*, and *L. infantum*, respectively. These binding free energy values indicate a robust binding affinity of malvidin to ARG across the three *Leishmania* species, with slight variations in energy values between the different pH levels.

Table 4. Average root mean square deviation (RMSD), radius of gyration (RG), and solvent accessible surface area (SASA) values obtained from molecular dynamics simulations.

	L. amazonensis	*L. braziliensis*	*L. infantum*
	RMSD (nm)		
pH 5	0.023 ± 0.001	0.023 ± 0.001	0.023 ± 0.001
pH 7	0.023 ± 0.001	0.023 ± 0.001	0.023 ± 0.001
	RG (nm)		
pH 5	3.017 ± 0.001	3.017 ± 0.001	3.017 ± 0.001
pH 7	3.016 ± 0.001	3.017 ± 0.001	3.017 ± 0.001
	SASA (nm^2)		
pH 5	332.244 ± 1.530	344.397 ± 1.517	335.716 ± 1.504
pH 7	331.772 ± 1.494	344.434 ± 1.552	334.429 ± 1.430

Table 5. Binding free energy calculation of malvidin–ARG complexes using molecular mechanics generalized born surface area (MM/GBSA).

Energy *	*L. amazonensis*		*L. braziliensis*		*L. infantum*	
	pH 5	**pH 7**	**pH 5**	**pH 7**	**pH 5**	**pH 7**
ΔE_{vDW}	−29.49	−28.82	−30.78	−34.12	−26.37	−26.96
ΔE_{EL}	148.22	−90.93	134.83	−157.23	177.00	−22.25
ΔE_{GB}	−124.69	110.87	−108.55	183.44	−159.78	42.08
ΔE_{SURF}	−3.77	−3.94	−3.91	−4.14	−3.68	−3.72
ΔG_{GAS}	118.72	−119.74	104.06	−191.35	150.63	−49.21
ΔG_{SOLV}	−128.46	106.92	−112.47	179.30	−163.46	38.36
ΔG_{TOTAL}	−9.73	−12.82	−8.41	−12.05	−12.83	−10.85

* All values are in kcal/mol.

The consistently negative ΔG_{TOTAL} values across all systems and pH conditions suggest that the binding interactions are thermodynamically favorable. The release of energy upon binding indicates a stable complex formation between ARG and malvidin. The absence of significant fluctuations in the binding energies across different species and pH levels reinforces the stability and consistency of these interactions.

The data in Table 5 indicate that at a pH of 5, the most significant energetic contributions arise from the electrostatic contribution to free energy calculated by generalized born (ΔE_{GB}) and polar solvation free energy (ΔG_{SOLV}). Conversely, at a pH of 7, the primary energetic contributions are from electrostatic energy (ΔE_{EL}) and gas-phase free energy (ΔG_{GAS}). These findings suggest that the overall charge distribution of the systems varies with the pH, as illustrated in Figure 2. At a pH of 5, the ΔG_{SOLV} contributes more favorably to the binding free energy than at a pH of 7. In contrast, at a pH of 7, the ΔG_{GAS} contributes more to the binding free energy.

These results underline the pH-dependent nature of binding interactions, where different energetic components dominate under different conditions. At a pH of 5, the solvation effects (ΔG_{SOLV}) are more prominent, suggesting that the solvation environment significantly influences binding. At a pH of 7, the dominance of gas-phase interactions

(ΔG_{GAS}) implies that the intrinsic electrostatic and van der Waals interactions play a more crucial role.

Understanding these pH-dependent energetic contributions is essential for optimizing ligand binding in drug design. It highlights the need to consider environmental factors such as pH when developing inhibitors targeting ARG in *Leishmania* species. Future experimental studies should aim to validate these computational insights and explore the implications for therapeutic efficacy under varying physiological conditions.

Our molecular docking, MD, and energetic contribution study investigated the potential of malvidin to inhibit ARG enzymes from the three *Leishmania* species of *L. amazonensis*, *L. braziliensis*, and *L. infantum*. This represents the first detailed analysis of malvidin's interactions with the active sites of these enzymes at two relevant pH levels (a pH of 5 and a pH of 7). The docking simulations revealed key amino acid residues involved in malvidin's binding, which varied slightly between species and pH levels but consistently interacted with critical functional groups. Additionally, electrostatic potential surface calculations confirmed that protonation states significantly influenced the overall charge of the ARG models. In Figure 3, at a pH of 5 (blue color), ARG exhibited a stronger positive electrostatic potential due to the protonation of basic residues, while at a pH of 7 (red color), a more pronounced negative charge was observed. Despite these differences, malvidin remained bound to the ARG enzyme at both pH levels, demonstrating stable interactions. In this comprehensive study, valuable insights are provided into malvidin's binding mechanisms and its potential as a therapeutic agent against different *Leishmania* species.

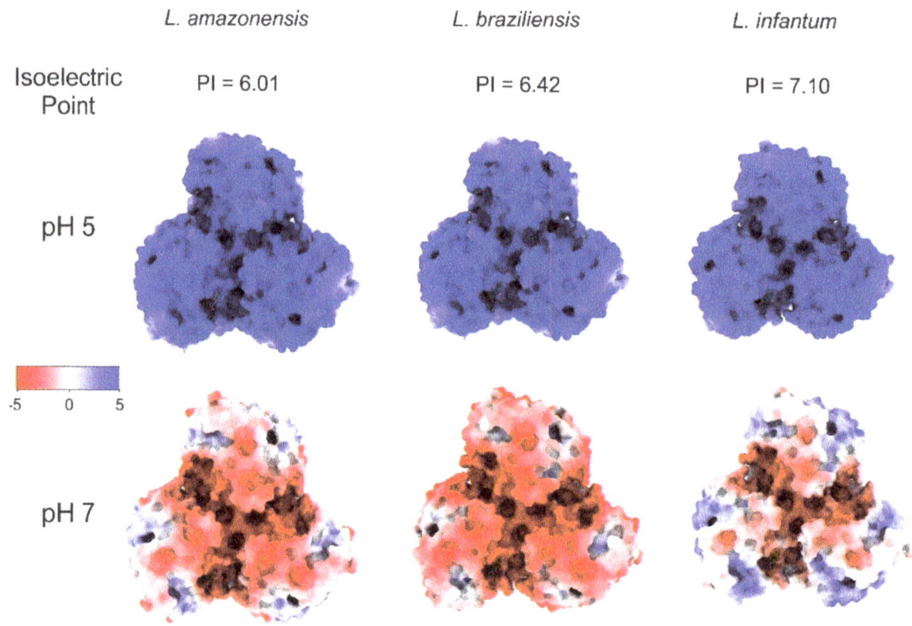

Figure 3. Representation of the electrostatic potential map of arginases (ARGs) for *L. amazonensis* (**left**), *L. braziliensis* (**middle**), and *L. infantum* (**right**) at a pH of 5 and a pH of 7 (in kBT/e). The electrostatic potential maps were calculated using the adaptive Poisson–Boltzmann solver (APBS) [63]. Blue represents positive potential, red represents negative potential, and white indicates neutral regions. Theoretical isoelectric point values were determined by ProtParam (https://web.expasy.org/protparam/ accessed on 15 October 2024), as part of the Expasy server [64].

For *L. amazonensis* (Figure S1), at a pH of 5, malvidin interacted with HIS 139, ASP 141, ASN 143 (carbon hydrogen bond), HIS 154, GLY 155 (carbon hydrogen bond), ALA 192, ASP 194, GLU 197, THR 257 (carbon hydrogen bond), PRO 258 (conventional hydrogen bond), VAL 259, ARG 260, and GLU 288. At a pH of 7, the interactions included HIS 139 (pi-pi stacked and pi alkyl), ASP 141 (carbon–hydrogen bond), ILE 142, ASN 143 (conventional and carbon–hydrogen bonds), SER 150, HIS 154 (pi alkyl and carbon–hydrogen bond), GLY 155 (carbon–hydrogen bond), ALA 192, VAL 193, ASP 194 (carbon–hydrogen bond), GLU 197, THR 257, PRO 258, VAL 259, and ARG 260. Notably, the residues HIS139, ASN143, SER150, GLY155, ALA192, GLU197, THR257, PRO258, and VAL259 are significant, with HIS139, GLY155, and THR257 displaying energy values above −1 kcal/mol.

In the case of *L. braziliensis* (Figure S2), at a pH of 5, the interacting amino acids were HIS 140 (conventional hydrogen bond and pi-pi stacked), ASP 142, ASN 144, HIS 155 (pi-alkyl), GLY 156, ASP 193, GLU 195, GLU 198, THR 258, PRO 259, VAL 260, and ARG 261 (unfavorable donor-donor). At a pH of 7, the interactions included HIS 140 (pi-pi stacked and conventional hydrogen bond), ALA 141, ASP 142 (carbon–hydrogen bond), ILE 143, ASN 144, SER 151, HIS 155 (pi alkyl), GLY 156 (carbon–hydrogen bond), ASP 193 (pi anion), GLU 195, GLU 198 (carbon and conventional hydrogen bonds), THR 248, PRO 259, VAL 260, ARG 261, and GLU 289. The residues HIS140, SER151, GLY156, GLU195, GLU198, THR258, PRO259, and VAL260, with HIS140 and THR258 showing energy values above −1 kcal/mol.

For *L. infantum* (Figure 4), at a pH of 5, the interacting amino acids were HIS 139 (conventional hydrogen bond and pi cation), ALA 140, ASP 141, ASN 143 (conventional hydrogen bond), SER 150, ASN 152, HIS 154 (pi-alkyl), GLY 155, ASP 194 (carbon–hydrogen bond), GLU 197 (conventional hydrogen bond), THR 257 (carbon–hydrogen bond), PRO 258 (conventional hydrogen bond), VAL 259, and ARG 260. At a pH of 7, the interactions included HIS 139 (pi cation), ASP 141 (carbon–hydrogen bond), ILE 142, ASN 143, SER 150, HIS 154 (pi alkyl), GLY 155 (carbon–hydrogen bond), ALA 192, VAL 193, ASP 194 (carbon–hydrogen bond), GLU 197 (conventional hydrogen bond), THR 257 (carbon–hydrogen bond), PRO 258 (conventional hydrogen bond), VAL 259, ARG 260 (conventional hydrogen bond), and GLU 288. The key residues contributing to the binding energy were HIS139, ASN143, SER150, GLY155, VAL193, ASP194, GLU197, THR257, PRO258, and VAL259, with HIS139 and PRO258 exhibiting the highest contributions (>−1 kcal/mol).

These findings highlight the specific amino acids that play significant roles in interacting with malvidin across different *Leishmania* species. The residues HIS139 and PRO258 were consistently significant in *L. amazonensis* and *L. infantum*, while HIS140 and THR258 were prominent in *L. braziliensis* (Figure S3).

These findings suggest that malvidin may inhibit *Leishmania* ARG activity by interacting with critical amino acid residues. Notably, HIS 139, a critical binding site for malvidin across all *Leishmania* species, is conserved in ARG family enzymes involved in the coordination of the binuclear manganese cluster (Mn^{2+}-Mn^{2+}) in the active site [65,66]. Blocking this metal coordination with other molecules has demonstrated the inhibition of ARG activity. Therefore, malvidin's interaction with HIS 139 is likely a pivotal mechanism for its inhibitory effect [67]. This is supported by other studies, such as involving fisetin, which also showed enhanced stability through similar interactions [68]. Another significant interaction involves HIS 154, which directly binds the L-arginine substrate in some *Leishmania* ARGs. Malvidin's interaction with HIS 154 could potentially hinder substrate binding, contributing to ARG inhibition [69]. Additionally, malvidin interacts with ALA 192, a residue previously targeted by other ARG inhibitors like phenylacetamide, ABH, and caryophyllene oxide. Since ALA 192 is not conserved in human ARG, this suggests a potential shared inhibitory mechanism between malvidin and these compounds [70–72]. Furthermore, some flavonoids, such as epicatechin, catechin, and gallic acid, have also been tested against ARG. Their interactions with additional residues, such as ASP 141, GLY 155, ASP 181, ASP 194, and THR 257, were observed [73]. Similar compounds, including polyphenols such as caffeic and rosmarinic acids, also interacted with HIS 139, ASP 194, and SER 150 [67].

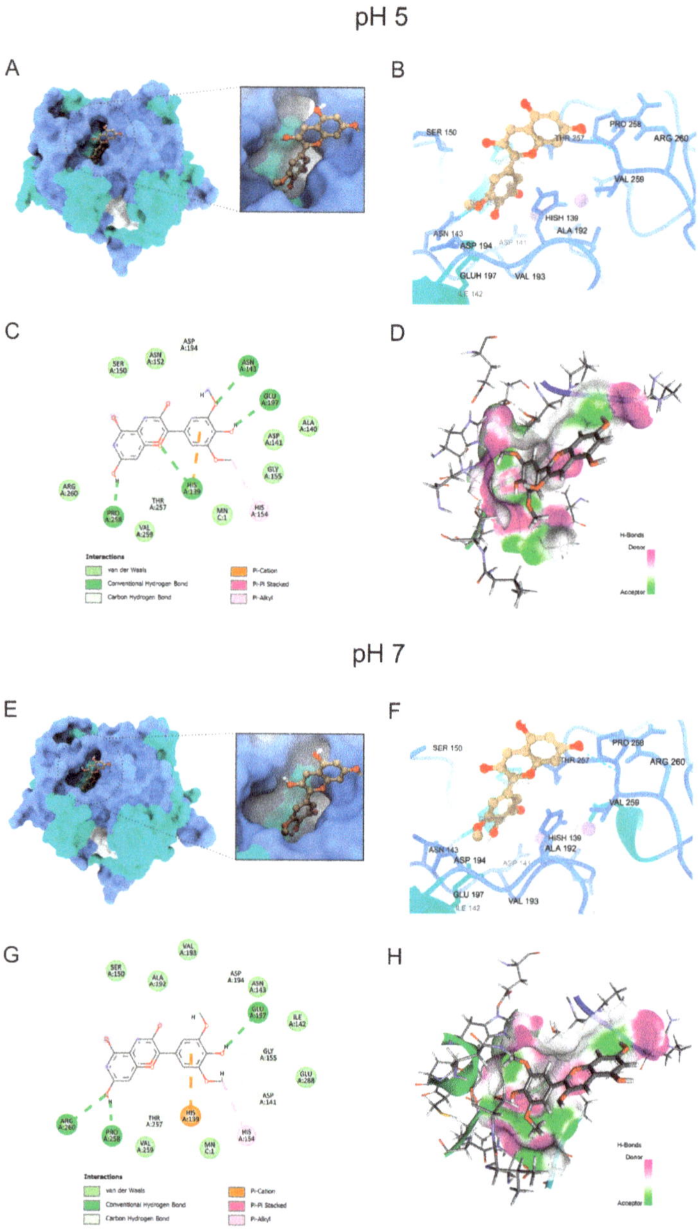

Figure 4. Docking of malvidin with arginase (ARG) from *Leishmania infantum* at a pH of 5. Panels (**A–D**) depict the 3-dimensional surface representation, the binding mode and molecular interactions of the interacting ligands, 2-dimensional view of ligand interactions with arginase residues, and hydrogen bonding interactions surface, respectively. Panels (**E–H**) reveal the corresponding docking analyses at a pH of 7.

These in silico findings provide a promising foundation for further research. In vitro and in vivo studies are necessary to confirm malvidin's ability to inhibit *Leishmania* ARG and evaluate its efficacy against the parasite. Additionally, exploring the structure–activity relationship of malvidin and its derivatives could lead to developing more potent ARG inhibitors for treating Leishmaniasis.

Network Analysis of Protein–Protein Interactions and Single-Cell RNAseq Analysis in Key Organ Systems for Potential Toxicity

The integration of the network analysis of protein–protein interactions with single-cell RNA sequencing (scRNAseq) is transforming our understanding of drug-induced toxicity across key organ systems. Network analysis offers a system-level view of complex protein interactions within cellular pathways [74]. This approach enables the identification of critical nodes and pathways that may serve as potential drug targets or toxicity markers, particularly those implicated in processes like inflammation, oxidative stress, and cellular metabolism, which are often linked to adverse effects. Meanwhile, scRNAseq analysis offers a high-resolution view of gene expression at the cellular level, uncovering cellular heterogeneity and specific responses to toxic agents that are often obscured in bulk tissue studies [75]. By integrating these methods, researchers can link alterations in protein interactions with gene expression changes at the cellular level, offering a comprehensive understanding of how toxicity develops across different organ systems. This combined approach improves the prediction of early toxicity prediction—a major challenge in pharmaceutical research. This method highlights both established and novel drug-pathway connections, leveraging big data to improve drug safety and guide the development of therapies with a reduced risk of adverse effects in clinical trials [76,77]. This integrative approach is illustrated in the network visualizations shown in Figure 5A which depict the intricate interplay between proteins within the specific tissues of the brain, heart, kidney, and liver. Each graph represents a tissue-specific protein interaction network, emphasizing key tissue-specific toxicity markers (represented as circles) and predicted targets (represented as squares) based on their centrality within the network. The color intensity scale (yellow to dark blue) indicates the relative importance of each node in terms of its role and action within the network.

In the brain network, critical markers are involved in neuronal function and implicated in neurodegenerative diseases. These include acetylcholinesterase (ACHE), apolipoprotein E (APOE), amyloid precursor protein (APP), brain-derived neurotrophic factor (BDNF), cyclin-dependent kinase 5 (CDK5), Huntingtin (HTT), nerve growth factor (NGF), Parkinson protein 7 (PARK7), prion protein (PRNP), and presenilin 1 (PSEN1). The intricate interactions among these markers influence cell survival and contribute to the complex pathology of neurodegenerative diseases such as Parkinson's, Alzheimer's, Huntington's, and Creutzfeldt-Jakob disease [78]. The heart network, on the other hand, features essential markers for maintaining cardiac function. Markers such as cysteine and glycine-rich protein 3 (CSRP3), myosin-binding protein C (MYBPC), telethonin (TCAP), tropomyosin 1 (TPM1), and titin (TTN) play crucial roles in maintaining cellular homeostasis within the heart muscle. Dysregulation of these genes can lead to various cardiac disorders, emphasizing their importance in negatively regulating cell death pathways [79].

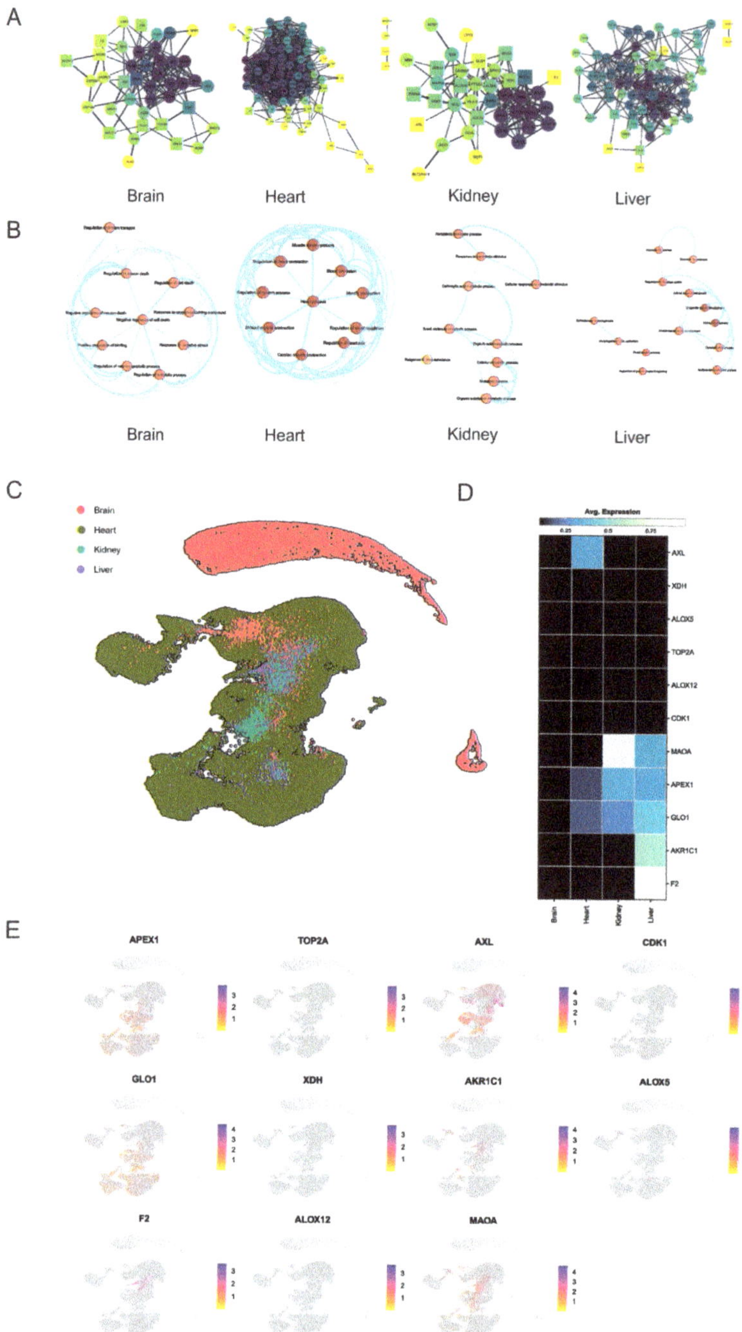

Figure 5. Single-cell RNA sequencing and comprehensive gene expression analysis across the brain, heart, kidney, and liver. (**A**) Network representations of gene interactions in the brain, heart, kidney, and liver, highlighting distinct patterns of connectivity and interaction hubs in each organ. Darker color represents maximal clique centrality (MCC). (**B**) Pathway analysis illustrating the major biological processes and pathways enriched in each organ, with significant nodes and connections

visualized. (**C**) UMAP plot showing the clustering of gene expression data from the brain, heart, kidney, and liver, with each color representing a different organ. (**D**) Heatmap of average expression levels for key genes across the brain, heart, kidney, and liver, indicating organ-specific expression patterns. (**E**) Spatial expression maps for selected genes (APEX1, TOP2A, AXL, GLO1, XDH, AKR1C1, ALOX5, F2, ALOX12, MAOA, CDK1) across the different organs, with color intensity representing expression levels.

The liver network emphasizes markers responsible for critical cellular defense mechanisms, neurotransmitter metabolism, and steroid hormone processing. Genes such as ATP-binding cassette subfamily C member 2 (ABCC2), ATP-binding cassette subfamily G member 2 (ABCG2), cytochrome P450 1A1 (CYP1A1), epoxide hydrolase 1 (EPHX1), UDP glucuronosyltransferase 1 family, polypeptide A10 (UGT1A10), UDP glucuronosyltransferase 1 family, polypeptide A3 (UGT1A3), and UDP glucuronosyltransferase 1 family, polypeptide A7 (UGT1A7) collectively contribute to liver detoxification, homeostasis, and metabolic regulation [80]. The kidney network highlights markers crucial for kidney development and function. Markers such as ataxia telangiectasia mutated (ATM), GATA binding protein 3 (GATA3), renin (REN), Ret proto-oncogene (RET), and tuberous sclerosis complex 2 (TSC2) play essential roles. Dysregulation of these genes can lead to structural abnormalities, impaired renal function, and electrolyte imbalances. Understanding their roles is vital for managing kidney health and developing targeted therapies [81,82].

In this context, several predicted malvidin targets were identified as central nodes within the protein interaction networks across key organ systems. These central nodes, due to their critical roles in maintaining the structural and functional integrity of key pathways, did not exhibit strong associations with potential toxic effects. Their centrality likely contributes to a stabilizing effect, where malvidin-induced perturbations are absorbed or buffered by the system, minimizing disruptions that could lead to toxicity. This observation highlights the significance of network topology when evaluating drug safety, as central proteins may play essential roles in cellular homeostasis without directly contributing to toxicity. These findings demonstrate the significance of integrating network analysis and scRNAseq to offer a comprehensive, system-level comprehension of drug-induced toxicity, hence guiding the development of safer therapies with reduced risks of side effects.

2.4.4. Pathway Analysis of Key Genes in Different Tissues

Figure 5B illustrates the functional pathways associated with key proteins in each tissue type. The brain's pathways involve proteins like monoamine oxidase A (MAOA), cyclin-dependent kinase 1 (CDK1), and xanthine dehydrogenase (XDH). These proteins play crucial roles in regulating cellular processes, impacting cell survival, development, and overall health. MAOA influences neurotransmitter levels [83], CDK1 orchestrates critical cell cycle events [84], and XDH generates reactive oxygen species [85]. The dysregulation of these proteins can lead to structural abnormalities, impaired function, and pathogenesis. Heart-specific pathways include interactions between thrombin (F2), AXL receptor tyrosine kinase, CDK1, endothelin-1 (EDN1), and XDH. These proteins play crucial roles in regulating the cellular processes within the heart. F2 plays a dual role in blood clotting and anticoagulation [86]. AXL influences cell survival and resistance to chemotherapy [87]. CDK1 orchestrates critical cell cycle events, maintaining genome stability [88]. EDN1 impacts vasoconstriction, inflammation, and tissue remodeling [89]. XDH generates oxidative stress and cellular health [90]. Kidney pathways feature proteins such as AXL, DNA topoisomerase 2 Alpha (TOP2A), BRCA1-associated protein 1 (BAP1), XDH, and F2. These proteins play crucial roles in regulating cellular processes within the kidney. AXL influences cell survival and tissue repair [91], TOP2A maintains genome stability [92], BAP1 is associated with cancer predisposition [93], XDH impacts purine metabolism and oxidative stress [94], and F2 plays a dual role in blood clotting and anticoagulation [95]. Liver pathways involve proteins like XDHs [96], aldo-keto reductase family 1 member C1 (AKR1C1) [97], and arachidonate 5-lipoxygenase (ALOX5) [98]. XL influences cell survival and tissue repair, TOP2A maintains genome stability [99], BAP1 is associated with cancer predisposition [100], XDH impacts purine metabolism and oxidative stress, and F2 plays a dual role in blood clotting and anticoagulation, highlighting the liver's role in detoxification and metabolic regulation [101].

The findings suggest that although the identified marker targets for malvidin are central within their respective networks, they are not involved in processes critical for cellular function and homeostasis that could lead to toxicity if interrupted. Their centrality indicates that while these targets play important roles in their networks, their interaction with malvidin is unlikely to disrupt critical biological pathways or cause adverse effects and toxicity.

2.4.5. Heatmap Analysis of Key Differentially Expressed Genes

The uniform manifold approximation and projection (UMAP) plot (Figure 5C) demonstrates distinct clustering of brain (red), heart (green), kidney (blue), and liver (purple) tissues based on their gene expression profiles. Noticeably, shifts in the clustering patterns of liver and kidney tissues suggest significant tissue-specific effects.

The heatmap (Figure 5D) displays the average expression levels of selected genes across different tissues following malvidin treatment. Interestingly, no significant overexpression of genes was observed in the brain. In the heart, malvidin treatment led to increased expression of AXL, a receptor tyrosine kinase known to be involved in cell survival, proliferation, migration, and angiogenesis [91]. However, AXL is often overexpressed in cancer cells and associated with poor prognosis [102]. Further investigation is needed to understand the specific role of AXL upregulation in the heart following malvidin treatment.

The kidney exhibited an increased expression of MAOA following malvidin treatment. Although primarily found in the brain, MAOA is also expressed at low levels in other tissues, including the kidneys. Studies have shown an association between lower levels of MAOA expression in the kidneys and chronic kidney disease (CKD) [103]. Increased MAOA expression in response to malvidin may offer potential benefits for kidney health, but further research is warranted.

In the liver, malvidin treatment resulted in the altered expressions of several genes, including MAOA, F2, and AKR1C1. F2, which can be interpreted in two ways within the context of the liver, plays a dual role in blood clotting. The liver produces F2 and its expression is monitored to assess the post-transplantation functionality of donated livers [104]. However, a higher level of F2 can also indicate stage 2 fibrosis, a condition characterized by scar tissue development in various liver diseases [105]. In this case, F2 likely reflects the state of the liver tissue rather than direct gene expression changes. AKR1C1 encodes an enzyme involved in the body's metabolic processes [106]. The research suggests that AKR1C1 might be overexpressed in liver cancer compared to healthy tissue, potentially serving as a disease marker or even contributing to its development [107].

Two genes, GLO1 and APEX1, were notably expressed across the heart, kidney, and liver. GLO1 acts as a detoxifying enzyme, specifically targeting a harmful molecule called methylglyoxal produced during normal cellular processes [108]. By removing methylglyoxal, GLO1 protects cells from damage and contributes to overall cellular health. APEX1, on the other hand, plays a critical role in DNA repair by removing damaged building blocks from DNA, allowing proper repair mechanisms to function [109]. Damaged DNA can lead to mutations and cell death, highlighting the importance of APEX1 in maintaining cellular health [110]. Similar to GLO1, APEX1 is expressed in various tissues and offers specific benefits to each organ. In the heart, APEX1 protects heart muscle cells from DNA damage linked to heart disease [111]. Similarly, it helps maintain kidney DNA integrity and potentially prevents kidney disease. Finally, in the liver, APEX1 is involved in DNA repair processes essential for detoxification and regeneration functions [111].

The combined analysis of the UMAP plot and heatmap provides valuable insights into the tissue-specific effects of malvidin treatment. The observed changes in gene expression patterns suggest potential benefits for various organs, including the reduction in oxidative stress in the liver and the upregulation of DNA repair mechanisms in the heart, kidney, and liver. However, further research is needed to fully understand the underlying mechanisms and potential therapeutic applications of malvidin.

It is important to acknowledge that UMAP, while a powerful tool for visualizing high-dimensional data, introduces some degree of distortion into the underlying data structure. Therefore, further bioinformatic analysis methods, such as principal component analysis (PCA) or t-distributed stochastic neighbor embedding (t-SNE), could better understand gene expression patterns across these tissues. These UMAP visualizations provide a valuable starting point for dissecting tissue-specific gene expression patterns. These findings can guide future studies utilizing techniques like RNA-seq and quantitative PCR (qPCR) to validate candidate genes and explore their functional roles in distinct organ systems [112]. Understanding these complex transcriptional landscapes is crucial for advancing our knowledge of organ development, physiology, and disease pathogenesis.

In the bottom row of Figure 5E, UMAP visualizations are presented for gene expression data from the brain (red), heart (green), kidney (blue), and liver (purple) tissues. UMAP, a dimensionality reduction technique, projects high-dimensional gene expression data into a two-dimensional space, allowing for the visualization of differentially expressed genes across these tissues. The UMAP plots demonstrate clear segregation between tissue types, suggesting distinct transcriptional landscapes across organs. This separation likely reflects the expression of genes critical for each tissue's specialized functions.

The heart UMAP might group genes encoding proteins involved in sarcomere function, electrical conduction, and cardiac rhythm regulation. Examples include genes for sarcomere function like myosin heavy chain 6 (MYH6) [113], alpha skeletal muscle actin (ACTA1) [114], troponin I type 2 (TNNI2) [115]; electrical conduction genes like sodium voltage-gated channel alpha subunit 5 (SCN5A) [116], potassium voltage-gated channel subunit (KCNQ1) [117], gap junction protein, alpha 1 (GJA1) [118]; and cardiac rhythm regulation genes like potassium channel, inwardly rectifying subfamily J member 11 (KCNJ11) [119], calcium channel, voltage-dependent, L-type, and alpha 1C subunit (CACNA1C) [120].

Kidney tissue may show enrichment for genes related to solute transport, electrolyte homeostasis, and waste product excretion. Examples include genes for solute transport like solute carrier family 2 (facilitated transporter) [121], aquaporin 1 (AQP1) [122], sodium–potassium–chloride cotransporter 1 (SLC12A1) [123]; electrolyte homeostasis genes like sodium–potassium-ATPase subunit alpha 1 (ATP1A1) [124], aquaporin 2 (AQP2) [125], chloride channel accessory 1 (CLCA1) [126]; and waste product excretion genes like solute carrier family 23 member 1 (SLC23A1) [127], multidrug resistance protein 1 (MDR1) [128], and uromodulin (UMOD) [129].

The liver UMAP may cluster genes implicated in xenobiotic detoxification, metabolic pathways, and bile acid synthesis. Examples include genes for xenobiotic detoxification like cytochrome P450 family 3 subfamily a member 4 (CYP3A4) [130], glutathione S-Transferase alpha 1 (GSTA1) [131], UDP glucuronosyltransferase 1 family member A1 (UGT1A1) [132]; metabolic pathway genes like glucokinase (GCK), fatty acid synthase (FASN), carnitine palmitoyltransferase 1A (CPT1A) [133]; and bile acid synthesis genes like cholesterol

7A-Hydroxylase (CYP7A1) [134], bile acid coenzyme A: amino acid N-Acetyltransferase (BAAT) [135], and sterol 12α-hydroxylase (CYP8B1) [136].

Further in-depth analysis of differentially expressed genes within each UMAP plot is warranted to identify tissue-specific markers and elucidate the underlying molecular mechanisms governing organ-specific functionalities. Functional enrichment analysis and gene ontology term association can be employed to identify biological processes and pathways significantly associated with each tissue cluster [137,138]. Additionally, co-expression network analysis can reveal potential interactions and regulatory relationships between differentially expressed genes within each tissue type. These UMAP visualizations provide a valuable starting point for dissecting tissue-specific gene expression patterns. These findings can guide future studies to unravel the complex interplay between genes and their functional roles in distinct organ systems. Understanding these complex transcriptional landscapes is crucial for advancing our knowledge of organ development, physiology, and disease pathogenesis.

3. Materials and Methods

3.1. Parasites

L. amazonensis (IFLA/BR/1967/PH-8), *L. braziliensis* (MHOM/BR/1975/M2903), and *L. infantum* (MHOM/BR/1970/BH46) were used as parasites. Stationary promastigotes were grown at a 24 °C in complete Schneider's medium (Sigma-Aldrich, St. Louis, MO, USA), which was composed of medium plus 20% (*v*/*v*) heat-inactivated fetal bovine serum (FBS; Sigma-Aldrich, USA) and 20 mM L-glutamine with a pH of 7.4 [139]. To obtain the axenic amastigotes, 10^9 stationary promastigotes were washed three times in sterile phosphate-buffered saline (PBS 1X and incubated in 5 mL FBS for 48 h (*L. amazonensis* and *L. braziliensis*) and 72 h (*L. infantum*), respectively, at 37 °C. Parasites were washed in cold PBS 1x, and their morphology was evaluated after staining using the Giemsa method in an optical microscope, as previously described [140].

3.2. Antileishmanial Activity

L. infantum IC_{50} was evaluated by incubating parasite stationary promastigotes or axenic amastigotes (10^6 cells, each) with malvidin (MolPort catalogue N° AA00E9KS, Latvia) (0 to 150.92 μM) or echioidinin (MolPort catalogue N°AK-693/21087015, Latvia) (0 to 178.3 μM) in 96-well culture plates (Nunc, Nunclon, Roskilde, Denmark) for 48 h at 24 °C. AmpB (0 to 10.8 μM; Sigma-Aldrich, USA) was used as control. Cell viability was assessed through 3-(4.5-dimethylthiazol-2-yl)-2.5-diphenyl tetrazolium bromide (Sigma-Aldrich, USA) method. Optical density (OD) values were measured in a microplate spectrophotometer (Molecular Devices, Spectra Max Plus, San Jose, CA, USA) at 570 nm. IC_{50} values were calculated by sigmoidal regression of dose–response curves in Microsoft Excel software (version 10.0) [141]. Two-tailed Student's *t*-tests ($p < 0.05$) were used to compare IC50 values between malvidin and echioidinin, assessing significant differences in inhibitory activity.

3.3. Cytotoxicity Assay

Cytotoxicity was evaluated ex vivo in murine macrophages and human red blood cells, for which concentrations inhibiting 50% of macrophages (CC_{50}) and red blood cells (RBC_{50}) were determined. Briefly, murine cells (5×10^5) or a 5% human red blood cells suspension were incubated in the presence of malvidin (0 to 301.84 μM) or echioidinin (0 to 356.76 μM) AmpB (0 to 10.82 μM) was used as control. Cells were incubated in RPMI 1640 medium for 48 h (murine macrophages) or 1 h (red blood cells) at 37 °C in 5% CO_2. Macrophage viability was assessed using the MTT method. The red blood cell suspension was centrifuged at 1000× *g* for 10 min at 4 °C, after which the percentage of cell lysis was evaluated spectrophotometrically at 570 nm. The hemolysis was determined by replacing the natural products with an equal volume of phosphate-buffered saline with a pH of 7.4 (PBS—negative control) or distilled water (positive control) [141]. CC_{50} and RBC_{50} values were calculated by sigmoidal regression of dose–response curves in Microsoft Excel software (version 10.0). SI was calculated as the ratio between CC_{50} and IC_{50} values.

3.4. *Treatment of Infected Macrophages*

To evaluate the efficacy of malvidin or echioidinin for treating infected macrophages, murine cells (5×10^5) were plated on round glass coverslips in 24-well plates in RPMI 1640 medium supplemented with 20% (*v*/*v*) FBS and 20 mM L-glutamine at a pH of 7.4 and incubated for 24 h at 37 °C in 5% CO_2. Stationary promastigotes were added to the wells at a ratio of 10 parasites per macrophage, and cultures were incubated further for 48 h at 37 °C for 5% CO_2. Free parasites were removed by extensive washing with RPMI 1640 medium, and infected macrophages were treated with malvidin (30.18, 60.37, and 120.74 μM), echioidinin (35.18, 70.36, and 140.72 μM), or AmpB (2.70 and 5.41 μM) for 48 h at 24 °C in 5% CO_2. After fixation with 4% paraformaldehyde, cells were washed and stained with Giemsa, and the infection percentage, the number of amastigotes per infected macrophage, and the reduction in the infection percentage were determined by counting 200 cells in triplicate, using an optical microscope [141].

3.5. *In Silico Analysis*

3.5.1. Molecular Geometry Optimization

Natural products and 2(S)-amino-6-boronohexanoic acid (ABH, CID 9793992) SDF files were selected from the PubChem database [142]. ABH was used as a control because it is known to have an inhibitory effect on arginase [36], allowing for a comparison of its effects. The molecular geometries of these compounds were optimized using the B3LYP/6-311++g(d,p) basis set, incorporating implicit water solvent effects (CPCM(WATER)) [143,144], as implemented in ORCA software version 5.0 [145,146]. Vibrational analyses were performed to ensure accurate ground-state geometries and confirm the absence of imaginary frequencies. Atomic charges were calculated using the Hirshfeld population analysis to account for the electrostatic effects of the ligands within the protein complexes [147]. These charges were then used to reparametrize the AMBER-ILDF force field, which was subsequently used for molecular dynamics simulations.

3.5.2. Global Reactivity Descriptors

Global reactivity descriptors were calculated using conceptual density functional theory (DFT) [148,149], which explains the relationships among structure, stability, and reactivity [150]. These descriptors were derived using Koopmans' theorem which evaluates the energies of the highest occupied molecular orbital (HOMO) and lowest unoccupied molecular orbital (LUMO) obtained from simple-point energy calculations using B3LYP/6-311++g(d,p) basis set [151,152] (see Table 6). This analysis provided insights into how the molecules might interact with other molecules.

Table 6. Global reactivity descriptors.

Reactivity Descriptors	Equation	References
Chemical Potential (μ)	$\mu = \frac{1}{2}(\varepsilon_L + \varepsilon_H)$	[153,154]
Global Hardness (η)	$\eta = \frac{1}{2}(\varepsilon_L - \varepsilon_H)$	[153,154]
Global Softness (*S*)	$S = \frac{1}{\eta}$	[153]

Table 6. *Cont.*

Reactivity Descriptors	Equation	References
Electronegativity (χ)	$\chi = -$	[153]
Electrophilicity (ω)	$\omega = \frac{\mu^2}{2\eta}$	[153,155]
HOMO–LUMO gap (ΔE gap)	$\Delta Egap = \varepsilon_L - \varepsilon_H$	

3.5.3. Homology Modeling, Docking, and Molecular Dynamics of Leishmania ARG

Homology modeling was performed to generate 3D models of ARG from protein sequences (https://www.uniprot.org/, accessed on 15 May 2024) of *L. amazonensis* (AAC95287.1), *L. braziliensis* (KAI5689219.1), and *Leishmania infantum* (CAC9543944.1), using the Swiss-Model server (https://swissmodel.expasy.org, accessed on 10 August 2023) with ARG from *L. mexicana* (PDB code: 4ITY) as the template [72]. Seven high-resolution X-ray crystal structures of ARG from *L. mexicana* (1.95 Å or better) were also referenced, including the inhibitorless form (PDB 4ITY), complexes with competitive inhibitors (PDB 4IU0, 4IU1, 4IU4, 5HJ9, 5HJA), and with the reaction product L-ornithine (PDB 4IU5) [54,71]. These structures provided insights into the active site, $Mn^{2+}{}_2$-binuclear cluster, metal ion coordination, and zones of surface electrostatic charge. Structural differences were also compared against human arginase I to gain further insight into developing Leishmania ARG-selective inhibitors [54]. Sequence alignments using BLAST (https://blast.ncbi.nlm.nih.gov/, accessed on 27 December 2024) revealed that ARG from *L. amazonensis*, *L. infantum*, and *L. braziliensis* differed by 2, 13, and 40 residues, respectively, compared to *L. mexicana* ARG (GenBank accession no. AAR06176.1). The Amber topology file for malvidin was generated with the Acpype server (https://www.bio2byte.be/acpype/, accessed on 21 August 2023) with a total charge of +1. High-resolution crystallographic data and sequence comparisons were used to strengthen computational docking predictions, particularly for species with high sequence similarity [156,157].

Molecular docking was performed using the DockThor server (https://www.dockthor.lncc.br/v2/, accessed on 7 September 2023), where grids were set in the active site of the chain A of each protein. Complexes with the best scores were selected for further analysis. Protonation states of ionizable residues were determined using the PROPKA v. 3 program [158,159]. The summary of pKa values was used to calculate the protonation states of Asp, Glu, Arg, Lys, and His residues, as well as the C-terminal and N-terminal ends. The semi-grand canonical Monte Carlo (SGCMC) procedure was used to calculate the different protonation states at a pH of 5 and a pH of 7, based on the free energy associated with the pKa of each titratable residue. This was calculated using the program developed by Barazorda et al., 2021, accessible online at https://github.com/smadurga/Protein-Protonation accessed on 10 September 2023 [160,161].

After the protein preparation, MD simulations were used to research the ARG-malvidin complexes. Simulations were conducted using the AMBER-ILDF force field in GROMACS v. 2023.4 software [162]. The systems were then solvated with the TIP3P water model, and sodium (Na^{+1}) or chloride (Cl^{-1}) ions were added to neutralize the system. The simulation box size was set to 12 Å × 12 Å × 12 Å. Before running the MD simulations, an energy minimization step was performed using the steep-descent algorithm for 20,000 steps. The MD simulation itself was conducted in the following two distinct stages: the first stage was in the canonical NVT ensemble considering position restraint of the backbone and the coordination of the metal in the active site with distance restraint, with a trajectory of 10 ns. The second stage was the production dynamics in the isothermal–isobaric NPT ensemble with a simulation time of 100 ns. The V-rescale thermostat maintained the temperature at a physiological value of 300 K, and the Parrinello–Rahman barostat maintained pressure at 1 bar, reflecting typical cellular pressure conditions.

Finally, the binding free energy estimation between ARG and malvidin was calculated using the molecular mechanics generalized born surface area (MM/GBSA) method. This calculation employed data from GROMACS simulation and was performed using a tool based on AMBER's MMPBSA.py script (gmx_MMPBSA v1.6.2 based on MMPBSA version 16.0 and AmberTools 20).

3.5.4. Prediction of Natural Product Targets and Retrieval of Tissue Toxicity Markers

The natural product's simplified molecular-input line-entry system (SMILES) codes were searched and retrieved from the PubChem server (https://pubchem.ncbi.nlm.nih.gov/, accessed on 21 March 2024) [163] and uploaded to the SwissTarget Prediction server (http://www.swisstargetprediction.ch/, accessed on 21 March 2024) [164], and to the Superpred server (https://prediction.charite.de/, accessed on 21 March 2024) [165], for target prediction on *Homo sapiens* species. To identify toxicity markers in the brain, heart, kidney, and liver tissues, the GeneCards database (https://www.genecards.org/, accessed on 21 March 2024) [166] was used, whereas the related protein/gene targets were retrieved using the keywords of "neurotoxicity", "cardiotoxicity", "nephrotoxicity", and "hepatotoxicity", respectively. The standard of proteins and gene names were obtained from the UniProt database (https://www.uniprot.org/, accessed on 15 April 2024) [167], and Venn diagrams were used to identify duplicate entries using the Draw Venn Diagram tool (http://bioinformatics.psb.ugent.be/webtools/Venn/, accessed on 15 April 2024).

3.5.5. Protein–Protein Interaction Network Analysis

The predicted targets and tissue toxicity-associated markers were used to construct protein–protein interaction (PPI) networks. The molecular networks were retrieved from the STRING database (https://string-db.org, accessed on 20 April 2024) [168] and analyzed using the Cytoscape platform v 3.10.3 [169]. The cytoHubba v 0.1plugin was used to score and rank network nodes based on the maximal clique centrality (MCC) topological analysis method [170]. Cytoscape default settings were considered to visualize the network, whereas node type and color were manually adjusted, considering the scores provided by the MCC analysis and the type of marker. Finally, the plugin StringApp v 2.2.0 [171] was used to retrieve functional enrichment for gene ontology (GO) terms regarding biological processes (BPs), molecular functions (MFs), and cellular components (CCs), with results displayed in circular radial layouts.

3.5.6. Single-Cell RNA Sequencing Analysis

The gene raw counts or normalized gene expression matrix of single-cell gene expression data of 04 relatively normal tissues were downloaded from the Gene Expression Omnibus (GEO) (https://www.ncbi.nlm.nih.gov/geo/, accessed on 30 October 2024). Kidney, GEO Accession No. GSE131685, from 3 normal controls [172] and liver, GEO Accession No. GSE115469, from 5 healthy patients [173]. The heart data are available on the Heart Cell Atlas database (https://www.heartcellatlas.org/) [174], 1 region was selected for download: the heart vascular datasets from 14 healthy donors. The brain data were searched at the scREAD database (https://bmblx.bmi.osumc.edu/scread/, accessed on 30 October 2024) [175], where the entorhinal cortex, GEO Accession No. GSE138852, from 4 healthy individuals [176]; the superior frontal gyrus, GEO Accession No. GSE147528, from 10 healthy individuals [177]; the prefrontal cortex, GEO Accession No. GSE129308, from 2 healthy controls [178], and the superior parietal lobe, GEO Accession No. GSE146639, from 5 non-demented elderly controls [179], were selected to constitute the organ dataset.

The data were loaded into R 4.1.0 and were preprocessed using standard parameters of the R packages 'Seurat' v.4.0.3 [180]. The expression matrix for each dataset was merged into one Seurat object using the "CreateSeuratObject" function. At the same time, cells with less than 100 expressed genes and higher than 25% mitochondrial genome transcript and genes expressed in less than three cells were removed. The gene expression data were normalized using the "NormalizeData" function, and the sources of cell–cell variation driven by batch were reverted out, using the number of detected UMI and mitochondrial gene expression by the "ScaleData" function. Highly variable genes were identified by the "FindVariableGenes" function and used for the principal component analysis (PCA) on the highly variant genes using the "RunPCA" function. The "JackStraw" function was implemented to remove the signal-to-noise ratio. Cells were then clustered utilizing the "FindClusters" function by embedding cells into a graph structure in PCA space. The clustered cells were then projected onto a two-dimensional space using the "RunUMAP" function. To create merged datasets of different organs, the "MergeSeurat" function was applied; the raw count matrices of two Seurat objects or more were merged into one, and a new Seurat object was created with the resulting combined raw count matrix.

4. Conclusions

In this study, the antileishmanial activity was evaluated of malvidin and echioidinin against *L. amazonensis*, *L. braziliensis*, and *L. infantum*. Malvidin demonstrated significantly higher in vitro efficacy, with lower IC50 values than echioidinin across both promastigote and axenic amastigote stages. Malvidin's lower IC_{50} values compared to echioidinin further emphasize its potent antileishmanial properties. For instance, malvidin's IC_{50} values for *L. amazonensis* were 197.71 ± 17 µM for stationary promastigotes and 258.07 ± 17 µM for axenic amastigotes, highlighting its potent antileishmanial properties

Cytotoxicity assays revealed that malvidin has a CC_{50} value of 2920.31 ± 80 µM, much higher than AmpB's CC_{50} value of 1.06 ± 0.12 µM, indicating lower toxicity to mammalian cells. Favorable selectivity index (SI) values and minimal hemolytic activity further support malvidin's therapeutic potential. Malvidin's effectiveness was further demonstrated in infected macrophages, where it reduced infection rates in a concentration-dependent manner, lowering *L. amazonensis* infection rate to 59.30% at 120.74 µM.

The promising malvidin in vitro results enabled further in silico analysis that predicted favorable interactions between malvidin and the ARG enzyme, particularly with critical residues such as HIS139 and PRO258. These interactions, with binding energy values of −12.82 kcal/mol at a pH of 7 for *L. amazonensis*, suggest a plausible mechanism for its antileishmanial activity. Malvidin's high electrophilicity and low global hardness enhance its adaptability within the ARG active site, reinforcing its ability to inhibit arginase and disrupt key metabolic pathways essential for parasite survival.

In this study, a detailed tissue-specific gene expression analysis was also conducted using single-cell RNA sequencing and network analysis. UMAP visualizations showed distinct brain, heart, kidney, and liver tissue clustering based on their gene expression profiles. Heatmap analysis revealed that malvidin treatment increased the expression of specific genes such as AXL in the heart and MAOA in the kidney, suggesting potential tissue-specific effects and benefits. Additionally, GLO1 and APEX1 were notably expressed across the heart, kidney, and liver, indicating malvidin's role in enhancing cellular health by reducing oxidative stress and improving DNA repair mechanisms.

The combined in vitro, in silico, and gene expression results highlight malvidin's potential as a low-toxicity antileishmanial candidate with specific ARG enzyme interactions. Its favorable selectivity index and macrophage infection rate reduction demonstrate therapeutic promise. Further in vivo studies and structure–activity relationship (SAR) explorations are needed to optimize its efficacy and fully develop its potential as a treatment for leishmaniasis.

Supplementary Materials: The following supporting information can be downloaded at: https://www.mdpi.com/article/10.3390/molecules30010173/s1, Figure S1. Docking of malvidin with arginase protein (ARG) from *Leishmania amazonensis*. Panels A, B, C, and D depict the 3D surface representation, the 3D structure of the binding mode and molecular interactions of the hit ligands, the 2D view of interaction types, and the H-bond interacting surface at a pH of 5, respectively. Panels E, F, G, and H show the corresponding visualizations at a pH of 7. Figure S2. Docking of malvidin with arginase protein (ARG) from *Leishmania braziliensis*. Panels A, B, C, and D depict the 3D surface representation, the 3D structure of the binding mode and molecular interactions of the hit ligands, the 2D view of interaction types, and the H-bond interacting surface at a pH of 5, respectively. Panels E, F, G, and H show the corresponding visualizations at a pH of 7. Figure S3. Decomposition of free energy per residue using the MM/GBSA method, showing the highest energy contribution from the ARG hot spot in interaction with malvidin.

Author Contributions: Conceptualization, L.D.G.-M. and M.A.C.-F.; methodology, L.D.G.-M., D.P.L., H.L.B.-C., M.P.-C., A.S.G., R.A.M.-d.-A. and E.A.F.C.; validation, D.P.L., H.L.B.-C., M.P.-C., R.C.G., E.A.F.C. and M.A.C.-F.; formal analysis, L.D.G.-M., D.P.L., H.L.B.-C., E.A.F.C. and M.A.C.-F.; investigation, L.D.G.-M., D.P.L., A.S.G., R.A.M.-d.-A. and R.C.G.; resources, M.A.C.-F.; writing—original draft preparation, L.D.G.-M. and D.P.L.; writing—review and editing, L.D.G.-M., H.L.B.-C., M.P.-C., M.A.C.-F., M.A.C.-P., G.D.-D.-C., A.S.G., R.A.M.-d.-A., R.C.G., E.L.D. and E.A.F.C.; supervision, M.A.C.-F.; funding acquisition, G.D.-D.-C. All authors have read and agreed to the published version of the manuscript.

Funding: This work was funded by the Consejo Nacional de Ciencia, Tecnología e Innovación Tecnológica (CONCYTEC), the Programa Nacional de Investigación Científica y Estudios Avanzados (PROCIENCIA), by the call "E067-2023-01 Proyectos Especiales: Proyectos de Incorporación de Investigadores Postdoctorales en Instituciones Peruanas" [número de contrato PE501084367-2023], and Grant PE501088204-2024. Also, it was funded by the Universidad Católica de Santa Maria (grants 27499-R-2020, 27574-R-2020, 7309-CU-2020, and 28048-R-2021) and by the Research Management Office from the Universidad Católica de Santa María. Furthermore, we were supported by grants APQ-02167-21 and RED-0067-23 (Rede Mineira de Imunobiológicos) from the Fundação de Amparo à Pesquisa do Estado de Minas Gerais (FAPEMIG) and by grant 402417/2023-2 from the Conselho Nacional de Desenvolvimento Científico e Tecnológico (CNPq).

Institutional Review Board Statement: Not applicable.

Informed Consent Statement: Not applicable.

Data Availability Statement: The software used in this study includes GROMACS and VMD v. 1.9.4, which are available to non-commercial users under specific distribution licenses, and CHARMM-GUI, which is free for academic use and requires a paid license for commercial purposes. All software tools are referenced throughout the Methods and Results sections. Additionally, structures, trajectories, and analysis scripts generated during this work are documented accordingly. All trajectories, structures, simulation scripts, analysis scripts, and data files will be shared with this work through a link hosted on the Amaro Lab Website (https://amarolab.ucsd.edu/covid19.php).

Acknowledgments: The authors express their gratitude for the financial support from the Programa Nacional de Investigación Científica y Estudios Avanzados—PROCIENCIA (PE501088204-2024 and PE501084367-2023). The authors would like to thank UFSJ, UFMG, UESC, USC, and UCSM for their support. The authors also thank CAPES (Code 001), CNPq and FAPEMIG for the financial support. ASG, RCG, RAMA, and EAFC would like to thank CNPq for their research PQ/DT fellowship.

Conflicts of Interest: The authors declare no conflicts of interest.

References

1. Tuon, F.F.; Amato Neto, V.; Sabbaga Amato, V. Leishmania: Origin, Evolution and Future since the Precambrian. *FEMS Immunol. Med. Microbiol.* **2008**, *54*, 158–166. [CrossRef] [PubMed]
2. Handler, M.Z.; Patel, P.A.; Kapila, R.; Al-Qubati, Y.; Schwartz, R.A. Cutaneous and Mucocutaneous Leishmaniasis: Differential Diagnosis, Diagnosis, Histopathology, and Management. *J. Am. Acad. Dermatol.* **2015**, *73*, 911–926. [CrossRef] [PubMed]
3. Herrera, G.; Barragán, N.; Luna, N.; Martínez, D.; De Martino, F.; Medina, J.; Niño, S.; Páez, L.; Ramírez, A.; Vega, L. An Interactive Database of *Leishmania* Species Distribution in the Americas. *Sci. Data* **2020**, *7*, 110. [CrossRef] [PubMed]
4. Ready, P.D. Epidemiology of Visceral Leishmaniasis. *Clin. Epidemiol.* **2014**, *6*, 147–154. [CrossRef] [PubMed]
5. Lainson, R. Ourpresent Knowledge of the Ecology and Controlof Leishmaniasis in the Amazon Region of Brazil. *Rev. Soc. Bras. Med. Trop.* **1985**, *18*, 47–56. [CrossRef]
6. Barral, A.; Pedral-Sampaio, D.; Grimaldi, G., Jr.; Almeida, R.B.; Barral-Netto, M. Leishmaniasis in Bahia, Brazil: Evidence That *Leishmania amazonensis* Produces a Wide Spectrum of Clinical Disease. *Am. J. Trop. Med. Hyg* **1991**, *44*, 536–546. [CrossRef] [PubMed]
7. Jones, T.C.; Johnson Jr, W.D.; Barretto, A.C.; Lago, E.; Badaro, R.; Cerf, B.; Reed, S.G.; Netto, E.M.; Tada, M.S.; Franca, F. Epidemiology of American Cutaneous Leishmaniasis Due to *Leishmania* Braziliensis Brasiliensis. *J. Infect. Dis.* **1987**, *156*, 73–83. [CrossRef]
8. De Brito, R.C.F.; Aguiar-Soares, R.D.d.O.; Cardoso, J.M.d.O.; Coura-Vital, W.; Roatt, B.M.; Reis, A.B. Recent Advances and New Strategies in Leishmaniasis Diagnosis. *Appl. Microbiol. Biotechnol.* **2020**, *104*, 8105–8116. [CrossRef] [PubMed]
9. Wamai, R.G.; Kahn, J.; McGloin, J.; Ziaggi, G. Visceral Leishmaniasis: A Global Overview. *J. Glob. Health Sci.* **2020**, *2*, e3. [CrossRef]
10. Dostálová, A.; Volf, P. *Leishmania* Development in Sand Flies: Parasite-Vector Interactions Overview. *Parasit. Vectors* **2012**, *5*, 276. [CrossRef]
11. Naderer, T.; McConville, M.J. The Leishmania–Macrophage Interaction: A Metabolic Perspective. *Cell. Microbiol.* **2008**, *10*, 301–308. [CrossRef] [PubMed]
12. Kima, P.E. The Amastigote Forms of *Leishmania* Are Experts at Exploiting Host Cell Processes to Establish Infection and Persist. *Int. J. Parasitol.* **2007**, *37*, 1087–1096. [CrossRef] [PubMed]
13. Damian, R.T. Molecular Mimicry: Parasite Evasion and Host Defense. In *Molecular Mimicry: Cross-Reactivity Between Microbes and Host Proteins as a Cause of Autoimmunity*; Springer: Berlin/Heidelberg, Germany, 1989; pp. 101–115.
14. Mukkada, A.J.; Meade, J.C.; Glaser, T.A.; Bonventre, P.F. Enhanced Metabolism of *Leishmania* Donovani Amastigotes at Acid p H: An Adaptation for Intracellular Growth. *Science* **1985**, *229*, 1099–1101. [CrossRef] [PubMed]
15. Lodge, R.; Descoteaux, A. Modulation of Phagolysosome Biogenesis by the Lipophosphoglycan of Leishmania. *Clin. Immunol.* **2005**, *114*, 256–265. [CrossRef]
16. da Silva Vieira, T.; Arango Duque, G.; Ory, K.; Gontijo, C.M.; Soares, R.P.; Descoteaux, A. *Leishmania* Braziliensis: Strain-Specific Modulation of Phagosome Maturation. *Front. Cell. Infect. Microbiol.* **2019**, *9*, 319. [CrossRef] [PubMed]
17. Liévin-Le Moal, V.; Loiseau, P.M. *Leishmania* Hijacking of the Macrophage Intracellular Compartments. *FEBS J.* **2016**, *283*, 598–607. [CrossRef]
18. Antoine, J.-C.; Prina, E.; Jouanne, C.; Bongrand, P. Parasitophorous Vacuoles of *Leishmania* Amazonensis-Infected Macrophages Maintain an Acidic PH. *Infect. Immun.* **1990**, *58*, 779–787. [CrossRef]
19. Monzote, L. Current Treatment of Leishmaniasis: A Review. *Open Antimicrob. Agents J.* **2009**, *1*, 9–19.
20. Kumari, S.; Kumar, V.; Tiwari, R.K.; Ravidas, V.; Pandey, K.; Kumar, A. Amphotericin B: A Drug of Choice for Visceral Leishmaniasis. *Acta Trop.* **2022**, *235*, 106661. [CrossRef]
21. Wilson, M.E.; Jeronimo, S.M.B.; Pearson, R.D. Immunopathogenesis of Infection with the Visceralizing *Leishmania* Species. *Microb. Pathog.* **2005**, *38*, 147–160. [CrossRef] [PubMed]
22. Croft, S.L.; Sundar, S.; Fairlamb, A.H. Drug Resistance in Leishmaniasis. *Clin. Microbiol. Rev.* **2006**, *19*, 111–126. [CrossRef] [PubMed]
23. Kropf, P.; Fuentes, J.M.; Fähnrich, E.; Arpa, L.; Herath, S.; Weber, V.; Soler, G.; Celada, A.; Modolell, M.; Müller, I. Arginase and Polyamine Synthesis Are Key Factors in the Regulation of Experimental Leishmaniasis in Vivo. *FASEB J.* **2005**, *19*, 1000–1002. [CrossRef] [PubMed]
24. Raj, S.; Sasidharan, S.; Balaji, S.N.; Saudagar, P. An Overview of Biochemically Characterized Drug Targets in Metabolic Pathways of *Leishmania* Parasite. *Parasitol. Res.* **2020**, *119*, 2025–2037. [CrossRef]
25. Mendes, L.d.C.; Dantas, T.B.V.; Severino, R.P.; da Silva, L.R.G.; de Souza, D.H.F.; Cass, Q.B.; Ramalho, S.D.; Vieira, P.C.; de Souza, G.E.; Guido, R.V.C. Using Molecular Networking and Docking to Explore Arginase Inhibitors among Drimys Brasiliensis Chemical Constituents. *Med. Chem. Res.* **2023**, *32*, 2208–2218. [CrossRef]
26. Camargo, P.G.; Dos Santos, C.R.; Girão Albuquerque, M.; Rangel Rodrigues, C.; da Silva Lima, C.H. Py-CoMFA, Docking, and Molecular Dynamics Simulations of *Leishmania (L.) amazonensis* Arginase Inhibitors. *Sci. Rep.* **2024**, *14*, 11575. [CrossRef]

27. Roberts, S.C.; Tancer, M.J.; Polinsky, M.R.; Gibson, K.M.; Heby, O.; Ullman, B. Arginase Plays a Pivotal Role in Polyamine Precursor Metabolism in Leishmania: Characterization of Gene Deletion Mutants. *J. Biol. Chem.* **2004**, *279*, 23668–23678. [CrossRef]
28. Pessenda, G.; da Silva, J.S. Arginase and Its Mechanisms in *Leishmania* Persistence. *Parasite Immunol.* **2020**, *42*, e12722. [CrossRef] [PubMed]
29. Barazorda-Ccahuana, H.L.; Goyzueta-Mamani, L.D.; Candia Puma, M.A.; Simões de Freitas, C.; de Sousa Vieria Tavares, G.; Pagliara Lage, D.; Ferraz Coelho, E.A.; Chávez-Fumagalli, M.A. Computer-Aided Drug Design Approaches Applied to Screen Natural Product's Structural Analogs Targeting Arginase in *Leishmania* Spp. *F1000Research* **2023**, *12*, 93. [CrossRef]
30. Damu, A.G.; Jayaprakasam, B.; Rao, K.V.; Gunasekar, D. A Flavone Glycoside from Andrographis Alata. *Phytochemistry* **1998**, *49*, 1811–1813. [CrossRef]
31. Ghule, B.; Kakad, P.; Shirke, A.; Kotagale, N.; Rathi, L. A Validated High-Performance Thin-Layer Chromatography Method for Quantification of Echioidin from Andrographis Echioides Plant. *JPC–J. Planar Chromatogr.–Mod. TLC* **2021**, *34*, 131–138. [CrossRef]
32. Mohammed, A.; Chiruvella, K.K.; Rao, Y.K.; Geethangili, M.; Raghavan, S.C.; Ghanta, R.G. In Vitro Production of Echioidinin, 7-O-Methywogonin from Callus Cultures of Andrographis Lineata and Their Cytotoxicity on Cancer Cells. *PLoS ONE* **2015**, *10*, e0141154. [CrossRef] [PubMed]
33. Hemalatha, P.; Sivakumar, V.; Parimaladevi, R.; Tilak, M. Studies on Anti-Fungal Activity of Leaf Extracts of *Andrographisechioides* (L.) Nees. *Int. J. Curr. Microbiol. App. Sci.* **2020**, *10*, 3853–3858. [CrossRef]
34. Umadevi, S.; Mohanta, G.P.; Chelladurai, V.; Manna, P.K.; Manavalan, R. Antibacterial and Antifungal Activity of *Andrographis echiodes*. *J. Nat. Remedies* **2003**, *3*, 185–188.
35. Roncaglioni, A.; Toropov, A.A.; Toropova, A.P.; Benfenati, E. In Silico Methods to Predict Drug Toxicity. *Curr. Opin. Pharmacol.* **2013**, *13*, 802–806. [CrossRef] [PubMed]
36. Raies, A.B.; Bajic, V.B. In Silico Toxicology: Computational Methods for the Prediction of Chemical Toxicity. *Wiley Interdiscip. Rev. Comput. Mol. Sci.* **2016**, *6*, 147–172. [CrossRef] [PubMed]
37. Burchmore, R.J.S.; Barrett, M.P. Life in Vacuoles–Nutrient Acquisition by *Leishmania* Amastigotes. *Int. J. Parasitol.* **2001**, *31*, 1311–1320. [CrossRef]
38. de Sousa, L.R.F.; Ramalho, S.D.; Burger, M.C.D.M.; Nebo, L.; Fernandes, J.B.; da Silva, M.F.D.G.F.; Iemma, M.R.D.C.; Correa, C.J.; Souza, D.H.F.D.; Lima, M.I.S.; et al. Isolation of Arginase Inhibitors from the Bioactivity-Guided Fractionation of Byrsonima Coccolobifolia Leaves and Stems. *J. Nat. Prod.* **2014**, *77*, 392–396. [CrossRef]
39. Abdullahi, S.A.; Unyah, N.Z.; Nordin, N.; Basir, R.; Nasir, W.M.; Alapid, A.A.; Hassan, Y.; Mustapha, T.; Majid, R.A. Phytochemicals and Potential Therapeutic Targets on Toxoplasma Gondii Parasite. *Mini Rev. Med. Chem.* **2020**, *20*, 739–753. [CrossRef]
40. Soto-Sánchez, J. Bioactivity of Natural Polyphenols as Antiparasitic Agents and Their Biochemical Targets. *Mini Rev. Med. Chem.* **2022**, *22*, 2661–2677. [CrossRef]
41. Chaniad, P.; Phuwajaroanpong, A.; Techarang, T.; Viriyavejakul, P.; Chukaew, A.; Punsawad, C. Antiplasmodial Activity and Cytotoxicity of Plant Extracts from the Asteraceae and Rubiaceae Families. *Heliyon* **2022**, *8*, e08848. [CrossRef] [PubMed]
42. Sifaoui, I.; López-Arencibia, A.; Martín-Navarro, C.M.; Reyes-Batlle, M.; Mejri, M.; Valladares, B.; Lorenzo-Morales, J.; Abderabba, M.; Piñero, J.E. Selective Activity of Oleanolic and Maslinic Acids on the Amastigote Form of *Leishmania* Spp. *Iran. J. Pharm. Res. IJPR* **2017**, *16*, 1190–1193.
43. Morais, L.S.; Dusi, R.G.; Demarque, D.P.; Silva, R.L.; Albernaz, L.C.; Báo, S.N.; Merten, C.; Antinarelli, L.M.R.; Coimbra, E.S.; Espindola, L.S. Antileishmanial Compounds from Connarus Suberosus: Metabolomics, Isolation and Mechanism of Action. *PLoS ONE* **2020**, *15*, e0241855. [CrossRef]
44. Girardi, C.; Fabre, N.; Paloque, L.; Ramadani, A.P.; Benoit-Vical, F.; González-Aspajo, G.; Haddad, M.; Rengifo, E.; Jullian, V. Evaluation of Antiplasmodial and Antileishmanial Activities of Herbal Medicine *Pseudelephantopus spiralis* (Less.) Cronquist and Isolated Hirsutinolide-Type Sesquiterpenoids. *J. Ethnopharmacol.* **2015**, *170*, 167–174. [CrossRef] [PubMed]
45. Al-Musayeib, N.M.; Mothana, R.A.; Al-Massarani, S.; Matheeussen, A.; Cos, P.; Maes, L. Study of the in Vitro Antiplasmodial, Antileishmanial and Antitrypanosomal Activities of Medicinal Plants from Saudi Arabia. *Molecules* **2012**, *17*, 11379–11390. [CrossRef] [PubMed]
46. Delgado-Altamirano, R.; Monzote, L.; Piñón-Tápanes, A.; Vibrans, H.; Rivero-Cruz, J.F.; Ibarra-Alvarado, C.; Rojas-Molina, A. In Vitro Antileishmanial Activity of Mexican Medicinal Plants. *Heliyon* **2017**, *3*, e00394. [CrossRef] [PubMed]
47. Ribeiro, T.G.; Chávez-Fumagalli, M.A.; Valadares, D.G.; Franca, J.R.; Lage, P.S.; Duarte, M.C.; Andrade, P.H.R.; Martins, V.T.; Costa, L.E.; Arruda, A.L.A.; et al. Antileishmanial Activity and Cytotoxicity of Brazilian Plants. *Exp. Parasitol.* **2014**, *143*, 60–68. [CrossRef] [PubMed]
48. Dykstra, C.E. Electrostatic Interaction Potentials in Molecular Force Fields. *Chem. Rev.* **1993**, *93*, 2339–2353. [CrossRef]
49. Bulat, F.A.; Murray, J.S.; Politzer, P. Identifying the Most Energetic Electrons in a Molecule: The Highest Occupied Molecular Orbital and the Average Local Ionization Energy. *Comput. Theor. Chem.* **2021**, *1199*, 113192. [CrossRef]

50. Aihara, J. Weighted HOMO-LUMO Energy Separation as an Index of Kinetic Stability for Fullerenes. *Theor. Chem. Acc.* **1999**, *102*, 134–138. [CrossRef]
51. Clemente, G.S.; van Waarde, A.; Antunes, I.F.; Dömling, A.; Elsinga, P.H. Arginase as a Potential Biomarker of Disease Progression: A Molecular Imaging Perspective. *Int. J. Mol. Sci.* **2020**, *21*, 5291. [CrossRef]
52. Clarke, R.J. Electrostatic Switch Mechanisms of Membrane Protein Trafficking and Regulation. *Biophys. Rev.* **2023**, *15*, 1967–1985. [CrossRef] [PubMed]
53. Shishova, E.Y.; Di Costanzo, L.; Emig, F.A.; Ash, D.E.; Christianson, D.W. Probing the Specificity Determinants of Amino Acid Recognition by Arginase. *Biochemistry* **2009**, *48*, 121–131. [CrossRef] [PubMed]
54. D'Antonio, E.L.; Christianson, D.W. Crystal Structures of Complexes with Cobalt-Reconstituted Human Arginase I. *Biochemistry* **2011**, *50*, 8018–8027. [CrossRef] [PubMed]
55. Srivastava, R. Chemical Reactivity Theory (CRT) Study of Small Drug-like Biologically Active Molecules. *J. Biomol. Struct. Dyn.* **2021**, *39*, 943–952. [CrossRef] [PubMed]
56. Boon, G.; De Proft, F.; Langenaeker, W.; Geerlings, P. The Use of Density Functional Theory-Based Reactivity Descriptors in Molecular Similarity Calculations. *Chem. Phys. Lett.* **1998**, *295*, 122–128. [CrossRef]
57. Pal, R.; Chattaraj, P.K. Chemical Reactivity from a Conceptual Density Functional Theory Perspective. *J. Indian Chem. Soc.* **2021**, *98*, 100008. [CrossRef]
58. Das, A.; Das, A.; Banik, B.K. Influence of Dipole Moments on the Medicinal Activities of Diverse Organic Compounds. *J. Indian Chem. Soc.* **2021**, *98*, 100005. [CrossRef]
59. Roos, G.; Geerlings, P.; Messens, J. Enzymatic Catalysis: The Emerging Role of Conceptual Density Functional Theory. *J. Phys. Chem. B* **2009**, *113*, 13465–13475. [CrossRef]
60. Erdoğan, M.; Serdar Çavuş, M.; Muğlu, H.; Yakan, H.; Türkeş, C.; Demir, Y.; Beydemir, Ş. Synthesis, Theoretical, in Silico and in Vitro Biological Evaluation Studies of New Thiosemicarbazones as Enzyme Inhibitors. *Chem. Biodivers.* **2023**, *20*, e202301063. [CrossRef]
61. Alyar, S.; Şen, T.; Özmen, Ü.Ö.; Alyar, H.; Adem, Ş.; Şen, C. Synthesis, Spectroscopic Characterizations, Enzyme Inhibition, Molecular Docking Study and DFT Calculations of New Schiff Bases of Sulfa Drugs. *J. Mol. Struct.* **2019**, *1185*, 416–424. [CrossRef]
62. Grillo, I.B.; Urquiza-Carvalho, G.A.; Rocha, G.B. Quantum Chemical Descriptors as a Modeling Framework for Large Biological Structures. In *Chemical Reactivity*; Elsevier: Amsterdam, The Netherlands, 2023; pp. 59–88.
63. Jurrus, E.; Engel, D.; Star, K.; Monson, K.; Brandi, J.; Felberg, L.E.; Brookes, D.H.; Wilson, L.; Chen, J.; Liles, K. Improvements to the APBS Biomolecular Solvation Software Suite. *Protein Sci.* **2018**, *27*, 112–128. [CrossRef] [PubMed]
64. Gasteiger, E.; Hoogland, C.; Gattiker, A.; Duvaud, S.; Wilkins, M.R.; Appel, R.D.; Bairoch, A. *Protein Identification and Analysis Tools on the ExPASy Server*; Springer: Berlin/Heidelberg, Germany, 2005; ISBN 1588293432.
65. da Silva, E.R.; Castilho, T.M.; Pioker, F.C.; de Paula Silva, C.H.T.; Floeter-Winter, L.M. Genomic Organisation and Transcription Characterisation of the Gene Encoding *Leishmania* (*Leishmania*) *amazonensis* Arginase and Its Protein Structure Prediction. *Int. J. Parasitol.* **2002**, *32*, 727–737. [CrossRef] [PubMed]
66. de Lima, E.C.; Castelo-Branco, F.S.; Maquiaveli, C.C.; Farias, A.B.; Renno, M.N.; Boechat, N.; Silva, E.R. Phenylhydrazides as Inhibitors of *Leishmania amazonensis* Arginase and Antileishmanial Activity. *Bioorg. Med. Chem.* **2019**, *27*, 3853–3859. [CrossRef]
67. da Silva, E.R.; Brogi, S.; Grillo, A.; Campiani, G.; Gemma, S.; Vieira, P.C.; do Carmo Maquiaveli, C. Cinnamic Acids Derived Compounds with Antileishmanial Activity Target *Leishmania amazonensis* Arginase. *Chem. Biol. Drug Des.* **2019**, *93*, 139–146. [CrossRef]
68. Manjolin, L.C.; dos Reis, M.B.G.; do Carmo Maquiaveli, C.; Santos-Filho, O.A.; da Silva, E.R. Dietary Flavonoids Fisetin, Luteolin and Their Derived Compounds Inhibit Arginase, a Central Enzyme in *Leishmania* (*Leishmania*) *amazonensis* Infection. *Food Chem.* **2013**, *141*, 2253–2262. [CrossRef] [PubMed]
69. Méndez-Cuesta, C.A.; Méndez-Lucio, O.; Castillo, R. Homology Modeling, Docking and Molecular Dynamics of the *Leishmania mexicana* Arginase: A Description of the Catalytic Site Useful for Drug Design. *J. Mol. Graph. Model.* **2012**, *38*, 50–59. [CrossRef] [PubMed]
70. Riley, E.; Roberts, S.C.; Ullman, B. Inhibition Profile of *Leishmania mexicana* Arginase Reveals Differences with Human Arginase I. *Int. J. Parasitol.* **2011**, *41*, 545–552. [CrossRef] [PubMed]
71. Hai, Y.; Christianson, D.W. Crystal Structures of *Leishmania mexicana* Arginase Complexed with α, α-Disubstituted Boronic Amino-Acid Inhibitors. *Acta Crystallogr. Sect. F Struct. Biol. Commun.* **2016**, *72*, 300–306. [CrossRef]
72. D'Antonio, E.L.; Ullman, B.; Roberts, S.C.; Dixit, U.G.; Wilson, M.E.; Hai, Y.; Christianson, D.W. Crystal Structure of Arginase from *Leishmania mexicana* and Implications for the Inhibition of Polyamine Biosynthesis in Parasitic Infections. *Arch. Biochem. Biophys.* **2013**, *535*, 163–176. [CrossRef]
73. dos Reis, M.B.G.; Manjolin, L.C.; Maquiaveli, C.D.C.; Santos-Filho, O.A.; da Silva, E.R. Inhibition of *Leishmania* (*Leishmania*) *amazonensis* and Rat Arginases by Green Tea EGCG, (+)-Catechin and (−)-Epicatechin: A Comparative Structural Analysis of Enzyme-Inhibitor Interactions. *PLoS ONE* **2013**, *8*, e78387. [CrossRef] [PubMed]

74. Ahmad, S.; Sharma, S.; Afjal, M.A.; Habib, H.; Akhter, J.; Goswami, P.; Parvez, S.; Akhtar, M.; Raisuddin, S. MRNA Expression and Protein-Protein Interaction (PPI) Network Analysis of Adrenal Steroidogenesis in Response to Exposure to Phthalates in Rats. *Environ. Toxicol. Pharmacol.* **2022**, *89*, 103780. [CrossRef]
75. Van de Sande, B.; Lee, J.S.; Mutasa-Gottgens, E.; Naughton, B.; Bacon, W.; Manning, J.; Wang, Y.; Pollard, J.; Mendez, M.; Hill, J.; et al. Applications of Single-Cell RNA Sequencing in Drug Discovery and Development. *Nat. Rev. Drug Discov.* **2023**, *22*, 496–520. [CrossRef] [PubMed]
76. Füzi, B.; Gurinova, J.; Hermjakob, H.; Ecker, G.F.; Sheriff, R. Path4Drug: Data Science Workflow for Identification of Tissue-Specific Biological Pathways Modulated by Toxic Drugs. *Front. Pharmacol.* **2021**, *12*, 708296. [CrossRef]
77. Wang, S.; Sun, S.-T.; Zhang, X.-Y.; Ding, H.-R.; Yuan, Y.; He, J.-J.; Wang, M.-S.; Yang, B.; Li, Y.-B. The Evolution of Single-Cell RNA Sequencing Technology and Application: Progress and Perspectives. *Int. J. Mol. Sci.* **2023**, *24*, 2943. [CrossRef] [PubMed]
78. Ahmad, K.; Baig, M.H.; Gupta, G.K.; Kamal, M.A.; Pathak, N.; Choi, I. Identification of Common Therapeutic Targets for Selected Neurodegenerative Disorders: An in Silico Approach. *J. Comput. Sci.* **2016**, *17*, 292–306. [CrossRef]
79. Gao, J.; Collyer, J.; Wang, M.; Sun, F.; Xu, F. Genetic Dissection of Hypertrophic Cardiomyopathy with Myocardial RNA-Seq. *Int. J. Mol. Sci.* **2020**, *21*, 3040. [CrossRef] [PubMed]
80. Vrana, M.; Whittington, D.; Nautiyal, V.; Prasad, B. Database of Optimized Proteomic Quantitative Methods for Human Drug Disposition-related Proteins for Applications in Physiologically Based Pharmacokinetic Modeling. *CPT Pharmacomet. Syst. Pharmacol.* **2017**, *6*, 267–276. [CrossRef] [PubMed]
81. Yao, T.; Wang, Q.; Zhang, W.; Bian, A.; Zhang, J. Identification of Genes Associated with Renal Cell Carcinoma Using Gene Expression Profiling Analysis. *Oncol. Lett.* **2016**, *12*, 73–78. [CrossRef]
82. Woolf, A.S.; Winyard, P.J.D.; Hermanns, M.H.; Welham, S.J.M. *Maldevelopment of the Human Kidney and Lower Urinary Tract: An Overview*; Vize, P.D., Woolf, A.S., Bard, J.B.L., Eds.; Academic Press: San Diego, CA, USA, 2003; pp. 377–393, ISBN 978-0-12-722441-1.
83. Handschuh, P.A.; Murgaš, M.; Vraka, C.; Nics, L.; Hartmann, A.M.; Winkler-Pjrek, E.; Baldinger-Melich, P.; Wadsak, W.; Winkler, D.; Hacker, M.; et al. Effect of MAOA DNA Methylation on Human in Vivo Protein Expression Measured by [11C]Harmine Positron Emission Tomography. *Int. J. Neuropsychopharmacol.* **2023**, *26*, 116–124. [CrossRef]
84. Łukasik, P.; Załuski, M.; Gutowska, I. Cyclin-Dependent Kinases (CDK) and Their Role in Diseases Development—Review. *Int. J. Mol. Sci.* **2021**, *22*, 2935. [CrossRef] [PubMed]
85. Danielyan, K.E.; Chailyan, S.G. Xanthine Dehydrogenase Inhibition Stimulates Growth and Development of Human Brain Derived Cells. *Am. J. Med. Biol. Res.* **2013**, *1*, 95–98. [CrossRef]
86. Davies, S.S.; Roberts II, L.J. F2-Isoprostanes as an Indicator and Risk Factor for Coronary Heart Disease. *Free Radic. Biol. Med.* **2011**, *50*, 559–566. [CrossRef] [PubMed]
87. Scaltriti, M.; Elkabets, M.; Baselga, J. Molecular Pathways: AXL, a Membrane Receptor Mediator of Resistance to Therapy. *Clin. Cancer Res.* **2016**, *22*, 1313–1317. [CrossRef] [PubMed]
88. Xie, B.; Wang, S.; Jiang, N.; Li, J.J. Cyclin B1/CDK1-Regulated Mitochondrial Bioenergetics in Cell Cycle Progression and Tumor Resistance. *Cancer Lett.* **2019**, *443*, 56–66. [CrossRef] [PubMed]
89. Gupta, R.M.; Hadaya, J.; Trehan, A.; Zekavat, S.M.; Roselli, C.; Klarin, D.; Emdin, C.A.; Hilvering, C.R.E.; Bianchi, V.; Mueller, C. A Genetic Variant Associated with Five Vascular Diseases Is a Distal Regulator of Endothelin-1 Gene Expression. *Cell* **2017**, *170*, 522–533. [CrossRef]
90. Berry, C.E.; Hare, J.M. Xanthine Oxidoreductase and Cardiovascular Disease: Molecular Mechanisms and Pathophysiological Implications. *J. Physiol.* **2004**, *555*, 589–606. [CrossRef]
91. Axelrod, H.; Pienta, K.J. Axl as a Mediator of Cellular Growth and Survival. *Oncotarget* **2014**, *5*, 8818. [CrossRef] [PubMed]
92. Pommier, Y.; Nussenzweig, A.; Takeda, S.; Austin, C. Human Topoisomerases and Their Roles in Genome Stability and Organization. *Nat. Rev. Mol. Cell Biol.* **2022**, *23*, 407–427. [CrossRef] [PubMed]
93. Popova, T.; Hebert, L.; Jacquemin, V.; Gad, S.; Caux-Moncoutier, V.; Dubois-d'Enghien, C.; Richaudeau, B.; Renaudin, X.; Sellers, J.; Nicolas, A. Germline BAP1 Mutations Predispose to Renal Cell Carcinomas. *Am. J. Hum. Genet.* **2013**, *92*, 974–980. [CrossRef]
94. Furuhashi, M. New Insights into Purine Metabolism in Metabolic Diseases: Role of Xanthine Oxidoreductase Activity. *Am. J. Physiol. Metab.* **2020**, *319*, E827–E834. [CrossRef] [PubMed]
95. Hertig, A.; Rondeau, E. Role of the Coagulation/Fibrinolysis System in Fibrin-Associated Glomerular Injury. *J. Am. Soc. Nephrol.* **2004**, *15*, 844–853. [CrossRef] [PubMed]
96. Stirpe, F.; Ravaioli, M.; Battelli, M.G.; Musiani, S.; Grazi, G.L. Xanthine Oxidoreductase Activity in Human Liver Disease. *Off. J. Am. Coll. Gastroenterol. ACG* **2002**, *97*, 2079–2085. [CrossRef] [PubMed]
97. Zeng, C.-M.; Chang, L.-L.; Ying, M.-D.; Cao, J.; He, Q.-J.; Zhu, H.; Yang, B. Aldo–Keto Reductase AKR1C1–AKR1C4: Functions, Regulation, and Intervention for Anti-Cancer Therapy. *Front. Pharmacol.* **2017**, *8*, 119. [CrossRef]
98. Titos, E.; Ferré, N.; Lozano, J.J.; Horrillo, R.; López-Parra, M.; Arroyo, V.; Clària, J. Protection from Hepatic Lipid Accumulation and Inflammation by Genetic Ablation of 5-Lipoxygenase. *Prostaglandins Other Lipid Mediat.* **2010**, *92*, 54–61. [CrossRef]

99. Cai, H.; Shao, B.; Zhou, Y.; Chen, Z. High Expression of TOP2A in Hepatocellular Carcinoma Is Associated with Disease Progression and Poor Prognosis. *Oncol. Lett.* **2020**, *20*, 1. [CrossRef] [PubMed]
100. Artegiani, B.; van Voorthuijsen, L.; Lindeboom, R.G.H.; Seinstra, D.; Heo, I.; Tapia, P.; López-Iglesias, C.; Postrach, D.; Dayton, T.; Oka, R. Probing the Tumor Suppressor Function of BAP1 in CRISPR-Engineered Human Liver Organoids. *Cell Stem Cell* **2019**, *24*, 927–943. [CrossRef]
101. Chinnaraj, M.; Planer, W.; Pozzi, N. Structure of Coagulation Factor II: Molecular Mechanism of Thrombin Generation and Development of next-Generation Anticoagulants. *Front. Med.* **2018**, *5*, 281. [CrossRef]
102. Li, Y.; Ye, X.; Tan, C.; Hongo, J.-A.; Zha, J.; Liu, J.; Kallop, D.; Ludlam, M.J.C.; Pei, L. Axl as a Potential Therapeutic Target in Cancer: Role of Axl in Tumor Growth, Metastasis and Angiogenesis. *Oncogene* **2009**, *28*, 3442–3455. [CrossRef]
103. Tu, H.-P.; Ko, A.M.-S.; Wang, S.-J.; Lee, C.-H.; Lea, R.A.; Chiang, S.-L.; Chiang, H.-C.; Wang, T.-N.; Huang, M.-C.; Ou, T.-T. Monoamine Oxidase A Gene Polymorphisms and Enzyme Activity Associated with Risk of Gout in Taiwan Aborigines. *Hum. Genet.* **2010**, *127*, 223–229. [CrossRef] [PubMed]
104. Parajuli, S.; Lockridge, J.B.; Langewisch, E.D.; Norman, D.J.; Kujovich, J.L. Hypercoagulability in Kidney Transplant Recipients. *Transplantation* **2016**, *100*, 719–726. [CrossRef] [PubMed]
105. Croquet, V.; Vuillemin, E.; Ternisien, C.; Pilette, C.; Oberti, F.; Gallois, Y.; Trossaert, M.; Rousselet, M.C.; Chappard, D.; Calès, P. Prothrombin Index Is an Indirect Marker of Severe Liver Fibrosis. *Eur. J. Gastroenterol. Hepatol.* **2002**, *14*, 1133–1141. [CrossRef] [PubMed]
106. Kiani, A.K.; Mor, M.; Bernini, A.; Fulcheri, E.; Michelini, S.; Herbst, K.L.; Buffelli, F.; Belgrado, J.-P.; Kaftalli, J.; Stuppia, L. Steroid-Converting Enzymes in Human Adipose Tissues and Fat Deposition with a Focus on AKR1C Enzymes. *Eur. Rev. Med. Pharmacol. Sci.* **2021**, *25*, 23–32. [PubMed]
107. Zhao, S.; Wang, S.; Zhao, Z.; Li, W. AKR1C1-3, Notably AKR1C3, Are Distinct Biomarkers for Liver Cancer Diagnosis and Prognosis: Database Mining in Malignancies. *Oncol. Lett.* **2019**, *18*, 4515–4522. [CrossRef]
108. Maessen, D.E.M.; Stehouwer, C.D.A.; Schalkwijk, C.G. The Role of Methylglyoxal and the Glyoxalase System in Diabetes and Other Age-Related Diseases. *Clin. Sci.* **2015**, *128*, 839–861. [CrossRef] [PubMed]
109. Kitsera, N.; Rodriguez-Alvarez, M.; Emmert, S.; Carell, T.; Khobta, A. Nucleotide Excision Repair of Abasic DNA Lesions. *Nucleic Acids Res.* **2019**, *47*, 8537–8547. [CrossRef] [PubMed]
110. Li, M.; Yang, X.; Lu, X.; Dai, N.; Zhang, S.; Cheng, Y.; Zhang, L.; Yang, Y.; Liu, Y.; Yang, Z. APE1 Deficiency Promotes Cellular Senescence and Premature Aging Features. *Nucleic Acids Res.* **2018**, *46*, 5664–5677. [CrossRef] [PubMed]
111. Guo, N.; Chen, Y.; Zhang, Y.; Deng, Y.; Zeng, F.; Li, X. Potential Role of APEX1 during Ferroptosis. *Front. Oncol.* **2022**, *12*, 798304. [CrossRef] [PubMed]
112. Guo, M.; Wang, H.; Potter, S.S.; Whitsett, J.A.; Xu, Y. SINCERA: A Pipeline for Single-Cell RNA-Seq Profiling Analysis. *PLoS Comput. Biol.* **2015**, *11*, e1004575. [CrossRef]
113. Granados-Riveron, J.T.; Ghosh, T.K.; Pope, M.; Bu'Lock, F.; Thornborough, C.; Eason, J.; Kirk, E.P.; Fatkin, D.; Feneley, M.P.; Harvey, R.P. α-Cardiac Myosin Heavy Chain (MYH6) Mutations Affecting Myofibril Formation Are Associated with Congenital Heart Defects. *Hum. Mol. Genet.* **2010**, *19*, 4007–4016. [CrossRef] [PubMed]
114. Gunning, P.; Ponte, P.; Blau, H.; Kedes, L. α-Skeletal and α-Cardiac Actin Genes Are Coexpressed in Adult Human Skeletal Muscle and Heart. *Mol. Cell. Biol.* **1983**, *3*, 1985–1995. [CrossRef]
115. Sandoval, Y.; Thordsen, S.E.; Smith, S.W.; Schulz, K.M.; Murakami, M.M.; Pearce, L.A.; Apple, F.S. Cardiac Troponin Changes to Distinguish Type 1 and Type 2 Myocardial Infarction and 180-Day Mortality Risk. *Eur. Heart J. Acute Cardiovasc. Care* **2014**, *3*, 317–325. [CrossRef]
116. Darbar, D.; Kannankeril, P.J.; Donahue, B.S.; Kucera, G.; Stubblefield, T.; Haines, J.L.; George Jr, A.L.; Roden, D.M. Cardiac Sodium Channel (SCN5A) Variants Associated with Atrial Fibrillation. *Circulation* **2008**, *117*, 1927–1935. [CrossRef]
117. Goldman, A.M.; Glasscock, E.; Yoo, J.; Chen, T.T.; Klassen, T.L.; Noebels, J.L. Arrhythmia in Heart and Brain: KCNQ1 Mutations Link Epilepsy and Sudden Unexplained Death. *Sci. Transl. Med.* **2009**, *1*, 2ra6. [CrossRef] [PubMed]
118. Yang, B.; Lin, H.; Xiao, J.; Lu, Y.; Luo, X.; Li, B.; Zhang, Y.; Xu, C.; Bai, Y.; Wang, H. The Muscle-Specific MicroRNA MiR-1 Regulates Cardiac Arrhythmogenic Potential by Targeting GJA1 and KCNJ2. *Nat. Med.* **2007**, *13*, 486–491. [CrossRef] [PubMed]
119. Choi, S.; Chung, J.H.; Nam, M.-H.; Bang, E.; Seo, J.B.; Chi, S.-G. KCNJ11 Downregulation Stimulates Cardiac Cell Apoptosis in Myocarditis. *Curr. Sci.* **2020**, *119*, 1106–1112. [CrossRef]
120. Napolitano, C.; Timothy, K.W.; Bloise, R.; Priori, S.G. CACNA1C-Related Disorders. In *GeneReviews*; University of Washington: Seattle, WA, USA, 2021.
121. Lewis, S.; Chen, L.; Raghuram, V.; Khundmiri, S.J.; Chou, C.-L.; Yang, C.-R.; Knepper, M.A. "SLC-Omics" of the Kidney: Solute Transporters along the Nephron. *Am. J. Physiol. Physiol.* **2021**, *321*, C507–C518. [CrossRef] [PubMed]
122. Hara-Chikuma, M.; Verkman, A.S. Aquaporin-1 Facilitates Epithelial Cell Migration in Kidney Proximal Tubule. *J. Am. Soc. Nephrol.* **2006**, *17*, 39–45. [CrossRef]

123. Acuña, R.; Martínez-de-la-Maza, L.; Ponce-Coria, J.; Vázquez, N.; Ortal-Vite, P.; Pacheco-Alvarez, D.; Bobadilla, N.A.; Gamba, G. Rare Mutations in SLC12A1 and SLC12A3 Protect against Hypertension by Reducing the Activity of Renal Salt Cotransporters. *J. Hypertens.* **2011**, *29*, 475–483. [CrossRef] [PubMed]
124. Li, Y.; Lu, X.; Yu, Z.; Wang, H.; Gao, B. Meta-Data Analysis of Kidney Stone Disease Highlights ATP1A1 Involvement in Renal Crystal Formation. *Redox Biol.* **2023**, *61*, 102648. [CrossRef]
125. Jung, H.J.; Kwon, T.-H. Molecular Mechanisms Regulating Aquaporin-2 in Kidney Collecting Duct. *Am. J. Physiol. Physiol.* **2016**, *311*, F1318–F1328. [CrossRef] [PubMed]
126. Liu, C.-L.; Shi, G.-P. Calcium-Activated Chloride Channel Regulator 1 (CLCA1): More than a Regulator of Chloride Transport and Mucus Production. *World Allergy Organ. J.* **2019**, *12*, 100077. [CrossRef] [PubMed]
127. Eck, P.; Kwon, O.; Chen, S.; Mian, O.; Levine, M. The Human Sodium-dependent Ascorbic Acid Transporters SLC23A1 and SLC23A2 Do Not Mediate Ascorbic Acid Release in the Proximal Renal Epithelial Cell. *Physiol. Rep.* **2013**, *1*, e00136. [CrossRef] [PubMed]
128. Ernest, S.; Rajaraman, S.; Megyesi, J.; Bello-Reuss, E.N. Expression of MDR1 (Multidrug Resistance) Gene and Its Protein in Normal Human Kidney. *Nephron* **1997**, *77*, 284–289. [CrossRef] [PubMed]
129. Devuyst, O.; Pattaro, C. The UMOD Locus: Insights into the Pathogenesis and Prognosis of Kidney Disease. *J. Am. Soc. Nephrol.* **2018**, *29*, 713–726. [CrossRef] [PubMed]
130. Kerr, B.M.; Thummel, K.E.; Wurden, C.J.; Klein, S.M.; Kroetz, D.L.; Gonzalez, F.J.; Levy, R. Human Liver Carbamazepine Metabolism: Role of CYP3A4 and CYP2C8 in 10, 11-Epoxide Formation. *Biochem. Pharmacol.* **1994**, *47*, 1969–1979. [CrossRef] [PubMed]
131. Coles, B.F.; Morel, F.; Rauch, C.; Huber, W.W.; Yang, M.; Teitel, C.H.; Green, B.; Lang, N.P.; Kadlubar, F.F. Effect of Polymorphism in the Human Glutathione S-Transferase A1 Promoter on Hepatic GSTA1 and GSTA2 Expression. *Pharmacogenet. Genom.* **2001**, *11*, 663–669. [CrossRef] [PubMed]
132. Girard, H.; Thibaudeau, J.; Fortier, L.; Villeneuve, L.; Caron, P.; Hao, Q.; von Moltke, L.L.; Greenblatt, D.J.; Guillemette, C. UGT1A1 Polymorphisms Are Important Determinants of Dietary Carcinogen Detoxification in the Liver. *Hepatology* **2005**, *42*, 448–457. [CrossRef]
133. Schlaepfer, I.R.; Joshi, M. CPT1A-Mediated Fat Oxidation, Mechanisms, and Therapeutic Potential. *Endocrinology* **2020**, *161*, bqz046. [CrossRef]
134. Zhang, L.; Huang, X.; Meng, Z.; Dong, B.; Shiah, S.; Moore, D.D.; Huang, W. Significance and Mechanism of CYP7a1 Gene Regulation during the Acute Phase of Liver Regeneration. *Mol. Endocrinol.* **2009**, *23*, 137–145. [CrossRef]
135. Fuchs, C.D.; Trauner, M. Role of Bile Acids and Their Receptors in Gastrointestinal and Hepatic Pathophysiology. *Nat. Rev. Gastroenterol. Hepatol.* **2022**, *19*, 432–450. [CrossRef]
136. Kaur, A.; Patankar, J.V.; de Haan, W.; Ruddle, P.; Wijesekara, N.; Groen, A.K.; Verchere, C.B.; Singaraja, R.R.; Hayden, M.R. Loss of Cyp8b1 Improves Glucose Homeostasis by Increasing GLP-1. *Diabetes* **2015**, *64*, 1168–1179. [CrossRef] [PubMed]
137. Chen, L.; Zhang, Y.-H.; Wang, S.; Zhang, Y.; Huang, T.; Cai, Y.-D. Prediction and Analysis of Essential Genes Using the Enrichments of Gene Ontology and KEGG Pathways. *PLoS ONE* **2017**, *12*, e0184129. [CrossRef] [PubMed]
138. Dezső, Z.; Nikolsky, Y.; Sviridov, E.; Shi, W.; Serebriyskaya, T.; Dosymbekov, D.; Bugrim, A.; Rakhmatulin, E.; Brennan, R.J.; Guryanov, A. A Comprehensive Functional Analysis of Tissue Specificity of Human Gene Expression. *BMC Biol.* **2008**, *6*, 49. [CrossRef] [PubMed]
139. Coelho, E.A.F.; Tavares, C.A.P.; Carvalho, F.A.A.; Chaves, K.F.; Teixeira, K.N.; Rodrigues, R.C.; Charest, H.; Matlashewski, G.; Gazzinelli, R.T.; Fernandes, A.P. Immune Responses Induced by the *Leishmania* (*Leishmania*) Donovani A2 Antigen, but Not by the LACK Antigen, Are Protective against Experimental *Leishmania* (*Leishmania*) *amazonensis* Infection. *Infect. Immun.* **2003**, *71*, 3988–3994. [CrossRef] [PubMed]
140. Valadares, D.G.; Duarte, M.C.; Oliveira, J.S.; Chávez-Fumagalli, M.A.; Martins, V.T.; Costa, L.E.; Leite, J.P.V.; Santoro, M.M.; Régis, W.C.B.; Tavares, C.A.P. Leishmanicidal Activity of the Agaricus Blazei Murill in Different *Leishmania* Species. *Parasitol. Int.* **2011**, *60*, 357–363. [CrossRef]
141. Tavares, G.d.S.V.; Mendonça, D.V.C.; Lage, D.P.; Granato, J.d.T.; Ottoni, F.M.; Ludolf, F.; Chávez-Fumagalli, M.A.; Duarte, M.C.; Tavares, C.A.P.; Alves, R.J. Antileishmanial Activity, Cytotoxicity and Mechanism of Action of Clioquinol against *Leishmania* Infantum and *Leishmania amazonensis* Species. *Basic Clin. Pharmacol. Toxicol.* **2018**, *123*, 236–246. [CrossRef] [PubMed]
142. Kim, S.; Chen, J.; Cheng, T.; Gindulyte, A.; He, J.; He, S.; Li, Q.; Shoemaker, B.A.; Thiessen, P.A.; Yu, B. PubChem in 2021: New Data Content and Improved Web Interfaces. *Nucleic Acids Res.* **2021**, *49*, D1388–D1395. [CrossRef] [PubMed]
143. Andersson, M.P.; Uvdal, P. New Scale Factors for Harmonic Vibrational Frequencies Using the B3LYP Density Functional Method with the Triple-ζ Basis Set 6-311+ G (d, P). *J. Phys. Chem. A* **2005**, *109*, 2937–2941. [CrossRef] [PubMed]
144. Johnson, B.G.; Gill, P.M.W.; Pople, J.A. The Performance of a Family of Density Functional Methods. *J. Chem. Phys.* **1993**, *98*, 5612–5626. [CrossRef]
145. Neese, F. The ORCA Program System. *Wiley Interdiscip. Rev. Comput. Mol. Sci.* **2012**, *2*, 73–78. [CrossRef]

146. Neese, F. Software Update: The ORCA Program System, Version 4.0. *Wiley Interdiscip. Rev. Comput. Mol. Sci.* **2018**, *8*, e1327. [CrossRef]
147. Hirshfeld, F.L. Bonded-Atom Fragments for Describing Molecular Charge Densities. *Theor. Chim. Acta* **1977**, *44*, 129–138. [CrossRef]
148. Geerlings, P.; De Proft, F.; Langenaeker, W. Conceptual Density Functional Theory. *Chem. Rev.* **2003**, *103*, 1793–1874. [CrossRef] [PubMed]
149. Parr, R.G.; Yang, W. Density Functional Approach to the Frontier-Electron Theory of Chemical Reactivity. *J. Am. Chem. Soc.* **1984**, *106*, 4049–4050. [CrossRef]
150. Domingo, L.R.; Sáez, J.A.; Pérez, P. A Comparative Analysis of the Electrophilicity of Organic Molecules between the Computed IPs and EAs and the HOMO and LUMO Energies. *Chem. Phys. Lett.* **2007**, *438*, 341–345. [CrossRef]
151. Koopmans, T. Über Die Zuordnung von Wellenfunktionen Und Eigenwerten Zu Den Einzelnen Elektronen Eines Atoms. *Physica* **1934**, *1*, 104–113. [CrossRef]
152. Teale, A.M.; De Proft, F.; Tozer, D.J. Orbital Energies and Negative Electron Affinities from Density Functional Theory: Insight from the Integer Discontinuity. *J. Chem. Phys.* **2008**, *129*, 044110. [CrossRef]
153. Chermette, H. Chemical Reactivity Indexes in Density Functional Theory. *J. Comput. Chem.* **1999**, *20*, 129–154. [CrossRef]
154. Balawender, R.; Geerlings, P. DFT-Based Chemical Reactivity Indices in the Hartree-Fock Method. II. Fukui Function, Chemical Potential, and Hardness. *J. Chem. Phys.* **2005**, *123*, 124103. [CrossRef]
155. Parr, R.G.; Laszlo, V. Szentpály & S. Liu, Electrophilicity Index. *J. Am. Chem. Soc.* **1999**, *121*, 1922–1924.
156. Du, H.; Brender, J.R.; Zhang, J.; Zhang, Y. Protein Structure Prediction Provides Comparable Performance to Crystallographic Structures in Docking-Based Virtual Screening. *Methods* **2015**, *71*, 77–84. [CrossRef] [PubMed]
157. Maveyraud, L.; Mourey, L. Protein X-Ray Crystallography and Drug Discovery. *Molecules* **2020**, *25*, 1030. [CrossRef] [PubMed]
158. Søndergaard, C.R.; Olsson, M.H.M.; Rostkowski, M.; Jensen, J.H. Improved Treatment of Ligands and Coupling Effects in Empirical Calculation and Rationalization of p K a Values. *J. Chem. Theory Comput.* **2011**, *7*, 2284–2295. [CrossRef] [PubMed]
159. Olsson, M.H.M.; Søndergaard, C.R.; Rostkowski, M.; Jensen, J.H. PROPKA3: Consistent Treatment of Internal and Surface Residues in Empirical p K a Predictions. *J. Chem. Theory Comput.* **2011**, *7*, 525–537. [CrossRef]
160. Barazorda-Ccahuana, H.L.; Nedyalkova, M.; Mas, F.; Madurga, S. Unveiling the Effect of Low PH on the SARS-CoV-2 Main Protease by Molecular Dynamics Simulations. *Polymers* **2021**, *13*, 3823. [CrossRef]
161. Barazorda-Ccahuana, H.L.; Gómez, B.; Mas, F.; Madurga, S. Effect of PH on the Supramolecular Structure of Helicobacter Pylori Urease by Molecular Dynamics Simulations. *Polymers* **2020**, *12*, 2713. [CrossRef] [PubMed]
162. Van Der Spoel, D.; Lindahl, E.; Hess, B.; Groenhof, G.; Mark, A.E.; Berendsen, H.J.C. GROMACS: Fast, Flexible, and Free. *J. Comput. Chem.* **2005**, *26*, 1701–1718. [CrossRef] [PubMed]
163. Kim, S.; Chen, J.; Cheng, T.; Gindulyte, A.; He, J.; He, S.; Li, Q.; Shoemaker, B.A.; Thiessen, P.A.; Yu, B. PubChem 2019 Update: Improved Access to Chemical Data. *Nucleic Acids Res.* **2019**, *47*, D1102–D1109. [CrossRef]
164. Daina, A.; Michielin, O.; Zoete, V. SwissTargetPrediction: Updated Data and New Features for Efficient Prediction of Protein Targets of Small Molecules. *Nucleic Acids Res.* **2019**, *47*, W357–W364. [CrossRef]
165. Gallo, K.; Goede, A.; Preissner, R.; Gohlke, B.-O. SuperPred 3.0: Drug Classification and Target Prediction—A Machine Learning Approach. *Nucleic Acids Res.* **2022**, *50*, W726–W731. [CrossRef]
166. Stelzer, G.; Rosen, N.; Plaschkes, I.; Zimmerman, S.; Twik, M.; Fishilevich, S.; Stein, T.I.; Nudel, R.; Lieder, I.; Mazor, Y. The GeneCards Suite: From Gene Data Mining to Disease Genome Sequence Analyses. *Curr. Protoc. Bioinform.* **2016**, *54*, 1.30.1–1.30.33. [CrossRef]
167. Lerat, E.; Fablet, M.; Modolo, L.; Lopez-Maestre, H.; Vieira, C. TEtools Facilitates Big Data Expression Analysis of Transposable Elements and Reveals an Antagonism between Their Activity and That of PiRNA Genes. *Nucleic Acids Res.* **2017**, *45*, e17. [CrossRef]
168. Szklarczyk, D.; Franceschini, A.; Wyder, S.; Forslund, K.; Heller, D.; Huerta-Cepas, J.; Simonovic, M.; Roth, A.; Santos, A.; Tsafou, K.P. STRING V10: Protein–Protein Interaction Networks, Integrated over the Tree of Life. *Nucleic Acids Res.* **2015**, *43*, D447–D452. [CrossRef]
169. Shannon, P.; Markiel, A.; Ozier, O.; Baliga, N.S.; Wang, J.T.; Ramage, D.; Amin, N.; Schwikowski, B.; Ideker, T. Cytoscape: A Software Environment for Integrated Models of Biomolecular Interaction Networks. *Genome Res.* **2003**, *13*, 2498–2504. [CrossRef] [PubMed]
170. Chin, C.-H.; Chen, S.-H.; Wu, H.-H.; Ho, C.-W.; Ko, M.-T.; Lin, C.-Y. CytoHubba: Identifying Hub Objects and Sub-Networks from Complex Interactome. *BMC Syst. Biol.* **2014**, *8* (Suppl. 4), S1. [CrossRef] [PubMed]
171. Gorodkin, J.; Jensen, L.; Doncheva, N.; Morris, J. Cytoscape StringApp: Network Analysis and Visualization of Proteomics Data. *J. Proteome Res.* **2019**, *18*, 623–632.
172. Barrett, T.; Wilhite, S.E.; Ledoux, P.; Evangelista, C.; Kim, I.F.; Tomashevsky, M.; Marshall, K.A.; Phillippy, K.H.; Sherman, P.M.; Holko, M. NCBI GEO: Archive for Functional Genomics Data Sets—Update. *Nucleic Acids Res.* **2012**, *41*, D991–D995. [CrossRef]

173. MacParland, S.A.; Liu, J.C.; Ma, X.-Z.; Innes, B.T.; Bartczak, A.M.; Gage, B.K.; Manuel, J.; Khuu, N.; Echeverri, J.; Linares, I. Single Cell RNA Sequencing of Human Liver Reveals Distinct Intrahepatic Macrophage Populations. *Nat. Commun.* **2018**, *9*, 4383. [CrossRef]
174. Litviňuková, M.; Talavera-López, C.; Maatz, H.; Reichart, D.; Worth, C.L.; Lindberg, E.L.; Kanda, M.; Polanski, K.; Heinig, M.; Lee, M. Cells of the Adult Human Heart. *Nature* **2020**, *588*, 466–472. [CrossRef]
175. Jiang, J.; Wang, C.; Qi, R.; Fu, H.; Ma, Q. ScREAD: A Single-Cell RNA-Seq Database for Alzheimer's Disease. *iScience* **2020**, *23*, 101769. [CrossRef]
176. Grubman, A.; Chew, G.; Ouyang, J.F.; Sun, G.; Choo, X.Y.; McLean, C.; Simmons, R.K.; Buckberry, S.; Vargas-Landin, D.B.; Poppe, D. A Single-Cell Atlas of Entorhinal Cortex from Individuals with Alzheimer's Disease Reveals Cell-Type-Specific Gene Expression Regulation. *Nat. Neurosci.* **2019**, *22*, 2087–2097. [CrossRef]
177. Leng, K.; Li, E.; Eser, R.; Piergies, A.; Sit, R.; Tan, M.; Neff, N.; Li, S.H.; Rodriguez, R.D.; Suemoto, C.K. Molecular Characterization of Selectively Vulnerable Neurons in Alzheimer's Disease. *Nat. Neurosci.* **2021**, *24*, 276–287. [CrossRef] [PubMed]
178. Otero-Garcia, M.; Xue, Y.-Q.; Shakouri, T.; Deng, Y.; Morabito, S.; Allison, T.; Lowry, W.E.; Kawaguchi, R.; Swarup, V.; Cobos, I. Single-Soma Transcriptomics of Tangle-Bearing Neurons in Alzheimer's Disease Reveals the Signatures of Tau-Associated Synaptic Dysfunction. *BioRxiv* **2020**, 2005–2020. [CrossRef]
179. Alsema, A.M.; Jiang, Q.; Kracht, L.; Gerrits, E.; Dubbelaar, M.L.; Miedema, A.; Brouwer, N.; Hol, E.M.; Middeldorp, J.; van Dijk, R. Profiling Microglia from Alzheimer's Disease Donors and Non-Demented Elderly in Acute Human Postmortem Cortical Tissue. *Front. Mol. Neurosci.* **2020**, *13*, 134. [CrossRef]
180. Stuart, T.; Butler, A.; Hoffman, P.; Hafemeister, C.; Papalexi, E.; Mauck, W.M.; Hao, Y.; Stoeckius, M.; Smibert, P.; Satija, R. Comprehensive Integration of Single-Cell Data. *Cell* **2019**, *177*, 1888–1902. [CrossRef] [PubMed]

molecules

MDPI

Article

In Silico Prediction of Alkaline Phosphatase Interaction with the Natural Inhibitory 5-Azaindoles Guitarrin C and D

Aleksandra Seitkalieva [1,2,†], Yulia Noskova [1], Marina Isaeva [1], Alla Guzii [1], Tatyana N. Makarieva [1], Sergey Fedorov [1] and Larissa Balabanova [1,2,*,†]

1 G.B. Elyakov Pacific Institute of Bioorganic Chemistry, Far Eastern Branch, Russian Academy of Sciences, Prospect 100-Letya Vladivostoka 152, 690022 Vladivostok, Russia; sasha0788@inbox.ru (A.S.); noskovaiulia@yandex.ru (Y.N.); issaeva@gmail.com (M.I.); gagry@rambler.ru (A.G.); makarieva@piboc.dvo.ru (T.N.M.); fedorov@piboc.dvo.ru (S.F.)

2 Youth Research Laboratory of Recombinant DNA Technologies, Advanced Engineering School, Institute of Biotechnology, Bioengineering and Food Systems, Far Eastern Federal University, 10 Ajax Bay, Russky Island, 690922 Vladivostok, Russia

* Correspondence: balaban@piboc.dvo.ru

† These authors contributed equally to this work.

Abstract: The natural 5-azaindoles, marine sponge guitarrin C and D, were observed to exert inhibitory activity against a highly active alkaline phosphatase (ALP) CmAP of the PhoA family from the marine bacterium *Cobetia amphilecti*, with IC_{50} values of 8.5 and 110 μM, respectively. The superimposition of CmAP complexes with *p*-nitrophenyl phosphate (*p*NPP), a commonly used chromogenic aryl substrate for ALP, and the inhibitory guitarrins C, D, and the non-inhibitory guitarrins A, B, and E revealed that the presence of a carboxyl group at C6 together with a hydroxyl group at C8 is a prerequisite for the inhibitory effect of 5-azaindoles on ALP activity. The 10-fold more active guitarrin C could compete with *p*NPP for binding sites in the ALP active site due to similarities in size, three-dimensional structure, and the orientation of the COOH group along the phosphate group. However, the inhibition of CmAP and calf intestinal ALP (CIAP) by guitarrin C was observed to occur via a non-competitive mode of action, as evidenced by a twofold decrease in V_{max} and an unchanged K_m. In contrast, the kinetic model with guitarrin D, with an additional OH group at C7, reflected a mixed type of inhibition, with a decrease in both values. The sensitivity of CIAP to guitarrins C and D was shown to be slightly lower than that of CmAP, with IC_{50} values of 195 and 230 μM, respectively. Nevertheless, these findings prompted the prediction of complexes of human ALP isoenzymes with guitarrins C and D.

Keywords: alkaline phosphatase; *Cobetia amphilecti*; marine sponge 5-azaindoles; guitarrins; non-competitive inhibitor; molecular docking

1. Introduction

Alkaline phosphatase (ALP) isoenzymes are of significant physiological and therapeutic interest due to their high expression in a number of severe metabolic disorders and diseases, including bone resorption (osteoporosis, Paget's disease), abnormal calcification (soft tissues, arteries, brain), congestive heart failure, liver and kidney dysfunction, sepsis, and malignant obstruction [1–9]. Together with specific ecto-enzymes, such as nucleoside triphosphate diphosphohydrolase CD39 (ENTPD), nucleotide pyrophosphatase/phosphodiesterase (eNPP1), and ecto-5′-nucleotidase CD73 (e5′-NT), the non-specific phosphomonoesterase ALP plays an essential role in cellular regulation [6]. It is responsible for the hydrolysis of extracellular signaling nucleotides to nucleosides, in particular adenosine, and for the control of the availability of these signaling molecules at their respective receptors, which trigger different cellular processes [6–8]. It has been demonstrated that aberrant expression of the tissue-nonspecific (TNAP) and intestinal

(IAP) isoenzymes is associated with a number of different cancers, including esophageal, breast, prostate, colon, ovarian, and liver cancer [9]. Thus, placental alkaline phosphatase (PLAP) was found to play a role in the formation of calcifications in the human breast cancer MDA-MB-231 cells through the PI3K-Akt signaling pathways, which is a process analogous to osteoblast differentiation [10]. Furthermore, ALP isoenzymes have been identified as key players in a complex anti-inflammatory mechanism due to their substrate non-specificity towards the terminal phosphate of cell-free signaling purines and pyrimidines, the cofactor nicotinamide adenine dinucleotide phosphate (NADPH) and its precursor nicotinamide mononucleotide (NMN) [6,8]. The aforementioned processes are also involved in the metabolism of DNA, RNA, and transmembrane phosphomonoesters, such as scavenger receptor fatty acid translocase B2/CB36, phosphoinositides, and protein kinases and their allosterically phosphorylated dual-specificity phosphatases [6–8,10–13].

The identification of effective natural or synthetic inhibitors of ALP represents a promising approach to drug discovery, particularly in the context of aberrant activity associated with TNAP, IAP and PLAP isoenzymes. Furthermore, this approach may facilitate a deeper understanding of the role of ALP in cellular processes, potentially offering new avenues for modulating these processes through the use of these inhibitors [3,4,14–19]. It has been shown that heterocyclic compounds, a purine derivative, theophylline (IC_{50} = 47 μM; K_i = 91 μM) (**1**), and an imidazothiazole derivative, levamisole (IC_{50} = 19 μM) (**2**), as well as L-phenylalanine (IC_{50} = 80 μM) (**3**), are well-known inhibitors of ALP and are used as standard inhibitory ligands [17–19] (Figure 1).

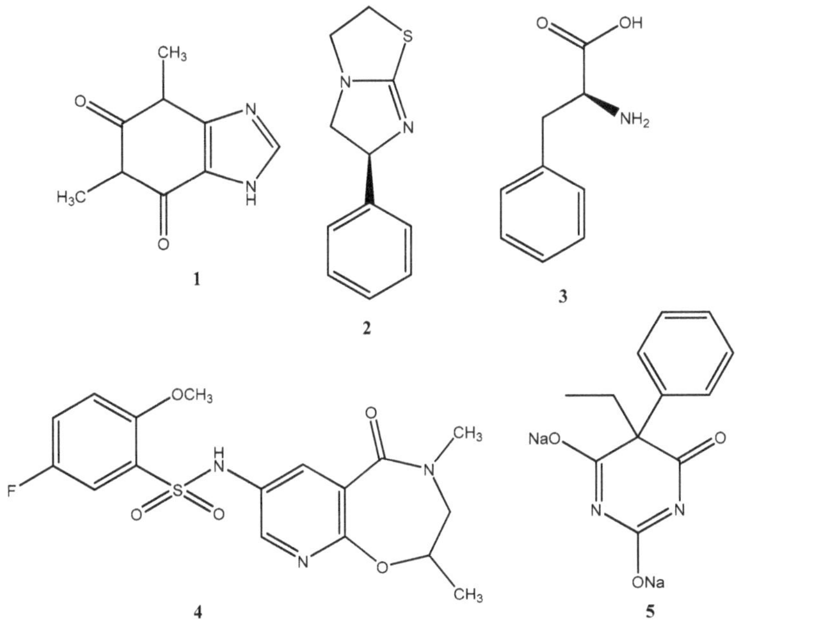

Figure 1. Conventional and newly discovered ALP inhibitors: theophylline (**1**); levamisole (**2**); L-phenylalanine (**3**); pyrido-oxazinone derivative DS-1211 (**4**); phenobarbital (**5**).

Recently, however, several groups have identified a number of pyrazole, benzo[b]thiophene, 1,2,4-triazole and 1,3,4-thiadiazole, 3,3′-carbonyl-bis(chromones), isonicotinohydrazide, sulfonamide, bisphosphonates, polyphenols, and quinoline derivatives, which are potent and selective inhibitors of ALP isoenzymes [5,9,18]. For example, the first clinical trial of a TNAP-specific inhibitor, DS-1211, a pyrido-oxazinone derivative (**4**), for the treatment of ectopic calcification has recently been completed [5]. A commonly used anticonvulsant barbiturate, phenobarbital-5-ethyl-5-phenyl-2,4,6(1H,3H,5H)-pyrimidinetrione (**5**), has been shown to be an effective drug against the ALP target [9] (Figure 1). It is important to note, however, that many of the ALP-specific ligands that have been identified may be involved in undesirable vital pathways, which could result in adverse effects. For example, the common ALP inhibitors, including vanadate, pervanadate, and okadaic acid, have been demonstrated to impact cell viability, functionality, growth, and uptake by inducing alterations in the redox state of mitochondria [20]. Given the lack of clarity surrounding the mechanisms of many pathophysiological conditions associated with increased plasma ALP activity, the study of specific and non-specific ALP binding is a valuable avenue for elucidating the ALP-specific role in metabolic pathways [2,4,13,14,21].

The interactions of small organic molecules with ALP proteins are currently being investigated through molecular docking in order to predict their mechanism of action, binding efficiency, and inactivation potency. Consequently, a series of novel tricyclic coumarin sulphonate esters, benzocoumarin thiazole–azomethines, bisthioureas of pimelic and 4-methylsalicylic acids, and quinoline-4-carboxylic acid and pyrazolo-oxothiazolidine derivatives have been predicted in silico to exert competitive, non-competitive, or uncompetitive inhibitory effects on human ALP isozymes [14,16,19,22]. To achieve this, the structures of TNAP and IAP were constructed using molecular modelling techniques based on the PLAP crystal structure, which was determined at 1.8 A resolution and has long been the only available experimental model for human ALPs [23–25]. However, the crystal structures of TNAP and its dysfunctional mutants responsible for hypophosphatasia, a metabolic bone disease that manifests as developmental abnormalities in bone and dental tissues, have been recently determined [26].

In the context of microorganisms, both inorganic and organic ALP inhibitors are valuable tools for investigating the mechanisms of enzymatic catalysis and the biogeochemical roles of these processes in the environment. Such applications include the regulation of microbiomes, induced mineralization in biofilms, and the remediation of heavy metals and organic contaminants [27–33]. The notable inhibitory activity of guitarrins, the inaugural structurally characterized natural 5-azaindoles extracted from the Northwest Pacific marine sponge *Guitarra fimbriata*, against the highly active alkaline phosphatase CmAP from the marine bacterium *C. amphilecti* KMM 296, has recently been documented [33,34]. In comparison with the chelator ethylenediamine tetraacetic acid (EDTA), the compounds guitarrins C and D exhibited greater activity against the enzyme, with a 1000-fold and 100-fold higher affinity, respectively [33]. Previously, azaindole derivatives have been evaluated as potential kinase inhibitors based on a number of different kinase targets, including adaptor-associated kinase 1 (AAK1), anaplastic lymphocyte kinase (ALK), AXL, cell division cycle 7 (Cdc7), cyclin-dependent kinases (CDKs), dual-specificity tyrosine (Y)-phosphorylation-regulated kinase 1A (DYRK1A), and so forth. It has been observed that azaindole heterocycles interact with the ATP binding sites of the fibroblast growth factor receptor 4 (FGFR4), phosphatidylinositol 3-kinase (PI3K), and other targets. This interaction is attributed to the ability of two nitrogen atoms from the heterocycles to interact with the aforementioned sites, as evidenced by previous studies [34,35].

This study provides data on the ligand structural features responsible for the inhibitory activity of the marine sponge-derived 5-azaindoles against the *C. amphilecti* KMM 296 alkaline phosphatase CmAP through kinetics and molecular docking. Experimental kinetics were applied to both CmAP and calf intestinal alkaline phosphatase (CIAP) to compare the inhibitory activity of guitarrins C and D and their mode of inhibition. In the absence of the solved crystal structure of CIAP, the amino acid residues involved in the interaction and their binding energy were predicted by docking guitarrins into the active centers of CmAP and human ALPs.

2. Results and Discussion

The screening of 5-azaindoles derived from the marine sponge *G. fimbriata* yielded a pair of effective inhibitors, namely guitarrin C and D (Figure 2), against the highly active alkaline phosphatase CmAP from the marine bacterium *C. amphilecti* KMM 296 [33], which belongs to the protein structural family PhoA, which includes animal and human ALP isoenzymes [36].

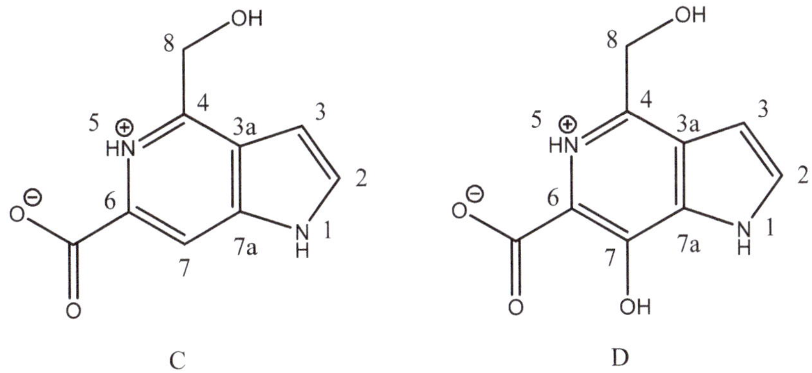

Figure 2. Chemical structures of guitarrin C ($C_9H_8N_2O_3$) and D ($C_9H_8N_2O_4$).

In the previous study [33], guitarrins A, B, and E were found to have no inhibitory activity, whereas guitarrins C and D had pronounced inhibitory activity against a highly active alkaline phosphatase, CmAP. This finding indicated that the marine bacterial ALP could serve as a model for the identification of efficacious inhibitors of clinically relevant mammalian ALPs. In addition, the high-quality homology model of the CmAP structure reported earlier [31] was almost identical to the solved structure of the *Vibrio* enzyme (PDB: 3E2D), with a Cα RMSD of 0.43 Å and identical active site metal coordination [27,31]. The superimposition of the molecular docking of the CmAP and VAP (PDB: 6T26) complexes (RMSD = 0.485 Å) (Figures S1–S3) with an aryl chromogenic substrate *p*NPP (Figure 3A), common to all ALPs, and the most effective CmAP inhibitor guitarrin C [33] demonstrated that their molecules compete for the CmAP binding site (Figure 4A).

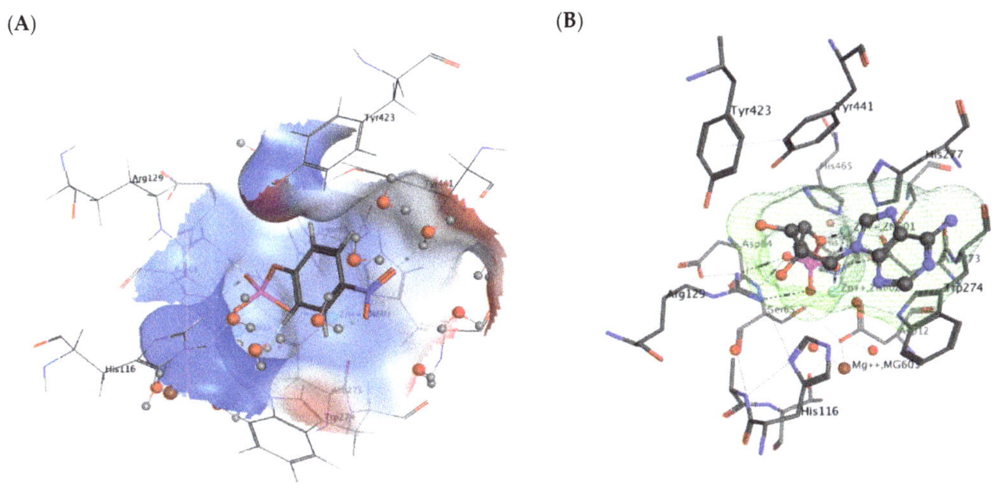

Figure 3. *C. amphilecti* KMM 296 alkaline phosphatase CmAP complexes with the substrates *p*NPP (**A**) and adenosine monophosphate (AMP) (**B**) in the active center, generated by MOE (v. 2020.09). The regions of the active center with an overall positive charge are indicated in blue, while those with

an overall negative charge are indicated in red (**A**). The water molecules are depicted as triatomic structures, with the oxygen atom rendered in red and the hydrogen atom in grey. The structures of the substrates and amino acid residues within the active center of the enzyme are illustrated as sticks, with the reactive groups highlighted in red and blue. The amino acid residues that form direct or water-mediated contacts with substrate and metal ions are indicated. The phosphate groups of the substrates are indicated in pink. The catalytic ion Zn^{2+} and the stabilizing ion Mg^{2+} are indicated in blue and brown, respectively. The invisible amino acid residues and metal ions important for catalysis, located behind or within the 3D image of the active site, shown as overlaps and indicated by shadow inscriptions.

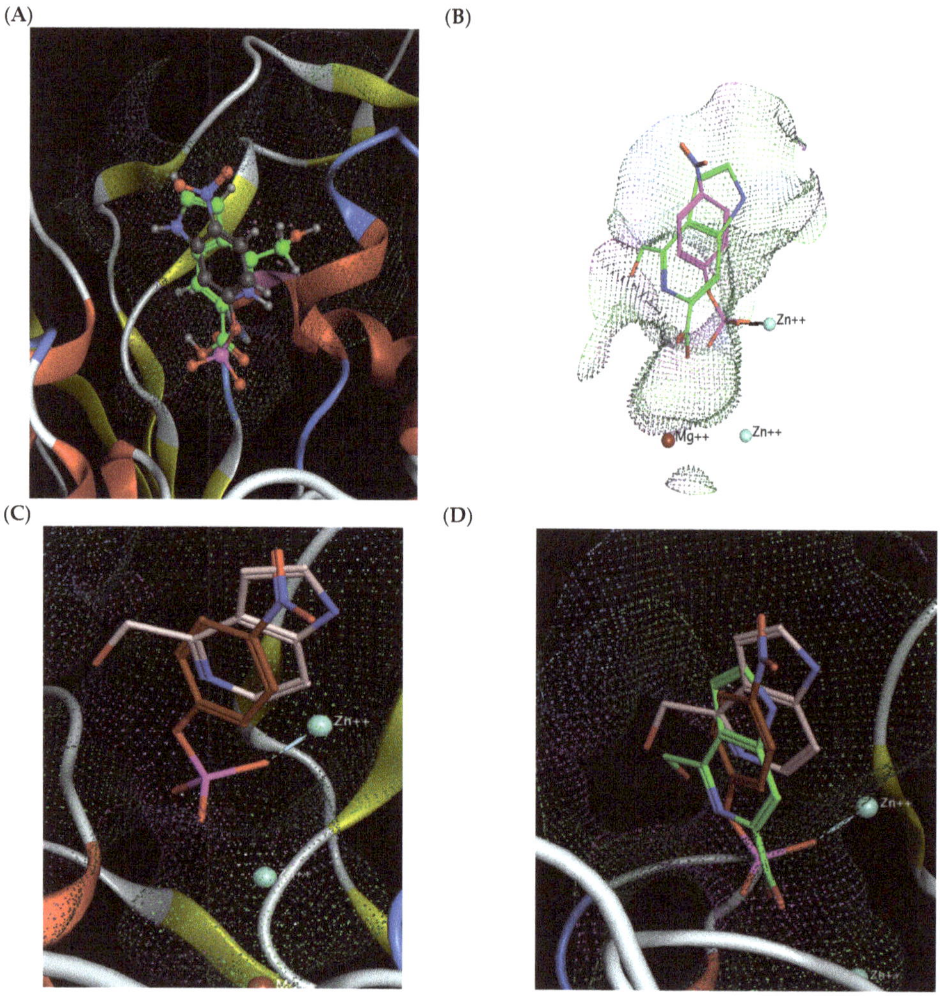

Figure 4. Superimposition of the molecular docking of natural 5-azaindoles into the active site of CmAP in a water molecule environment: (**A**) Guitarrin C (green) and *p*NPP (grey); (**B**) Guitarrin C (green) and *p*NPP (pink); (**C**) Guitarrin B (pink) and *p*NPP (brown); (**D**) Guitarrin C (green), B (pink) and *p*NPP (brown). The reactive end groups are indicated in red and the phosphate group of the substrate *p*NPP is highlighted in pink. Catalytic Zn^{2+} ions are represented by blue spheres, with Mg^{2+} ions represented by brown spheres. The 3D molecular electrostatic potential contour map of the CmAP active site is presented as a network.

As a consequence of the similarities in size, three-dimensional structure, orientation, and precise superposition of the carbon atom from the terminal reactive group COOH of the guitarrin C with the substrate POOH group, hydrogen bonds are formed with the catalytic nucleophile Ser 65 via the oxygen atom (Figure 4A).

However, the introduction of solvent molecules (H_2O) into the surrounding environment of the enzyme resulted in the guitarrin C molecule becoming integrated into the common hydrogen bond network formed by the atoms of CmAP and H_2O (Figure 4B). This resulted in a modification of the guitarrin's interactions within the active site, leading to a transition from precise to approximate superposition with *p*NPP (Figures 4B, 5A,B and S2).

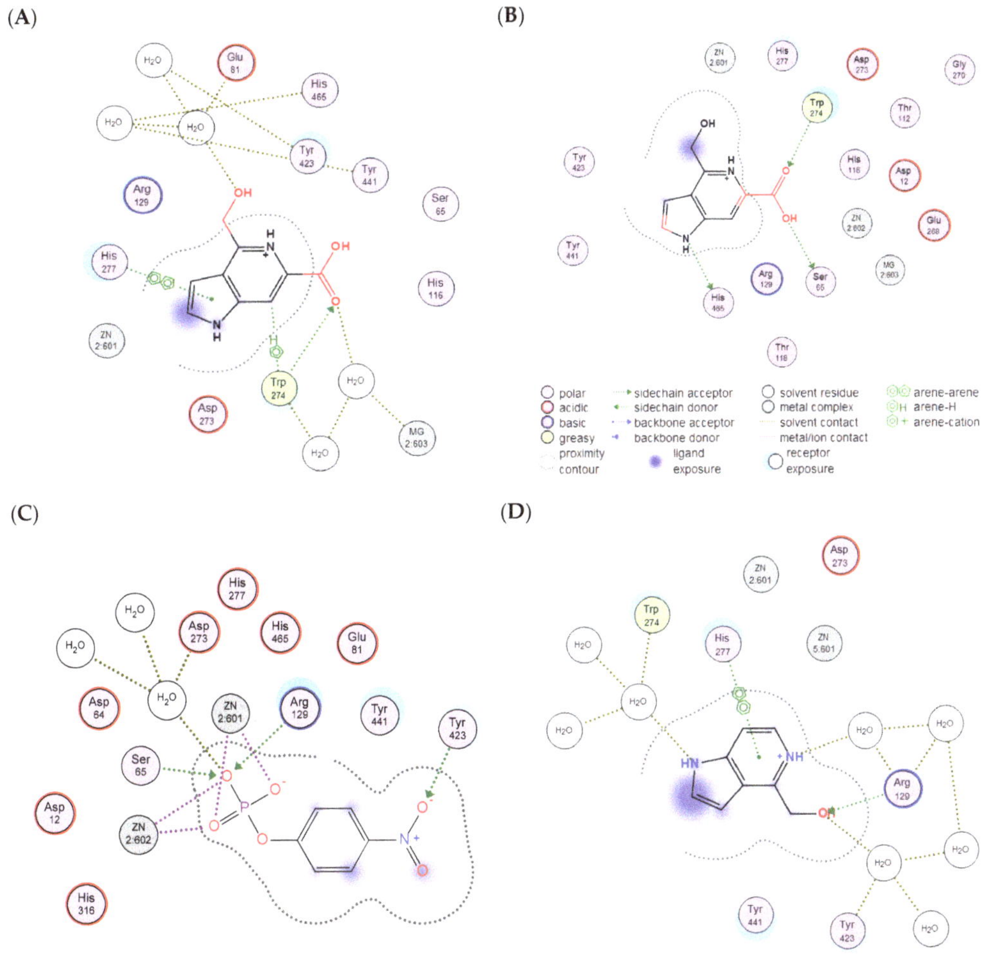

Figure 5. Comparative 2D diagrams of the contacts of guitarrin C (**A**,**B**), *p*NPP (**C**), and guitarrin B (**D**) with amino acid residues of the CmAP active site in a water molecule environment (**A**,**C**,**D**) and without H_2O in the environment (**B**). Contact symbols are in (**B**) (generated in MOE, v. 2020.09).

In the presence of H_2O molecules, the interaction with guitarrin C does not affect the nucleophile Ser65 and the two catalytic ions Zn^{2+} in the active site of CmAP. This is in contrast to the substrate *p*NPP (Figures 4B and 5A,C). While all ALPs interact with *p*NPP by forming contacts with the phosphate ion in the active site via Zn^{2+} to coordinate it with the nucleophilic Ser residue (Figure 5C), it is believed that the Mg^{2+} ion regulates the protonation state of Ser via the water molecules and provides an octahedral geometry to stabilize the most active conformation of the catalytic residues in the ALP dimer [37–39]. Furthermore, Mg^{2+} is more likely to be involved in allosteric interactions between subunits of the ALP dimer [37,38].

It can be proposed that in the presence of water and a guitarrin, the catalytic residues of ALP may remain in contact with the substrate *p*NPP, with the limiting step being the release of the reaction product by non-competitive or uncompetitive inhibition [27]. To illustrate, the crystal structure of the marine bacterium *Vibrio splendidus* alkaline phosphatase (VAP) was described in complex with an inhibitory cyclohexylamine, which was employed as a template to model the structures of CmAP (Figure S1). This revealed that the contacts with the inhibitor were formed in the vicinity of the substrate *p*NPP within the active site. However, the kinetic model of catalysis exhibited a non-competitive type of inhibition, as evidenced by the unchanging value of the Michaelis–Menten constant (195 ± 4 μM). This was accompanied by a K_i equal to IC_{50} (35.7 mM), with a corresponding decrease in V_{max} [27].

The CmAP complexes exhibited distinctive interactions when the active center was bound to the highly inhibitory guitarrin C and the guitarrin B, which lacks inhibitory activity [33] (Figures 4C,D and 5D). The inhibitor guitarrin C forms contacts with amino acid residues of the active center via a hydrogen network and a key residue, W274, which is involved in Mg^{2+} ion binding and in subunit dimerization and substrate coordination prior to catalysis [27]. The absence of the COOH group on the C6 atom in guitarrin B results in a reduction in the molecular size, which provides a different orientation in relation to the aromatic residues that surround the active center of CmAP. Furthermore, the active center residues that form hydrogen bonds with the heterocycle are also distinct (Figures 4C,D and 5D). It would appear that the formation of hydrogen bonds between guitarrin B and the substrate-binding residue Arg129 is insufficient to inhibit CmAP (Figure 5D). Indeed, the introduction of mutations in the ALP substrate-binding Arg residue did not result in a notable alteration in enzyme activity, as evidenced by the findings of the study [40]. Moreover, the replacement of the residue with Ser provided additional stabilization of the transition state. Similarly, guitarrins A and E, which lack the COOH group and are found in the same marine sponge, also demonstrated no inhibitory activity against CmAP alkaline phosphatase [33].

It is noteworthy that the compound guitarrin D exhibited an order of magnitude lower inhibitory activity towards CmAP than guitarrin C (Table 1). This is likely due to the presence of an additional OH group at carbon atom C7 [33], which impedes the inhibitor molecule from reaching the active site as deeply as guitarrin C (Figures 2–4).

Table 1. Kinetic and inhibition values for alkaline phosphatases CIAP and CmAP.

Value	CIAP	CmAP
IC_{50} (Guit C) *	195 ± 2.2 μM	8.5 ± 0.08 μM
IC_{50} (Guit D) **	230 ± 1.6 μM	110 ± 0.8 μM
K_i (Guit C) *	38 ± 0.1 μM	1.58 ± 0.04 μM
K_i (Guit D) *	45 ± 0.2 μM	20.56 ± 0.06 μM
V_{max}	0.35 ± 0.012 U mL^{-1}	5.98 ± 0.03 U mL^{-1}
$V_{1/2}$	0.175 ± 0.007 U mL^{-1}	2.99 ± 0.01 U mL^{-1}
K_m	0.49 ± 0.015 mM	0.46 ± 0.2 mM

*, guitarrin C; **, guitarrin D.

The IC_{50} values for guitarrin C and D were 8.5 and 110 μM, respectively, at 50% of the maximum inhibition of the activity of CmAP (85%), resulting in a residual activity of 15% (Table 1, Figure 6). For CIAP, the comparable inhibition constants for guitarrin C and D were determined, and a maximum inhibition of only 57% of the CmAP activity was observed (Table 1, Figure 6).

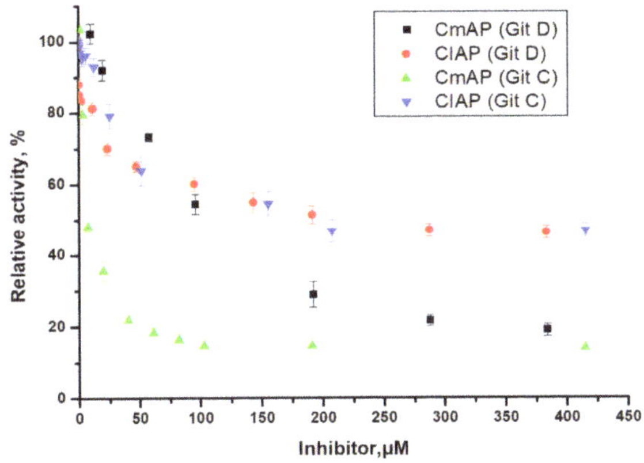

Figure 6. Comparative inhibitory effect of guitarrin C and D on CmAP and CIAP activity. Activity was measured using 2 mM *p*NPP at pH 10.2 and 37 °C in a 100 mM Tris-HCl buffer containing 0.2 M KCL.

The marine bacterial ALPs from the Antarctic TAB5 bacterium and *V. splendidus*, for which the enzyme serves as the prototype for the CmAP homology model, were also distinguished from calf intestinal alkaline phosphatases (CAP) and *Escherichia coli* (ECAP) by the inhibitory potency of cyclohexylamine with a non-competitive mode of action [27].

The determination of the type of inhibition of the CmAP enzyme by the highly specific inhibitor guitarrin C demonstrated that the K_i value is less than the IC_{50} value, with a decrease in V_{max} from 5.98 ± 0.03 to 2.62 ± 0.02 U mL^{-1} and an unaltered K_m value (0.46 ± 0.2 mM). This is consistent with a non-competitive kinetic model (Table 1, Figure 7). However, guitarrin D exhibited a mixed type of inhibition, with a more than 2-fold decrease in V_{max} from 5.98 ± 0.03 to 3.98 ± 0.02 U mL^{-1} and a 1.6-fold increase in K_m from 0.46 ± 0.2 to 0.72 ± 0.1 mM (Table 1, Figure 7).

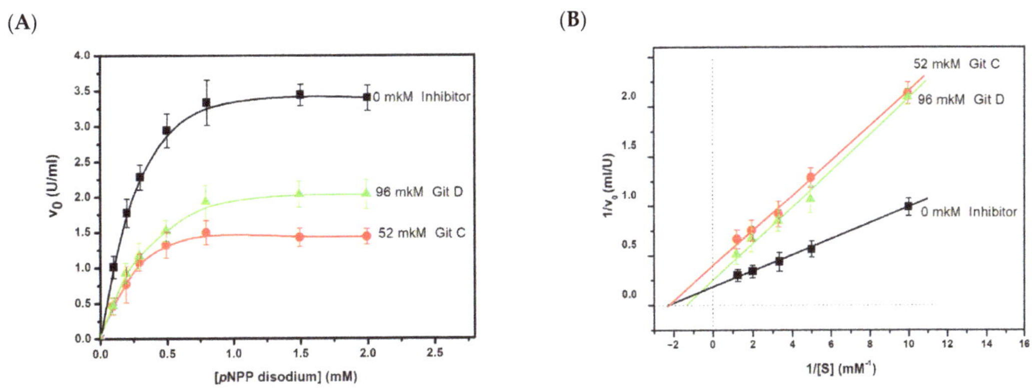

Figure 7. Effect of guitarrin C and D on the V_{max} of *p*NPP (2 mM) catalytic hydrolysis under CmAP (0.003 mg mL^{-1}) (**A**). The Linweaver–Burk reciprocal analysis (**B**). The experiments were performed in triplicate.

Furthermore, an analysis of the BRENDA database indicated that mixed-type inhibition is more likely to occur by binding to the active site of the enzyme rather than to the allosteric site of the substrate–enzyme complex, which is a commonly held belief [41]. This type of inhibition is observed in multi-substrate reactions or in reactions involving allosterically regulated enzymes. In their natural environment, marine sponges may regulate ALP activity via 5-azaindoles in order to control cell density and the synthesis of biologically active secondary metabolites, including those produced by the sponges themselves and those produced by bacterial symbionts and pathogens [33,36]. For example, transcriptome analysis of the alkaline phosphatase PhoA in diatoms revealed increased expression of this enzyme coinciding with decreased activity levels of numerous biosynthetic pathways, including iron transport, tetrapyrrole metabolism (particularly chlorophyll), and ATP-dependent proteins [42]. These findings indicate that PhoA is involved in the regulation of the cell cycle and redox processes. It is of interest to note that the CmAP alkaline phosphatase gene is located within the same marine bacterial gene cluster as the electron transport chain proteins, including ATP-dependent NAD^+-transhydrogenase [36]. This observation indicates that PhoA may perform a similar function in bacteria and eukaryotic cells. Moreover, 5-azaindoles are known to exhibit inhibitory activity against protein kinases involved in cell signaling pathways. This approach has been successfully employed in the development of agents that can be used to switch off aberrant activities of these pathways during oncogenesis [35].

It can be stated that the presence of a COOH group on the C6 atom and an OH group on the C8 atom is a prerequisite for the inhibitory effect of 5-azoindoles on ALP activity (Figure 2). This assertion is also applicable to mammalian enzymes belonging to the PhoA structural family (Table 1). Nevertheless, guitarrin D revealed that the inhibitory activity against CIAP exhibited a mere 1.2-fold difference compared to that of guitarrin C (Table 1). It seems reasonable to posit that the active site of mammalian ALPs has a larger entrance diameter than that of bacterial CmAP [31] (Figures 3, 8 and S4). However, within 4 Å of the two inhibitory ligands L-phenylalanine (L-Phe) and cyclohexylamine in the alignment of human placental (PLAP) and marine bacterial (VAP) ALPs, respectively, were the conserved metal coordination residues [27]. Therefore, a similar binding mode of the ligands in the presence of bound inorganic phosphate could be expected in VAP and CmAP (Figures S1–S3).

As the crystal structure of calf intestinal ALP has yet to be determined, the binding contacts of guitarrins C and D were analyzed by molecular docking using recently resolved human tissue-nonspecific ALP isoenzyme TNAP [26] and the PLAP-based modelled IAP. The oligomeric interfaces of PLAP and TNAP were found to exhibit significant differences, particularly in the presence of reverse charged or missing residues in PLAP [26]. This indicates that the N-terminal helix performs a supplementary function in TNAP, in addition to that observed in PLAP. These interfacial differences result in the formation of the functional octamer, which exerts a stabilizing effect on the dimer interfaces. This is likely to be of great importance in the context of osteogenesis. Nevertheless, given that PLAP and TNAP have a 55% sequence identity, they display considerable similarity in terms of their overall folding and metal binding/catalytic sites [26].

Following the minimization of energy during the molecular docking of guitarrin D with the TNAP active site, the contacts of the reactive group COO- remain with the same key residues of the active site (Asp 294, Glu 342, Arg 335, and His 171) that are involved in the interaction with guitarrin C (Figures 8 and S5). However, the orientation of the molecule undergoes a slight alteration due to the presence of an additional hydroxyl group at the C7 atom, resulting in an enhancement of the calculated binding energy from -8 to -6 kcal/mol (Figure 8). It is notable that the residues Asp 294 and His 171 have been identified as coordinating with both metal ions and the substrate. Single mutations at these residues have been demonstrated to significantly reduce the relative enzymatic activity of TNAP [26].

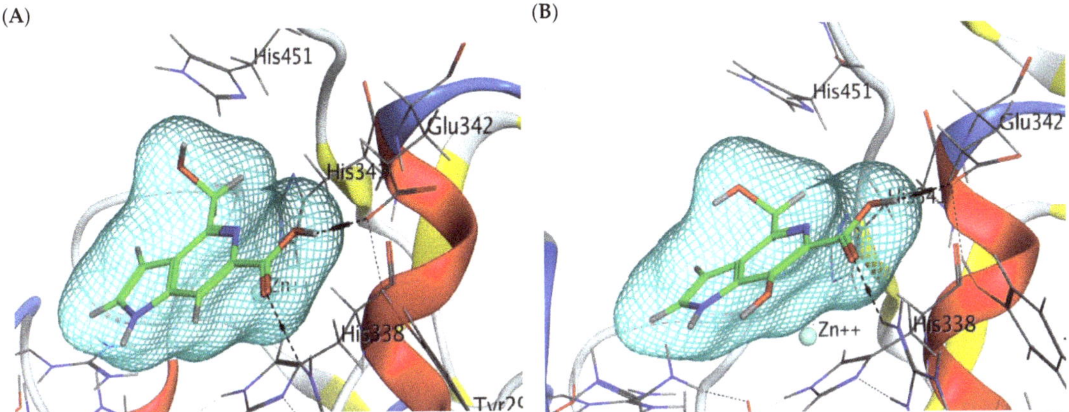

Ligand	Receptor	Residues	Interaction	Distance	E (kcal/mol)
N7 7	OD2	ASP 294	H-donor	2.99	–6.2
O13 13	OE2	GLU 342	H-donor	2.54	–5.4
O14 14	NH2	ARG 335	H-acceptor	2.72	–4.9
5-ring	NE2	HIS 171	pi-H	4.08	–1.5

Figure 8. Molecular docking of TNAP alkaline phosphatase with guitarrin C (**A**) and D (**B**). The amino acid residues of the active centre (not all fit in the figure) involved in the interaction with the ligand and coordinating zinc ions are signed and shown as sticks. The remaining enzyme residues are shown as ribbons. The molecules of gittarins C and D are shown as sticks (in green). Interactions between the enzyme and the reactive groups (in blue, red and grey) of the ligands are shown as dashed lines. The 3D molecular electrostatic potential contour map of the ligand binding site is shown as a network (in light blue). The modes of ligand–receptor interactions, distance, and calculated binding energy are presented below (generated by MOE, v. 2020.09).

It may, therefore, be proposed that further investigation of both guitarrins C and D as inhibitory regulators of ALP isoenzymes in animals and humans may prove beneficial. Moreover, both bacterial and mammalian alkaline phosphatase (ALP) have been observed to demonstrate the highest affinity for their natural purinergic substrates, particularly adenosine monophosphate (AMP) (Figure 3B) [7,43,44]. The substrate preference of bacterial PhoA enzymes remains uncertain [43,44]. However, the physiological role of mammalian ALP is to reduce the expression of pro-inflammatory signaling metabolites through the hydrolysis of purine nucleotides and the production of anti-inflammatory adenosine [7,8,26,45]. As illustrated in Figures 2–5, due to their structural similarity, guitarrin C and D can directly compete with AMP for binding sites in the active site of ALP, thereby causing competitive inhibition of the enzymes and the limiting the process of the formation of the substrate–enzyme complex [27]. Nevertheless, the inhibitory activity of the 5-azoindoles, guitarrin C and D, and the kinetic model for human ALP isoenzymes remain to be determined experimentally in order to ascertain their practical efficacy.

3. Materials and Methods

3.1. *Homology Modelling and Molecular Docking*

Three-dimensional structures of the guitarrins C and D were constructed and optimized using the molecular modelling software Molecular Operating Environment (MOE), version 2020.09, developed by the Chemical Computing Group Inc. (1010 Sherbooke St. West, Suite #910, Montreal, QC, Canada, H3A 2R7, 2020) [46]. A homology model of alkaline phosphatase CmAP from the marine bacterium *C. amphilecti* KMM 296 was constructed using the high-quality crystal structures (1.4 Å) of *Vibrio* alkaline phosphatase (VAP) (PDB: 3E2D; 6T26), with the co-crystallized inhibitory ligands sulphate and cyclohexylamine, respectively [27,31]. The homology model of the intestinal-type alkaline phosphatase (IAP, Uniprot ID P09923) was obtained from the Swiss-Model server [47,48] using the crystal structure of placental alkaline phosphatase (PLAP, PDB: 3mk1) as a template. Molecular docking was conducted using the Dock module of the MOE program. Docking was performed for the active sites of the alkaline phosphatases CmAP, TNAP (PDB: 7yiv), and IAP using the Score London dG function, with 30 poses initially generated and subsequently refined within 4.5 Å of the placed ligand using the GBVI/WSA dG function [49]. Analysis of the ligand contacts in the complexes was performed using the Ligand Interaction module of the MOE program.

3.2. Production of Recombinant Phosphatase CmAP

The *E. coli* Rosetta strain (DE3) was transformed by the plasmid pET40 (b+) containing the gene encoding for the mature alkaline phosphatase CmAP from the marine bacterium *C. amphilecti* KMM 296 [31]. The transformed cells were grown in 25 mL of liquid LB medium containing 50 μg/mL kanamycin at 200 rpm at 37 °C for 16 h. Subsequently, the cells were transferred to fresh LB medium (1 L) containing 50 μg/mL kanamycin and incubated at 37 °C on a shaker at 200 rpm until they reached an optical density at 600 nm of 0.6–0.8. Subsequently, 0.2 mM IPTG was added to induce recombinant protein CmAP expression, and incubation was continued at 18 °C for 18 h at 200 rpm. The cells were precipitated by centrifugation at 4000 rpm for 15 min at 8 °C. They were then suspended in 20 mL of 50 mM Tris-HCl buffer (pH 8.0) and subjected to ultrasonic treatment on a Bandeline ultrasonic disintegrator (Berlin, Germany) at 100% power (22 kHz) and 0–4 °C in 40-s pulses until the suspension was clear. Subsequently, the suspension was subjected to centrifugation at 11,000 rpm for 30 min at 8 °C. The precipitate was discarded, and the CmAP phosphatase activity was determined in the resulting extract. The protein concentration was determined by the Bradford method, with bovine serum albumin (BSA) employed as the reference standard [50].

3.3. Isolation and Purification of Recombinant Phosphatase CmAP

For the isolation of CmAP, the resulting supernatant was applied to a 25 × 3.2 cm Ni-IMAC-Sepharose column (Cytiva (GE Healthcare) Life Sciences, Buckinghamshire, UK) that had been equilibrated with 50 mM Tris-HCl, pH 8.0 (buffer A), and washed with five volumes of the same buffer. The recombinant protein was eluted with a linear gradient of 0–0.5 M imidazole in 50 mM Tris-HCl buffer, pH 8.0, and 0.5 M NaCl (6 column volumes) at a rate of 3 mL/min. The fraction containing CmAP was purified on a 10 × 1.4 cm Source 15 Q column (Cytiva (GE Healthcare) Life Sciences, Buckinghamshire, UK) that had been equilibrated with buffer A containing 2 mM $MgCl_2$ (buffer B). The protein was then eluted with a linear gradient of 0–0.5 M NaCl in buffer B. Ion exchange chromatography was performed at a rate of 1 mL/min, with the volume of the fractions being 1 mL. The fractions containing CmAP were collected and treated with L-HEP enterokinase at a final concentration of 1 U per 1 mg protein for 18 hs at 25 °C with constant stirring in a vortex. Subsequently, the protein solution was applied to a HisTrap™ high-performance column (Cytiva (GE Healthcare) Life Sciences, Buckinghamshire, UK) that had been pre-equilibrated with buffer A. The recombinant protein was eluted with 10 volumes of buffer B. Fractions containing CmAP were collected and subjected to concentration by ion-exchange chromatography using a Mono-Q HR column (4 × 0.8 cm) (Cytiva (GE Healthcare) Life Sciences, Buckinghamshire, UK). The column was then equilibrated with buffer B and washed with 10 volumes of buffer B, and the target protein was eluted with a linear gradient of 0–0.5 M NaCl in buffer B at a rate of 0.5 mL/min, resulting in 1 mL fractions. The specific activity of CmAP was determined in the fractions obtained.

3.4. Alkaline Phosphatase Activity Assay

The enzymatic activity was evaluated through the addition of 5 μL of the protein solution sample to a solution of 495 μL buffer, comprising 0.1 M Tris-HCl (pH 10.0), 0.2 M KCl, and 2 mM *p*NPP. The reaction mixture was incubated at 37 °C for a period of 30 min. To halt the reaction, 2 mL of chilled 0.5 M NaOH was introduced. The quantity of p-nitrophenol (*p*NP) produced was quantified spectrophotometrically at a wavelength of 400 nm, with the control sample, which lacked the enzyme, serving as a reference point. The specific activity was calculated using the following formula: (2.5 × OD400)/(18.6 × 0.005 × t × C), where OD400 is the optical density at a wavelength of 400 nm, 2.5 is the volume of the reaction mixture in mL, 18.6 is the extinction coefficient of *p*NP (M^{-1}/cm^{-1}), t is the incubation time at 37 °C in minutes, 0.005 is the sample volume in mL, and C is the protein concentration in the sample (mg/mL). The quantity of enzyme required for the generation of 1 μM *p*NP within a one-minute period was established as the unit of activity.

3.5. Inhibitory Activity Assay

The effect of the natural inhibitors guitarrins C and D [33] on the activity of the alkaline phosphatases CmAP from *C. amphilecti* KMM 296 and calf intestinal alkaline phosphatase (CIAP, Invitrogen™ 18009019, Thermo Fisher Scientific, Waltham, MA, USA) was investigated in a buffer containing 0.1 M tris-HCl (pH 10.0) with 0.2 KCl. For this purpose, 5 μL of enzyme, 20 μL of inhibitor of different concentrations (0–10.4 μM), and 470 μL of the indicated buffer were added to an incubation mixture totaling 500 μL and incubated for 60 min, and then the enzymatic reaction was initiated by adding 5 μL of *p*NPP substrate. The activity of ALPs was determined using the standard method described above. CmAP (3.22 mg/mL, 2929 U/mg) and CIAP (0.99 mg/mL, 647 U/mg) were dissolved in 50 mM tris-HCl (pH 8.0) buffer with 2 mM $MgCl_2$. Results were presented as the percentage of inhibition relative to control activity. The IC_{50} for each compound tested was defined as the concentration of compound that inhibited the enzyme activity by 50%. IC_{50} values are given as the mean of three experiments and standard deviation.

3.6. Determination of Inhibition Type and Constant (K_i)

The kinetic parameters (K_m and V_{max}) were determined for alkaline phosphatases CmAP and CIAP. The mean K_m values and standard deviations were presented. In order to ascertain the nature of the inhibition, the K_m was determined in the presence of guitarrin C (52.0 μM) and guitarrin D (96.0 μM) by preincubating CmAP with the inhibitor for 60 min at room temperature (23 °C). The inhibition constant, K_i, was determined using the Cheng–Prusoff formula [51]. The binding activity of the inhibitor to the reactive substrate (Ki) can be calculated using the following formula: $Ki = IC_{50}/(1 + [S]/Km)$, where IC_{50} is the functional activity of the inhibitor, [S] is the concentration of the reactive substrate, and KM is the Michaelis constant. The results were presented in the Origin 7.0 software.

4. Conclusions

The kinetic parameters and chemical nature of the inhibitory effect of natural 5-azaindoles, guitarrin C and D derived from the marine sponge, on the alkaline phosphatase CmAP from the marine bacterium were determined. Guitarrin C was identified as a highly selective non-competitive inhibitor of the marine enzyme due to the presence of a carboxyl group at C6 and a hydroxyl group at C8, with an IC_{50} of 8.5 ± 0.08 μM and a K_i of 1.58 ± 0.04 μM. It was hypothesized that this compound acts as an allosteric regulator of enzymes present in both resident and pathogenic sponge microflora. The non-competitive inhibitor would form contacts with amino acid residues of the active center via a hydrogen network and a key residue, W274, which is involved in Mg^{2+} ion binding and in subunit dimerization and substrate coordination prior to catalysis. However, the less active inhibitor, guitarrin D, which features an additional hydroxyl group at C7, demonstrated a mixed type of inhibition, which may also have regulatory implications.

It was proposed that both 5-azaindoles may interact with the active site of human ALPs, thereby modulating their function by forming hydrogen bonds with the nitrogen and oxygen atoms of the guitarrin C and D heterocycles. This hypothesis is based on the relatively low calculated binding energy between the key human ALP residues Asp 294, Glu 342, Arg 335, and His 171, which are situated in proximity to and coordinating with the catalytic metal ions, as well as the inhibitory activity of both guitarrins towards the animal ALP isoenzyme. The IC_{50} value for the bovine ALP isoenzyme in the presence of guitarrin C was found to be 195 ± 2.2 μM, with a K_i value of 38 ± 0.1 μM. Similarly, the IC_{50} value for guitarrin D was determined to be 230 ± 1.6 μM, with a K_i value of 45 ± 0.2 μM.

Supplementary Materials: The following supporting information can be downloaded at: https://www.mdpi.com/article/10.3390/molecules29235701/s1, Figure S1: Molecular docking of *Vibrio* alkaline phosphatase VAP (PDB: 6T26) and the inhibitor cyclohexylamine using MOE v. 2020.09. The key residue interactions with the ligand are consistent with those of experimental VAP model described in [27]; Figure S2: Ligand interactions in the superimposed complexes of CmAP with *p*NPP substrate and guitarrin C inhibitor (MOE v. 2020.09). Ligand interaction report is below; Figure S3: Active sites binding of *p*NPP (dark yellow sticks) and the inhibitory ligands guitarrin C (green sticks), and cyclohexylamine (light yellow sticks) to the bacterial alkaline phosphatases CmAP and VAP (PDB: 6T26.A, [27]) aligned (shown as ribbons). The active site zinc ions are shown as blue spheres. The superposition report for CmAP (1: ap) and VAP (3: 6T26.A) is below; Figure S4: Electrostatic surface potentials of the TNAP entrance to the active site bound with guitarrin C. Positively charged amino acid residues are indicated by blue color. Negatively charged amino acid residues are indicated by red color (generated in MOE, v. 2020.09); Figure S5: 2-D diagram of the contacts of guitarrin C and human alkaline phosphatase TNAP. The mode of ligand-receptor interactions, the distance and the calculated binding energy are presented in the table below (generated in MOE, v. 2020.09).

Author Contributions: Conceptualization, L.B.; methodology and investigation, A.S., Y.N., A.G. and T.N.M.; resources, M.I. and S.F.; writing—review and editing, L.B. All authors have read and agreed to the published version of the manuscript.

Funding: This research was funded by the Ministry of Science and Higher Education of the Russian Federation, FZNS-2022-0015.

Institutional Review Board Statement: Not applicable.

Informed Consent Statement: Not applicable.

Data Availability Statement: Data are contained within the article and Supplementary Materials.

Acknowledgments: The authors would like to express their sincerest gratitude to the recently deceased Likhatskaya G.N. for providing the molecular modelling results.

Conflicts of Interest: The authors declare no conflicts of interest.

References

1. Nargis, W.; Ahamed, B.U.; Hossain, M.A.; Biswas, S.; Ibrahim, M. Pattern of elevated serum alkaline phosphatase (ALP) levels in hospitalized patients: A single centre study. *Pulse* **2013**, *6*, 20–26. [CrossRef]
2. Levitt, M.D.; Hapak, S.M.; Levitt, D.G. Alkaline phosphatase pathophysiology with emphasis on the seldom-discussed role of defective elimination in unexplained elevations of serum alp—A case report and literature review. *Clin. Exp. Gastroenterol.* **2022**, *15*, 41–49. [CrossRef]
3. Bassi, G.; Favalli, N.; Pellegrino, C.; Onda, Y.; Scheuermann, J.; Cazzamalli, S.; Manz, M.G.; Neri, D. Specific inhibitor of placental alkaline phosphatase isolated from a dna-encoded chemical library targets tumor of the female reproductive tract. *Med. Chem.* **2021**, *64*, 15799–15809. [CrossRef]
4. Baqi, Y. Ecto-nucleotidase inhibitors: Recent developments in drug discovery. *Rev. Med. Chem.* **2015**, *15*, 21–33. [CrossRef]
5. Maruyama, S.; Visser, H.; Ito, T.; Limsakun, T.; Zahir, H.; Ford, D.; Tao, B.; Zamora, C.A.; Stark, J.G.; Chou, H.S. Phase I studies of the safety, tolerability, pharmacokinetics, and pharmacodynamics of DS-1211, a tissue-nonspecific alkaline phosphatase inhibitor. *Clin. Transl. Sci.* **2022**, *15*, 967–980. [CrossRef]
6. Jablonska, P.; Kutryb-Zajac, B.; Mierzejewska, P.; Jasztal, A.; Bocian, B.; Lango, R.; Rogowski, J.; Chlopicki, S.; Smolenski, R.T.; Slominska, E.M. The new insight into extracellular NAD^+ degradation-the contribution of CD38 and CD73 in calcific aortic valve disease. *J. Cell Mol. Med.* **2021**, *25*, 5884–5898. [CrossRef]

7. Gao, C.; Koko, M.Y.F.; Ding, M.; Hong, W.; Li, J.; Dong, N.; Hui, M. Intestinal alkaline phosphatase (IAP, IAP Enhancer) attenuates intestinal inflammation and alleviates insulin resistance. *Front. Immunol.* **2022**, *13*, 927272. [CrossRef] [PubMed]
8. Balabanova, L.; Bondarev, G.; Seitkalieva, A.; Son, O.; Tekutyeva, L. Insights into alkaline phosphatase anti-inflammatory mechanisms. *Biomedicines* **2024**, *12*, 2502. [CrossRef]
9. Ejaz, S.A.; Zain-ul-Abideen, M.; Channar, P.A.; Saeed, A.; Ahmed, A.; Alsaiari, N.S.; Katubi, K.M.; Abbas, Q.; Dahlous, K.A.; Raza, H.; et al. Synthesis, biochemical characterization and molecular modeling studies of 5-(substituted benzylidene) pyrimidine-2,4,6-trione: Potential inhibitors of alkaline phosphatase. *J. Mol. Struct.* **2023**, *1282*, 135225. [CrossRef]
10. Fushimi, A.; Takeyama, H.; Tachibana, T.; Manome, Y. Osteogenic cocktail induces calcifications in human breast cancer cell line via placental alkaline phosphatase expression. *Sci. Rep.* **2020**, *10*, 12669. [CrossRef]
11. Briolay, A.; Bessueille, L.; Magne, D. TNAP: A New Multitask Enzyme in Energy Metabolism. *Int. J. Mol. Sci.* **2021**, *22*, 10470. [CrossRef] [PubMed]
12. Zhang, Z.; Nam, H.K.; Crouch, S.; Hatch, N.E. Tissue Nonspecific Alkaline Phosphatase Function in Bone and Muscle Progenitor Cells: Control of Mitochondrial Respiration and ATP Production. *Int. J. Mol. Sci.* **2021**, *22*, 1140. [CrossRef]
13. Sato, M.; Saitoh, I.; Kiyokawa, Y.; Iwase, Y.; Kubota, N.; Ibano, N.; Noguchi, H.; Yamasaki, Y.; Inada, E. Tissue-nonspecific alkaline phosphatase, a possible mediator of cell maturation: Towards a new paradigm. *Cells* **2021**, *10*, 3338. [CrossRef]
14. Channar, P.A.; Irum, H.; Mahmood, A.; Shabir, G.; Zaib, S.; Saeed, A.; Ashraf, Z.; Larik, F.A.; Lecka, J.; Sévigny, J.; et al. Design, synthesis and biological evaluation of trinary benzocoumarin-thiazoles-azomethines derivatives as effective and selective inhibitors of alkaline phosphatase. *Bioorg. Chem.* **2019**, *91*, 103137. [CrossRef]
15. Khurshid, A.; Saeed, A.; Ashraf, Z.; Abbas, Q.; Hassan, M. Understanding the enzymatic inhibition of intestinal alkaline phosphatase by aminophenazone-derived aryl thioureas with aided computational molecular dynamics simulations: Synthesis, characterization, SAR and kinetic profiling. *Mol. Divers.* **2021**, *25*, 1701–1715. [CrossRef]
16. Salar, U.; Khan, K.M.; Iqbal, J.; Ejaz, S.A.; Hameed, A.; Al-Rashida, M.; Perveen, S.; Tahir, M.N. Coumarin sulfonates: New alkaline phosphatase inhibitors; in vitro and in silico studies. *Eur. J. Med. Chem.* **2017**, *131*, 29–47. [CrossRef]
17. al-Rashida, M.; Iqbal, J. Inhibition of alkaline phosphatase: An emerging new drug target. *Med. Chem.* **2015**, *15*, 41–51. [CrossRef]
18. Jassas, R.S.; Naeem, N.; Sadiq, A.; Mehmood, R.; Alenazi, N.A.; Al-Rooqi, M.M.; Mughal, E.U.; Alsantali, R.I.; Ahmed, S.A. Current status of N-, O-, S-heterocycles as potential alkaline phosphatase inhibitors: A medicinal chemistry overview. *RSC Adv.* **2023**, *13*, 16413–16452. [CrossRef] [PubMed]
19. Hosseini Nasab, N.; Raza, H.; Shim, R.S.; Hassan, M.; Kloczkowski, A.; Kim, S.J. Potent Alkaline Phosphatase Inhibitors, Pyrazolo-Oxothiazolidines: Synthesis, Biological Evaluation, Molecular Docking, and Kinetic Studies. *Int. J. Mol. Sci.* **2022**, *23*, 13262. [CrossRef]
20. Le-Vinh, B.; Akkuş-Dağdeviren, Z.B.; Le, N.-M.N.; Nazir, I.; Bernkop-Schnürch, A. Alkaline phosphatase: A reliable endogenous partner for drug delivery and diagnostics. *Adv. Therap.* **2022**, *5*, 2100219. [CrossRef]
21. Singh, S.B.; Carroll-Portillo, A.; Coffman, C.; Ritz, N.L.; Lin, H.C. Intestinal Alkaline Phosphatase Exerts Anti-Inflammatory Effects Against Lipopolysaccharide by Inducing Autophagy. *Sci. Rep.* **2020**, *10*, 3107. [CrossRef]
22. Mumtaza, A.; Saeeda, K.; Mahmoodb, A. Bisthioureas of pimelic acid and 4-methylsalicylic acid derivatives as selective inhibitors of tissue-nonspecific alkaline phosphatase (TNAP) and intestinal alkaline phosphatase (IAP): Synthesis and molecular docking studies. *Bioorg. Chem.* **2020**, *101*, 10399. [CrossRef]
23. Le Du, M.H.; Stigbrand, T.; Taussig, M.J.; Ménez, A.; Stura, E.A. Crystal structure of alkaline phosphatase from human placenta at 1.8 A resolution. Implication for a substrate specificity. *J. Biol. Chem.* **2001**, *276*, 9158–9165. [CrossRef]
24. Akdel, M.; Pires, D.E.V.; Pardo, E.P.; Jänes, J.; Zalevsky, A.O.; Mészáros, B.; Bryant, P.; Good, L.L.; Laskowski, R.A.; Pozzati, G.; et al. A structural biology community assessment of AlphaFold2 applications. *Nat. Struct. Mol. Biol.* **2022**, *29*, 1056–1067. [CrossRef] [PubMed]
25. Burke, D.F.; Bryant, P.; Barrio-Hernandez, I.; Memon, D.; Pozzati, G.; Shenoy, A.; Zhu, W.; Dunham, A.S.; Albanese, P.; Keller, A.; et al. Towards a structurally resolved human protein interaction network. *Nat. Struct. Mol. Biol.* **2023**, *30*, 216–225. [CrossRef]
26. Yu, Y.; Rong, K.; Yao, D.; Zhang, Q.; Cao, X.; Rao, B.; Xia, Y.; Lu, Y.; Shen, Y.; Yao, Y.; et al. The structural pathology for hypophosphatasia caused by malfunctional tissue non-specific alkaline phosphatase. *Nat. Commun.* **2023**, *14*, 4048. [CrossRef]
27. Asgeirsson, B.; Markússon, S.; Hlynsdóttir, S.S.; Helland, R.; Hjörleifsson, J.G. X-ray crystal structure of *Vibrio* alkaline phosphatase with the non-competitive inhibitor cyclohexylamine. *Biochem. Biophys. Rep.* **2020**, *24*, 100830. [CrossRef]
28. Minai-Tehrani, D.; Soheili, Z.; Yahyavi, E. Inhibition of microbial alkaline phosphatase by cimetidine; kinetics and molecular model of binding. *Curr. Enzyme. Inhib.* **2015**, *11*, 39–45. [CrossRef]
29. Srivastava, A.; Saavedra, D.E.M.; Thomson, B.; García, J.A.L.; Zhao, Z.; Patrick, W.M.; Herndl, G.J.; Baltar, F. Enzyme promiscuity in natural environments: Alkaline phosphatase in the ocean. *ISME J.* **2021**, *15*, 3375–3383. [CrossRef] [PubMed]
30. Dong, H.; Huang, L.; Zhao, L.; Zeng, Q.; Liu, X.; Sheng, Y.; Shi, L.; Wu, G.; Jiang, H.; Li, F.; et al. A critical review of mineral–microbe interaction and co-evolution: Mechanisms and applications. *Natl. Sci. Rev.* **2022**, *9*, nwac128. [CrossRef] [PubMed]
31. Golotin, V.; Balabanova, L.; Likhatskaya, G.; Rasskazov, V. Recombinant production and characterization of a highly active alkaline phosphatase from marine bacterium *Cobetia marina*. *Mar. Biotechnol.* **2015**, *17*, 130–143. [CrossRef] [PubMed]
32. Zorzetto, L.; Scoppola, E.; Raguin, E.; Blank, K.G.; Fratzl, P.; Bidan, C.M. Induced mineralization of hydroxyapatite in escherichia coli biofilms and the potential role of bacterial alkaline phosphatase. *Chem. Mater.* **2023**, *35*, 2762–2772. [CrossRef]

33. Guzii, A.G.; Makarieva, T.N.; Denisenko, V.A.; Gerasimenko, A.V.; Udovenko, A.A.; Popov, R.S.; Dmitrenok, P.S.; Golotin, V.A.; Fedorov, S.N.; Grebnev, B.B.; et al. Guitarrins A–E and Aluminumguitarrin A: 5-azaindoles from the Northwestern Pacific marine sponge *Guitarra fimbriata*. *J. Nat. Prod.* **2019**, *82*, 1704–1709. [CrossRef]
34. Motati, D.R.; Amaradhi, R.; Ganesh, T. Azaindole therapeutic agents. *Bioorg. Med. Chem.* **2020**, *28*, 115830. [CrossRef]
35. Fang, G.; Chen, H.; Cheng, Z.; Tang, Z.; Wan, Y. Azaindole derivatives as potential kinase inhibitors and their SARs elucidation. *Eur. J. Med. Chem.* **2023**, *258*, 115621. [CrossRef]
36. Balabanova, L.; Bakholdina, S.; Buinovskaya, N.; Noskova, Y.; Kolpakova, O.; Vlasova, V.; Bondarev, G.; Seitkalieva, A.; Son, O.; Tekutyeva, L. LPS-Dephosphorylating *Cobetia amphilecti* alkaline phosphatase of PhoA family divergent from the multiple homologues of *Cobetia* spp. *Microorganisms* **2024**, *12*, 631. [CrossRef]
37. Chen, S.-L.; Liao, R.-Z. Phosphate monoester hydrolysis by trinuclear alkaline phosphatase; DFT study of transition states and reaction mechanism. *ChemPhysChem* **2014**, *11*, 2321–2330. [CrossRef]
38. Borosky, G.L. Quantuchanical study on the catalytic mechanism of alkaline phosphatases. *J. Chem. Inf. Model.* **2017**, *3*, 540–549. [CrossRef]
39. Orhanović, S.; Pavela-Vrančič, M. Dimer asymmetry and the catalytic cycle of alkaline phosphatase from *Escherichia coli*. *Eur. J. Biochem.* **2003**, *21*, 4356–4364. [CrossRef]
40. O'Brien, P.J.; Lassila, J.K.; Fenn, T.D.; Zalatan, J.G.; Herschlag, D. Arginine coordination in enzymatic phosphoryl transfer: Evaluation of the effect of Arg166 mutations in *Escherichia coli* alkaline phosphatase. *Biochemistry* **2008**, *47*, 7663–7672. [CrossRef] [PubMed]
41. Pesaresi, A. Mixed and non-competitive enzyme inhibition: Underlying mechanisms and mechanistic irrelevance of the formal two-site model. *J. Enzyme. Inhib. Med. Chem.* **2023**, *38*, 2245168. [CrossRef] [PubMed]
42. Zhang, K.; Li, J.; Zhou, Z.; Huang, R.; Lin, S. Roles of Alkaline Phosphatase PhoA in Algal Metabolic Regulation under Phosphorus-replete Conditions. *J. Phycol.* **2021**, *57*, 703–707. [CrossRef]
43. Plisova, E.Y.; Balabanova, L.A.; Ivanova, E.P.; Kozhemyako, V.B.; Mikhailov, V.V.; Agafonova, E.V.; Rasskazov, V.A. A highly active alkaline phosphatase from the marine bacterium cobetia. *Mar. Biotechnol.* **2005**, *7*, 173–178. [CrossRef]
44. Reid, T.W.; Wilson, I.B. 17 *E. coli* Alkaline Phosphatase. *Enzymes* **1971**, *4*, 373–415. [CrossRef]
45. Rosin, D.L.; Perry Hall, J.; Zheng, S.; Huang, L.; Campos-Bilderback, S.; Sandoval, R.; Bree, A.; Beaumont, K.; Miller, E.; Larsen, J.; et al. Human Recombinant Alkaline Phosphatase (Ilofotase Alfa) Protects against Kidney Ischemia-Reperfusion Injury in Mice and Rats through Adenosine Receptors. *Front. Med.* **2022**, *9*, 931293. [CrossRef]
46. Labute, P. *Molecular Operating Environment*; Chemical Computing Group, Inc.: Montreal, QC, Canada, 2008.
47. Waterhouse, A.; Bertoni, M.; Bienert, S.; Studer, G.; Tauriello, G.; Gumienny, R.; Heer, F.T.; de Beer, T.A.P.; Rempfer, C.; Bordoli, L.; et al. SWISS-MODEL: Homology modelling of protein structures and complexes. *Nucleic Acids Res.* **2018**, *46*, W296–W303. [CrossRef]
48. Bienert, S.; Waterhouse, A.; de Beer, T.A.; Tauriello, G.; Studer, G.; Bordoli, L.; Schwede, T. The SWISS-MODEL Repository-new features and functionality. *Nucleic Acids Res.* **2017**, *45*, D313–D319. [CrossRef] [PubMed]
49. Corbeil, C.R.; Williams, C.I.; Labute, P. Variability in docking success rates due to dataset preparation. *J. Comput. Aided Mol. Des.* **2012**, *26*, 775–786. [CrossRef]
50. Bradford, M.M. A rapid and sensitive method for the quantitation of microgram quantities of protein utilizing the principle of protein-dye binding. *Anal. Biochem.* **1976**, *72*, 248–254. [CrossRef]
51. Cheng, Y.; Prusoff, W.H. Relationship between the inhibition constant (K1) and the concentration of inhibitor which causes 50 per cent inhibition (I50) of an enzymatic reaction. *Biochem. Pharmacol.* **1973**, *22*, 3099–3108. [CrossRef]

molecules

MDPI

Article

Exploring Radioiodinated Anastrozole and Epirubicin as AKT1-Targeted Radiopharmaceuticals in Breast Cancer: In Silico Analysis and Potential Therapeutic Effect with Functional Nuclear Imagining Implications

Mazen Abdulrahman Binmujlli

Department of Internal Medicine, College of Medicine, Imam Mohammad Ibn Saud Islamic University (IMSIU), P.O. Box 90950, Riyadh 11623, Saudi Arabia; dr1mazen@gmail.com

Abstract: This study evaluates radio-iodinated anastrozole ([^{125}I]anastrozole) and epirubicin ([^{125}I]epirubicin) for AKT1-targeted breast cancer therapy, utilizing radiopharmaceutical therapy (RPT) for personalized treatment. Through molecular docking and dynamics simulations (200 ns), it investigates these compounds' binding affinities and mechanisms to the AKT1 enzyme, compared to the co-crystallized ligand, a known AKT1 inhibitor. Molecular docking results show that [^{125}I]epirubicin has the highest ΔG_{bind} (−11.84 kcal/mol), indicating a superior binding affinity compared to [^{125}I] anastrozole (−10.68 kcal/mol) and the co-crystallized ligand (−9.53 kcal/mol). Molecular dynamics (MD) simulations confirmed a stable interaction with the AKT1 enzyme, with [^{125}I]anastrozole and [^{125}I]epirubicin reaching stability after approximately 68 ns with an average RMSD of around 2.2 Å, while the co-crystallized ligand stabilized at approximately 2.69 Å after 87 ns. RMSF analysis showed no significant shifts in residues or segments, with consistent patterns and differences of less than 2 Å, maintaining enzyme stability. The [^{125}I]epirubicin complex maintained an average of four H-bonds, indicating strong and stable interactions, while [^{125}I]anastrozole consistently formed three H-bonds. The average Rg values for both complexes were ~16.8 ± 0.1 Å, indicating no significant changes in the enzyme's compactness, thus preserving structural integrity. These analyses reveal stable binding and minimal structural perturbations, suggesting the high potential for AKT1 inhibition. MM-PBSA calculations confirm the potential of these radio-iodinated compounds as AKT1 inhibitors, with [^{125}I]epirubicin exhibiting the most favorable binding energy (−23.57 ± 0.14 kcal/mol) compared to [^{125}I]anastrozole (−20.03 ± 0.15 kcal/mol) and the co-crystallized ligand (−16.38 ± 0.14 kcal/mol), highlighting the significant role of electrostatic interactions in stabilizing the complex. The computational analysis shows [^{125}I]anastrozole and [^{125}I]epirubicin may play promising roles as AKT1 inhibitors, especially [^{125}I]epirubicin for its high binding affinity and dynamic receptor interactions. These findings, supported by molecular docking scores and MM-PBSA binding energies, advocate for their potential superior inhibitory capability against the AKT1 enzyme. Nevertheless, it is crucial to validate these computational predictions through in vitro and in vivo studies to thoroughly evaluate the therapeutic potential and viability of these compounds for AKT1-targeted breast cancer treatment.

Keywords: breast cancer; AKT1; molecular docking; molecular dynamic; MM-PBSA

1. Introduction

Breast cancer is one of the most prevalent malignancies affecting women worldwide, accounting for a significant number of cancer-related deaths each year [1]. According to the World Health Organization (WHO), breast cancer is the most common cancer among women, with an estimated 2.3 million new cases diagnosed in 2020 alone [2,3]. The global burden of breast cancer continues to rise, partly due to increasing life expectancy, urbanization, and the adoption of Western lifestyles [4]. Despite advances in therapeutic strategies, early and accurate diagnoses remain a critical challenge [5,6]. The five-year survival rate for breast cancer patients is significantly higher when the disease is detected at an early stage, underscoring the importance of effective diagnostic tools [7].

Mammography has been the gold standard for breast cancer screening; however, its sensitivity and specificity can be limited, particularly in women with dense breast tissue [8]. In recent years, molecular imaging techniques have gained prominence in the field of oncology, offering enhanced sensitivity and specificity for detecting malignancies [9,10]. Radiopharmaceuticals, which are radioactive compounds used for diagnosis and therapy, have emerged as valuable tools for imaging and targeting tumors [11,12]. These compounds can provide detailed information about the biological processes underlying cancer, facilitating early detection and personalized treatment strategies [11,12].

The integration of radiopharmaceuticals into clinical practice holds significant potential to transform breast cancer management [13]. Radiopharmaceuticals are employed in advanced imaging techniques, such as positron emission tomography (PET) and single-photon emission computed tomography (SPECT), to capture and visualize the metabolic and molecular processes occurring within tumors [14–16]. These techniques provide functional imaging capabilities, enabling clinicians to assess tumor metabolism, proliferation, and receptor status, which are essential for developing personalized treatment strategies [14–16]. In breast cancer, hormonal receptors, like estrogen (ER), progesterone (PR), and androgens (ARs), serve as crucial targets for the non-invasive whole-body evaluation of hormonal status, helping to predict cancer response to endocrine treatments [13,17]. A notable example is 16α-[^{18}F]-fluoro-17β-estradiol (FES), a radiopharmaceutical widely used for the PET imaging of ER expression [18]. Additionally, radiolabeled monoclonal antibody trastuzumab, applicable in both SPECT and PET, facilitates the imaging of the human epidermal growth factor receptor 2 (HER2) [19–21].

Molecular imaging with [^{18}F]-fluoroestradiol positron emission tomography ([^{18}F]-FES-PET) offers non-invasive insights into ER status across all metastases within a patient's body [22]. This technique is valuable for assessing the ER binding of endocrine modulators or down-regulators by comparing ER status before and during therapy [17]. Preclinical studies have demonstrated a strong correlation between [^{18}F]-FES uptake and ER expression as determined by immunohistochemistry, with [^{18}F]-FES uptake levels predicting the response to hormonal therapy (Figure 1) [23,24]. Recent studies have reported an overall sensitivity of 84% and specificity of 98% for [^{18}F]-FES-PET in breast cancer patients, with a Standard Uptake Value max (SUVmax) threshold of 1.5 used to predict therapeutic response; values below this threshold generally indicate the lack of a response to endocrine therapy [25–27]. Moreover, [^{18}F]-FES-PET proves useful for clinical dilemmas where conventional imaging and biopsies are inconclusive or impractical, aiding in therapy decisions by enhancing diagnostic clarity [28,29]. This imaging modality is also beneficial for monitoring ER status in progressive disease, thereby informing the continuation of or alteration in anti-hormonal treatments [18]. For therapy evaluation, [^{18}F]-FES-PET can detect treatment response significantly earlier than traditional CT scans, potentially guiding the timely adjustments to therapy [30–32]. Although primarily visualizing ERα, [^{18}F]-FES-PET has also shown efficacy in assessing ER status in gynecological tumors predominantly expressing ERβ, with ongoing efforts to develop specific ERβ tracers [33,34].

In addition to these advancements, radiopharmaceuticals have shown potential utility in managing triple-negative breast cancer (TNBC), an aggressive subtype characterized by the absence of ER, PR, and HER2 receptors, which limits the efficacy of conventional targeted therapies [35–37]. Radiopharmaceuticals targeting other molecular markers or metabolic pathways specific to TNBC could provide new therapeutic avenues, especially in cases where traditional treatments are ineffective [35–37]. This highlights the broader applicability of radiopharmaceuticals beyond receptor-positive breast cancers, underscoring their potential for addressing challenging cancer subtypes, like TNBC [35–37].

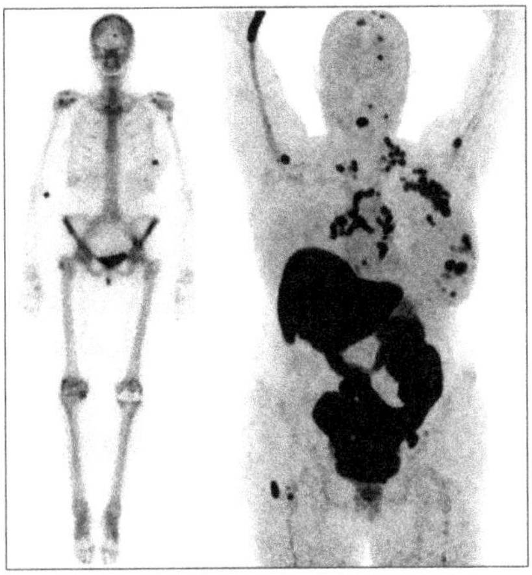

Figure 1. Female patient with the estrogen receptor (ER)-positive breast cancer in her left breast. Left image: bone scan for dissemination: uptake in the primary tumor in the left breast and uptake in the skull [13]. Right image: [^{18}F]-FES-PET with multiple lesions in the left breast, multiple ER-positive lymph node metastases (axillae, clavicular regions, neck, mediastinum, and hili), and multiple ER-positive bone metastases (skull, spine, left humerus, and right femur).

Targeting key signaling pathways in cancer cells has shown promise for improving diagnostic and therapeutic outcomes [38]. The PI3K/AKT/mTOR signaling pathway is a critical regulator of cell growth, proliferation, survival, and metabolism, and its dysregulation is commonly observed in breast cancer [39,40]. AKT1, a serine/threonine kinase, is a pivotal component of this pathway and represents a compelling target for anticancer therapy [41,42]. Aberrant activation of AKT1 has been associated with cancer progression, metastasis, and resistance to therapy, particularly in breast cancer [43,44]. Inhibiting AKT1 can suppress tumor growth, induce apoptosis, and enhance the efficacy of existing treatments by sensitizing cancer cells to chemotherapy and radiotherapy [45]. A notable example is ARQ092 (Miransertib), a selective allosteric inhibitor of AKT1, which has shown promise in clinical trials for patients with PI3K/Akt-driven tumors or Proteus syndrome [46]. Therefore, targeting AKT1 could have a significant impact on improving the treatment outcomes for breast cancer patients. Recent studies have highlighted the potential of radio-iodinated compounds, such as anastrozole and epirubicin, in the field of oncology [47,48]. Radioactive iodine attached to an sp3 carbon atom is a rare feature in radiopharmaceutical compounds; however, some have proven effective in treating solid tumors [49]. For instance, iododoxorubicin has demonstrated significant advancements, showcasing the potential of radio-iodinated derivatives in cancer treatment [50–52]. Studies on iododoxorubicin have demonstrated enhanced tumor uptake and an improved pharmacokinetic profile compared to its non-radio-iodinated variants [50–52]. In this context, radio-iodinated anastrozole ([^{125}I]anastrozole) and radio-iodinated epirubicin ([^{125}I]epirubicin) (Figure S1) have been studied as potential targeting radiopharmaceuticals for solid tumor imaging [47].

Although [^{125}I]anastrozole and [^{125}I]epirubicin demonstrate significant potential in targeting solid tumors, elucidating their interactions with enzymes remains a complex challenge. Traditional methodologies for evaluating enzyme inhibition are often both intricate and resource intensive [53,54]. In this regard, computational approaches, such as molecular docking and molecular dynamics simulations, offer indispensable alternatives [55]. These techniques provide cost-effective and precise methods to explore the interaction dynamics between radio-iodinated compounds and target enzymes, like AKT1 [56]. The main objective of this study is to leverage these computational strategies to assess the binding affinity and structural integrity of [^{125}I]anastrozole and [^{125}I]epirubicin in complex with the AKT1 enzyme. By elucidating the binding mechanisms involved, this research aims to underscore the potential of computational methodologies in expediting the development of targeted imaging agents for solid tumors implicating the AKT1 pathway.

2. Results

2.1. Molecular Docking

In this study, molecular docking simulations were conducted for [^{125}I]anastrozole and [^{125}I]epirubicin, along with the co-crystallized ligand ((2S)-2-(4-chlorobenzyl)-3-oxo-3-[4-(7H-pyrrolo [2,3-d]pyrimidin-4-yl)piperazin-1-yl]propan-1-amine), targeting the human AKT1 enzyme (PDB ID: 3OCB). The results are summarized in Figure 2 and Table 1.

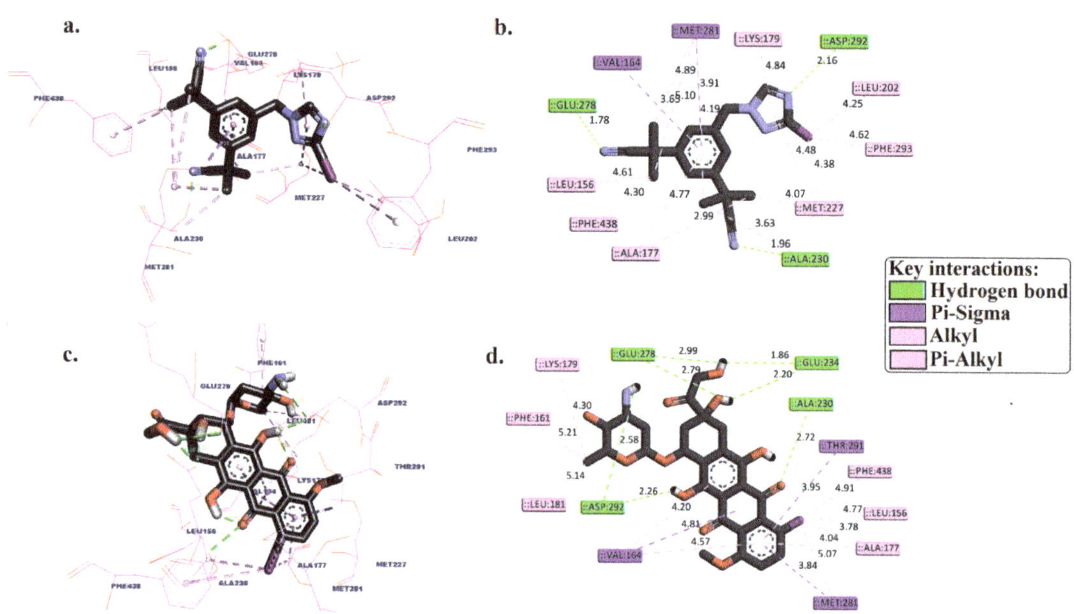

Figure 2. Three-dimensional and 2D binding pose interactions between [^{125}I]anastrozole (**a**,**b**) and [^{125}I]epirubicin (**c**,**d**) into the active binding site of the human AKT1 enzyme (PDB ID: 3OCB).

Table 1. Molecular docking scores, expressed as free binding energy (kcal/mol), for [^{125}I]anastrozole (a & b), [^{125}I]epirubicin, and the co-crystallized ligand, targeting the human AKT1 enzyme. The table also provides a detailed analysis of the 2D molecular interactions between these compounds and the key residues within the AKT1 active site (PDB ID: 3OCB).

Compounds	Free Binding Energy (kcal/mol)	Molecular Interactions Analysis within the AKT1 Active Binding Site			
		H-bond	Distance (Å)	Pi-Sigma	Hydrophobic Interaction
[^{125}I]anastrozole	−10.68	ALA230, GLU278, and ASP292	1.96, 1.78, and 2.16	VAL164 and MET281	LEU156, ALA177, LYS179, LEU202, MET227, MET281, PHE293, and PHE438
[^{125}I]epirubicin	−11.84	ALA230, GLU234, GLU234, GLU278, GLU278, ASP292, and ASP292	2.72, 1.86, 2.20, 2.79, 2.99, 2.26, and 2.58	VAL164, MET281, and THR291	LEU156, PHE161, VAL164, ALA177, LYS179, LEU181, PHE281, and PHE348
* Co-crystalized ligand (original pose)	−9.53	ALA230 and GLU278	1.98 and 2.67	-	VAL164, ALA177, LYS179, ALA230, and MET281

* Co-crystalized ligand: (2S)-2-(4-chlorobenzyl)-3-oxo-3-[4-(7H-pyrrolo [2,3-d]pyrimidin-4-yl)piperazin-1-yl]propan-1-amine.

2.2. Molecular Dynamic Simulation

The interactions between [^{125}I]anastrozole and [^{125}I]epirubicin with the human AKT1 enzyme were examined through a 200 ns molecular dynamics (MD) simulation using the GROMACS 2016 software suite. This analysis also included the co-crystallized ligand for comparison. Critical metrics, including the root mean square deviation (RMSD), root mean square fluctuation (RMSF), radius of gyration (Rg), and hydrogen bond profiles, were thoroughly analyzed. Furthermore, MM-PBSA calculations were conducted to evaluate the interactions between the ligands and the backbone atoms of the enzyme. The detailed results are illustrated in Figures 3–6 and are summarized in Table 2.

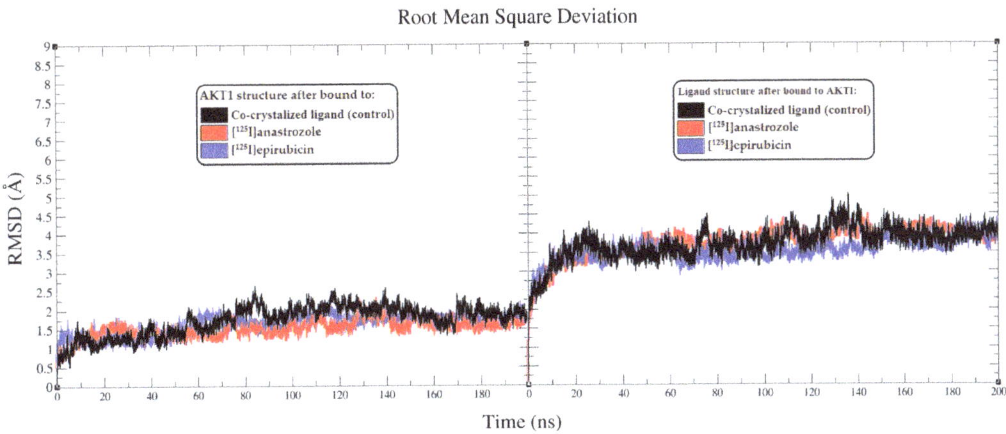

Figure 3. Analysis of root mean square deviation (RMSD) from the molecular dynamics (MD) simulations conducted over a 200 ns period. The left panel presents RMSD plots showing the molecular fluctuations in the AKT1 enzyme backbone when interacting with the co-crystallized ligand (black), [^{125}I]anastrozole (red), and [^{125}I]epirubicin (blue). The RMSD plots further illustrate the conformational shifts of the co-crystallized ligand (black), [^{125}I]anastrozole (red), and [^{125}I]epirubicin (blue) as they bind to the AKT1 enzyme.

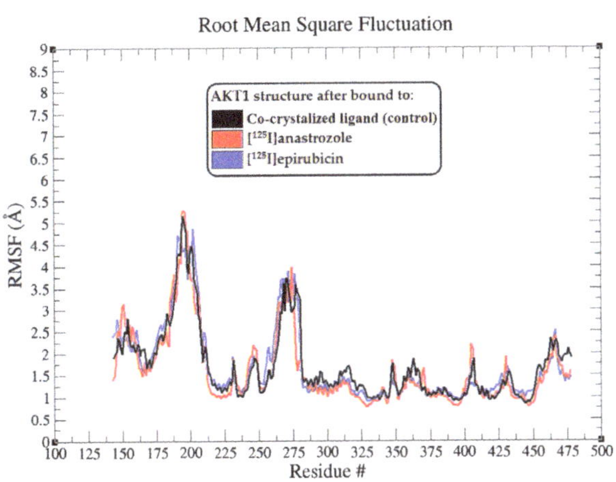

Figure 4. The root mean square fluctuation (RMSF) plots illustrate the fluctuations in the AKT1 backbone atoms over a 200 ns MD simulation for each system. These RMSF values capture the extent of atomic movement for each enzyme residue as its interact with the ligands throughout the simulation.

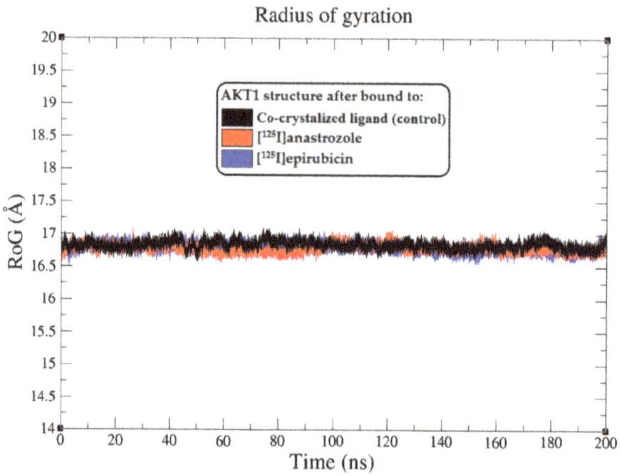

Figure 5. Radius of gyration (RoG) plots showing the AKT1 enzyme backbone atoms throughout the 200 ns MD simulation, illustrating the interactions with the co-crystallized ligand (black), [^{125}I]anastrozole (red), and [^{125}I]epirubicin (blue).

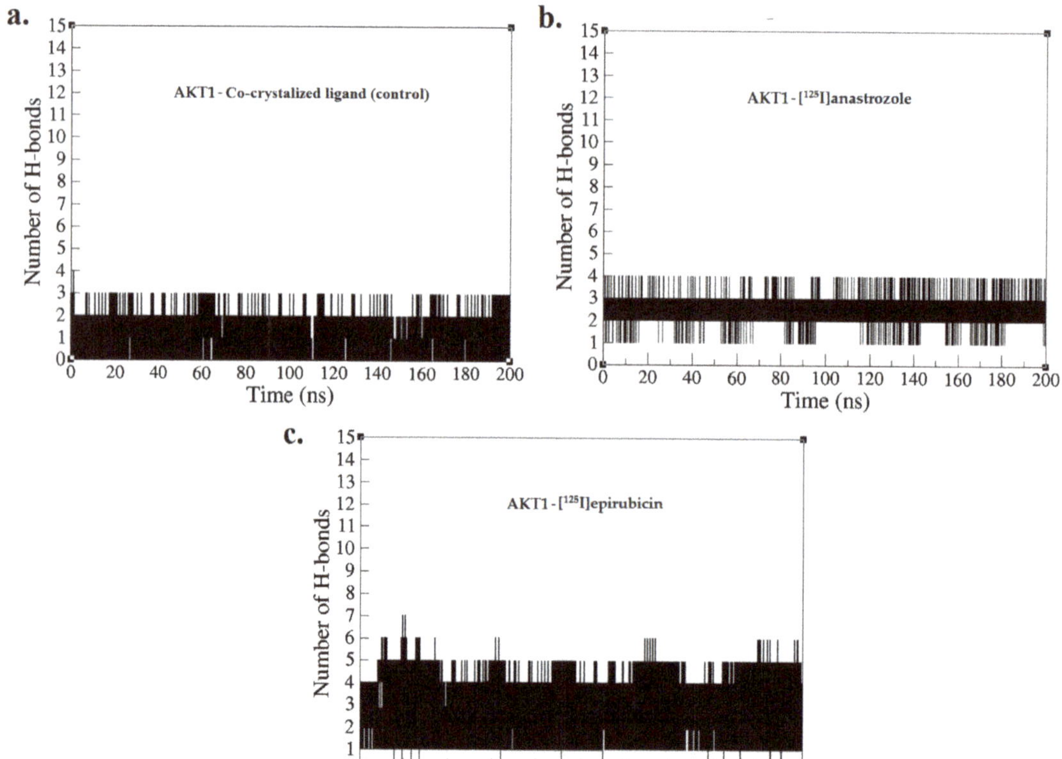

Figure 6. Hydrogen bond profiles obtained from the MD simulation over a time period of 0–200 ns for (**a**) AKT1-co-crystallized ligand, (**b**) AKT1-[^{125}I]anastrozole, and (**c**) AKT1-[^{125}I]epirubicin.

Table 2. MM-PBSA binding energies (ΔG_{bind}) of the co-crystallized ligand (control), [^{125}I]anastrozole, and [^{125}I]epirubicin at the active binding site of the AKT1 enzyme (PDB ID: 3OCB). Energy values are expressed in kcal/mol.

System	ΔG_{bind} (kcal/mol)	Electrostatic (kcal/mol)	Van der Waal (kcal/mol)	Polar Salvation (kcal/mol)	Non-Polar Salvation (kcal/mol)
AKT1-Co-crystallized ligand	−16.38 ± 0.14	−11.28 ± 0.11	−13.52 ± 0.12	19.11 ± 0.12	−10.69 ± 0.13
AKT1-[^{125}I]anastrozole	−20.03 ± 0.15	−12.86 ± 0.13	−14.69 ± 0.11	19.84 ± 0.13	−12.32 ± 0.12
AKT1-[^{125}I]epirubicin	−23.57 ± 0.14	−14.73 ± 0.12	−15.84 ± 0.14	19.86 ± 0.12	−12.86 ± 0.13

3. Discussion

3.1. Molecular Docking Simulation

Molecular docking is an essential computational technique for predicting the interaction between small molecules and their target proteins, forming stable complexes [57]. This method provides critical insights into binding orientations and affinities, playing a pivotal role in drug discovery by enabling the investigation of potential inhibitory effects on specific target enzymes [58–62]. In our study, the redocking of the co-crystallized served as a validation step for the docking parameters, ensuring that the computational model faithfully replicated the experimentally observed orientation of the ligand within the AKT1 active site [63].

Figure S1 demonstrates the effectiveness of the docking process, with panel (a) displaying the overlay of the co-crystallized ligand and the redocked ligand, highlighting the precision with which the redocking process replicated the original ligand orientation within the binding pocket [58–60,64,65]. The root mean square deviation (RMSD) of 0.98 Å serves as a quantitative measure of this comparison, indicating a high degree of conformational similarity between the two poses. Panels (b) and (c) in Figure S1 dissect the interaction profiles of the co-crystallized and redocked ligands, respectively. The diagrams detail key interactions, including hydrogen bonds, pi–sigma interactions, and hydrophobic interactions, which contribute to the stabilization of the ligand within the enzymatic cleft. Notably, the hydrogen bond between the ligand and ALA230 is maintained at a distance of 1.98 Å in the co-crystallized ligand, and a slightly longer distance of 2.67 Å in the redocked ligand, indicating a minor deviation in binding interaction post-redocking. The preservation of critical interactions, such as the hydrophobic interactions with VAL164 and MET281, demonstrates the redocking's ability to emulate the original binding mode. The interaction distances in the redocked ligand, ranging from 2.16 Å to 2.99 Å, are within acceptable limits compared to the co-crystallized ligand, affirming the reliability of the software used for the docking process [64–70]. Given the RMSD value of 0.98 Å, the redocking results fall within the generally accepted range for high-quality docking simulations, as established by the literature standards [64–70]. This degree of precision suggests that the docking software can be confidently applied to similar molecular investigations, such as the study of [^{125}I]anastrozole and [^{125}I]epirubicin, and the co-crystallized ligand against the AKT1 enzyme.

Figure 2 and Table 1 present a detailed analysis of the interactions involving [^{125}I] anastrozole, [^{125}I]epirubicin, and the co-crystallized ligand within the active binding site of AKT1. This analysis examines each molecule's binding affinity and interaction profile, focusing on key interactions, such as hydrogen bonds, hydrophobic interactions, and specifically, the interactions of the ^{125}I atom with the enzyme, which is crucial for understanding the mode of binding. [^{125}I]anastrozole exhibits a binding affinity with a ΔG_{bind} of −10.68 kcal/mol. It forms hydrogen bonds with ALA230, GLU278, and ASP292 at distances of 1.96 Å, 1.78 Å, and 2.16 Å, respectively. Additionally, it engages in pi–sigma interactions with VAL164 and MET281. Extensive hydrophobic interactions are noted with residues such as LEU156, ALA177, LYS179, LEU202, MET227, MET281, PHE293, and PHE438, which are crucial for non-polar binding complementarity. The interaction of ^{125}I with specific residues further suggests a competitive and reversible binding mode, likely contributing to the stabilization of the AKT1-ligand complex.

[^{125}I]epirubicin shows the most favorable binding affinity with a ΔG_{bind} of −11.84 kcal/mol. It forms a network of hydrogen bonds with ALA230, GLU234, GLU278, and ASP292 at distances ranging from 1.86 Å to 2.99 Å. In addition, it engages in pi–sigma interactions with VAL164, MET281, and THR291, alongside extensive hydrophobic interactions involving residues such as LEU156, PHE161, VAL164, ALA177, LYS179, LEU181, and PHE348. The interactions of ^{125}I with the enzyme suggest a highly specific binding mode, potentially leading to effective inhibition through competitive and reversible binding mechanisms. The co-crystallized ligand, with a ΔG_{bind} of −9.53 kcal/mol, forms hydrogen bonds with ALA230 and GLU278 at distances of 1.98 Å and 2.67 Å, respectively, but does not form pi–sigma interactions. It also engages in hydrophobic interactions with VAL164, ALA177, LYS179, ALA230, and MET281. The absence of iodine in this ligand precludes any direct comparison regarding iodine-specific interactions, but the overall binding profile highlights its role as a control for comparison.

Comparatively, [^{125}I]anastrozole and [^{125}I]epirubicin establish additional hydrogen bonds and specific ^{125}I interactions when compared to the co-crystallized ligand, potentially indicating stronger or additional points of enzyme engagement. This evidence suggests that [^{125}I]anastrozole and [^{125}I]epirubicin are promising candidates as AKT1 enzyme inhibitors, likely acting through competitive and reversible binding. However, while the binding energies and static interaction profiles are informative, they provide only a snapshot of a dynamic process. To gain a comprehensive understanding of the behavior and stability of these molecules at the molecular level, molecular dynamics simulations are essential. Such simulations will provide temporal insights into the flexibility, conformational changes, and resilience of the binding interactions under physiological conditions, which are critical parameters for the successful development of therapeutic agents.

3.2. Molecular Dynamic Simulation

The stability and conformational dynamics of the AKT1 enzyme when complexed with [^{125}I]anastrozole, [^{125}I]epirubicin, and the co-crystallized ligand were explored over a 200 ns simulation period. To assess structural dynamics, root mean square deviation (RMSD) analyses were performed, focusing on how each ligand influenced the conformational stability of AKT1 [71,72]. The RMSD trajectories for the AKT1 enzyme backbone displayed distinct stability patterns depending on the ligand involved (Figure 3). The complex with the co-crystallized ligand stabilized at an average RMSD value of approximately 2.69 Å after around 87 ns, reflecting consistent enzyme structure throughout the simulation. This stability provides a benchmark for evaluating the investigational ligands. Comparatively, the AKT1 enzyme exhibited greater stability in complexes with [^{125}I]anastrozole and [^{125}I]epirubicin, with both systems achieving stability after approximately 68 ns and maintaining an RMSD value near 2.2 Å. This suggests that these radio-iodinated ligands contribute to a more stable interaction within the enzyme's active site compared to the co-crystallized ligand. The RMSD fluctuations for the enzyme in these complexes were within a narrower range, less than 1.3 Å, indicating robust structural alignment and reduced conformational variability.

The RMSD values of the ligands themselves provide further insights into their conformational stability within the AKT1 binding site. Both [^{125}I]anastrozole and [^{125}I]epirubicin exhibited minimal conformational changes, with RMSD fluctuations averaging less than 1 Å. This consistent stability underscores their potential fit and effectiveness within the active site, suggesting that these radio-iodinated compounds maintain a high degree of structural integrity when bound to AKT1. The stable RMSD values reflect a strong and specific interaction, which is crucial for effective inhibition [73]. These RMSD data underscore the varying stability profiles of the AKT1-ligand complexes. Both [^{125}I]anastrozole and [^{125}I]epirubicin demonstrated significant conformational stability within the AKT1 binding site, indicating a strong potential to sustain interactions essential for effective enzyme inhibition. The stability observed with the co-crystallized ligand, while slightly less robust, still reinforces its potential as an AKT1 inhibitor, though with more variability in interaction dynamics compared to the radio-iodinated compounds.

The root mean square fluctuation (RMSF) analysis extends our understanding of the dynamic interplay between the AKT1 enzyme and the bound ligands, highlighting the flexibility of particular amino acid residues upon ligand binding [74]. The RMSF profiles, as depicted in Figure 4, are essential for pinpointing regions within the protein that show notable mobility, which could be critical for AKT1's function and interaction with ligands. Interestingly, the RMSF plots showed that there were no significant shifts in the residues or segments of the enzyme, and all ligands affected the mobility of the residues in a consistent pattern, with differences of less than 2 Å. This suggests that the overall flexibility of the AKT1 enzyme is preserved, regardless of the ligand binding, indicating a stable interaction profile across all tested ligands.

In this comprehensive analysis, the RoG was evaluated as an indicator of the enzyme's structural compactness upon ligand binding [74]. The Rg plots, illustrated in Figure 5, depict the Rg values of the AKT1 backbone atoms throughout the 0–200 ns MD simulation. The average Rg values offer a quantitative assessment of the overall shape and compactness of the protein–ligand complexes. The AKT1 enzyme, when complexed with the co-crystallized ligand, maintained an average RoG of approximately 16.8 ± 0.1 Å, suggesting a tightly packed protein structure. Similarly, the AKT1 complexes with [^{125}I]anastrozole and [^{125}I]epirubicin exhibited consistent average RoG values of around 16.8 ± 0.1 Å, indicating that these ligands do not significantly alter the compactness of the enzyme structure. The RMSF and RoG analyses provide additional evidence supporting the high stability and strong interaction profiles of [^{125}I]anastrozole and [^{125}I]epirubicin within the AKT1 binding site. These findings contribute to the growing body of data suggesting that these radio-iodinated compounds could serve as potent and specific inhibitors of the AKT1 enzyme, with potential applications in targeted breast cancer therapy.

Figure 6 depicts the number of hydrogen bonds formed between the ligands and the AKT1 enzyme during the simulation, providing a quantitative assessment of interaction strength and stability, which is essential for evaluating the efficacy of these compounds as inhibitors [75]. The AKT1 complex with the co-crystallized ligand consistently maintained between two and three hydrogen bonds throughout the duration of the simulation, with two bonds being consistently formed. This result confirms the predicted binding pose and indicates a stable interaction, consistent with the initial docking results presented in Figure 2. [^{125}I]anastrozole showed greater consistency in forming hydrogen bonds, fluctuating between three and four bonds, with three bonds being consistently maintained throughout the 200 ns MD simulation. This implies a potentially more effective interaction with AKT1, suggesting that [^{125}I]anastrozole may provide enhanced inhibitory effects due to its robust and stable hydrogen bonding. Conversely, [^{125}I]epirubicin exhibited the highest number of hydrogen bonds, fluctuating between four and five, with four bonds consistently maintained. The higher number of consistent hydrogen bonds suggests an even stronger and more stable interaction with the AKT1 enzyme, potentially leading to more effective inhibition due to improved positional integrity within the enzyme's active site. These hydrogen bond profiles from the MD simulations underscore the varying interaction strengths and stabilities across the different ligand-AKT1 complexes. [^{125}I]epirubicin, with its greater number of consistent hydrogen bonds, emerges as a particularly promising candidate for further development. The variations observed with the co-crystallized ligand highlight the critical role of optimizing hydrogen bond interactions to enhance ligand efficacy. [^{125}I]anastrozole, with its ability to maintain a stable number of hydrogen bonds, also demonstrates potential as an effective AKT1 inhibitor. These results suggest that radio-iodinated compounds, especially [^{125}I]epirubicin, could be effective targeted radiopharmaceuticals for breast cancer therapy.

To further support the previous analyses, Molecular Mechanics Poisson–Boltzmann Surface Area (MM-PBSA) calculations were conducted to obtain deeper insights into the binding free energies of the AKT1 complexes with the radio-iodinated compounds [76]. These calculations are crucial for validating the significance of the observed hydrogen bond interactions and for understanding the energetic contributions within the AKT1 enzyme's active binding site [76]. Table 2 summarizes the MM-PBSA binding energies for the co-crystallized ligand (control), [^{125}I]anastrozole, and [^{125}I]epirubicin, providing a quantitative perspective on the various energetic factors contributing to the overall binding affinity at the AKT1 active site.

The binding free energy (ΔG_{bind}) for the AKT1-co-crystallized ligand complex is -16.38 ± 0.14 kcal/mol, which serves as a reference point for comparing the other ligands. The electrostatic contributions for the co-crystallized ligand are substantial at -11.28 ± 0.11 kcal/mol, indicating that electrostatic forces play a significant role in the binding affinity. These are balanced by van der Waals interactions at -13.52 ± 0.12 kcal/mol and an unfavorable polar solvation energy of 19.11 ± 0.12 kcal/mol, offset by a favorable non-polar solvation energy of -10.69 ± 0.13 kcal/mol. The non-polar solvation energy is crucial for stabilizing the ligand within the enzyme's hydrophobic pockets. [^{125}I]anastrozole has a more favorable binding free energy of -20.03 ± 0.15 kcal/mol, indicating a stronger binding affinity than the co-crystallized ligand. This stronger affinity is due to higher electrostatic interactions at -12.86 ± 0.13 kcal/mol and more substantial van der Waals forces at -14.69 ± 0.11 kcal/mol. The elevated electrostatic interaction scores indicate an improved alignment within the active site, thereby enhancing the ligand's interaction with crucial residues. While the polar solvation penalty is slightly higher at 19.84 ± 0.13 kcal/mol, the favorable non-polar solvation energy of -12.32 ± 0.12 kcal/mol suggests effective accommodation within the enzyme's non-polar regions. [^{125}I]epirubicin shows the most favorable ΔG_{bind} of -23.57 ± 0.14 kcal/mol, with significant electrostatic interactions at -14.73 ± 0.12 kcal/mol and notable van der Waals forces at -15.84 ± 0.14 kcal/mol. These contributions indicate a well-balanced interaction dynamic within the active site, resulting in a robust binding affinity. Despite the slightly higher polar solvation energy at 19.86 ± 0.12 kcal/mol, the favorable non-polar solvation energy of -12.86 ± 0.13 kcal/mol supports the ligand's stable binding. The MM-PBSA binding energy analysis indicates that all ligands possess potential as AKT1 inhibitors, with [^{125}I]epirubicin exhibiting the highest binding affinity, closely followed by [^{125}I]anastrozole. The favorable binding free energies and substantial electrostatic contributions suggest that these investigational ligands hold promise as effective AKT1 inhibitors.

Overall, the computational analysis demonstrated that [^{125}I]epirubicin exhibited the most favorable binding affinity to AKT1, followed closely by [^{125}I]anastrozole, surpassing the affinity of the co-crystallized ligand. To contextualize these findings, it is imperative to compare them with the binding affinities and interaction profiles of other known AKT1 inhibitors documented in the literature. Prominent AKT1 inhibitors, such as capivasertib (AZD5363) [77,78] and afuresertib (GSK2110183) [79–81], have been extensively studied and have shown significant inhibitory effects on AKT1. For instance, capivasertib has been reported to exhibit binding affinities in the range of -9.0 to -10.0 kcal/mol [82,83], while afuresertib has shown affinities ranging from -8.5 to -9.5 kcal/mol [84,85]. When compared to these established inhibitors, the binding affinity of [^{125}I]epirubicin at -11.84 kcal/mol indicates a potentially stronger interaction with the AKT1 kinase domain, suggesting an enhanced therapeutic potential. The interaction profiles of capivasertib and afuresertib commonly involve critical residues within the AKT1 active site, such as THR308, SER473, and ASP292, which are integral to the enzyme's catalytic function. The findings indicate that [^{125}I]epirubicin and [^{125}I]anastrozole engage with similar residues, particularly ASP292, thereby implying that these radio-iodinated compounds could effectively inhibit AKT1 by interfering with its phosphorylation and subsequent activation. Notably, the MM-PBSA calculations revealed that [^{125}I]epirubicin forms more stable hydrogen bonds and van der Waals interactions compared to traditional AKT1 inhibitors, which may contribute to its superior binding stability.

However, it is important to recognize that the findings presented in this study are based on computational predictions, which, while informative, are inherently limited by the approximations and assumptions underlying molecular docking and dynamics simulations [58–60,64,65]. These methods do not fully capture the complexity of biological systems, including the influence of cellular environments, off-target effects, and metabolic stability. Thus, it is essential to validate these computational predictions with in vitro and in vivo studies to confirm the therapeutic efficacy and safety profiles of these compounds. Experimental studies will be crucial to verifying the investigational ligands' inhibitory action on AKT1 and determining their suitability for clinical development. Such studies will provide a more comprehensive understanding of the pharmacodynamics, pharmacokinetics, and potential toxicity of [^{125}I]anastrozole and [^{125}I]epirubicin, which are critical for assessing their viability as therapeutic agents.

4. Methodology

4.1. Molecular Docking Simulation

In this research, the focus was on the AKT1 enzyme, analyzed for its interaction with the radio-iodinated compounds anastrozole and epirubicin. The selection of AKT1 as the target for this study was based on its critical role in the PI3K/AKT/mTOR signaling pathway, which is frequently dysregulated in breast cancer and contributes to tumor growth and survival [39,40]. The crystallographic details of the AKT1 enzyme in complex with a pyrrolopyrimidine inhibitor ((2S)-2-(4-chlorobenzyl)-3-oxo-3-[4-(7H-pyrrolo [2,3-d]pyrimidin-4-yl)piperazin-1-yl]propan-1-amine) were obtained from the Protein Data Bank (PDB) [86], with the accession number 3OCB [87], (https://www.rcsb.org/structure/3OCB) accessed on 29 April 2024.

In preparation for molecular docking, the enzyme target underwent preprocessing to ensure its suitability for computational analysis. This involved the removal of non-essential water molecules and heteroatoms using Biovia Discovery Studio Visualizer [88], yielding a refined enzyme structure in a PDB format. Missing residues within the enzyme structure were reconstructed using the YASARA web server [71,89,90]. The protonation states of titratable amino acids were calculated at a physiological pH of 7.4 via the H^{++} web server [91]. Subsequently, polar hydrogen atoms were added, and Kollman charges were assigned, facilitating the conversion of the structure to a PDBQT format for subsequent docking simulations using AutoDock Tools version 1.5.6 [92].

The preparation and optimization of ligands for molecular docking simulations were conducted with the objective of maximizing the accuracy and reliability of the resulting computational predictions [93]. Radio-iodinated molecules ([^{125}I]anastrozole and [^{125}I]epirubicin) were drawn using the ChemDraw JS (v7.1) web server [71], inspired by structures outlined in previous studies [47], and saved in a structural data file (SDF) format. The ligands were subjected to energy minimization utilizing the universal force field and a conjugate gradient optimization algorithm [94]. This procedure, executed over a thousand iterations, was performed using Open Babel software (https://openbabel.org/) [95], resulting in final structures saved in the PDB format. Subsequently, Gasteiger charges were assigned using AutoDock Tools version 1.5.6, preparing the structures for docking simulations in the PDBQT format.

To elucidate the interactions between the ligands and the active site of AKT1, molecular docking was performed using AutoDock 4.2 [92]. A Lamarckian genetic algorithm was employed to optimize ligand configurations within the binding site, allowing flexibility in the ligand while keeping the macromolecule fixed [96]. The grid-box dimensions were set to $40 \times 40 \times 40$ Å for the X, Y, and Z axes, respectively, with central grid point coordinates at 12.916, 0.179, and 17.863. The docking process was conducted over 100 trials with a maximum of 25,000,000 evaluations per run, adhering to default parameters to ensure consistency in the analyses.

4.2. Molecular Dynamic Simulation

This study primarily investigated the interaction and stability of [^{125}I]anastrozole and [^{125}I]epirubicin within the active sites of the AKT1 enzyme using molecular dynamics (MD) simulations over a span of 200 ns. The crystallographic structure utilized for these simulations was the human AKT1 kinase domain complexed with a pyrrolopyrimidine inhibitor ((2S)-2-(4-chlorobenzyl)-3-oxo-3-[4-(7H-pyrrolo [2,3-d]pyrimidin-4-yl)piperazin-1-yl]propan-1-amine), obtained from the Protein Data Bank (PDB ID: 3OCB). The simulations were conducted using the GROMACS 2016.3 software package with the Gromos96 54a7 force field to perform an in-depth analysis of the ligand–protein interactions [97]. Over the 200 ns simulation period, valuable data on binding dynamics and molecular stability within the enzyme's active sites were gathered.

To initiate the simulations, topology files for both the ligands and the protein were prepared. Ligand topology was generated using the GROMACS 'pdb2gmx' function, while protein topology was created via the PRODRG server, accessed on 5 May 2024 [http://davapc1.bioch.dundee.ac.uk/cgi-bin/prodrg]. The systems were subsequently solvated in a TIP3P water model and neutralized with counterions to establish a balanced environment conducive to interaction studies. Energy minimization was then performed using the steepest descent method for 50,000 steps, with each step measured at 0.01 energy units, to ensure optimal system stability [71,72].

The system equilibration process was executed in two phases: initially using the NVT ensemble for 100 ps at a temperature of 310 K with the v-rescale thermostat, followed by the NPT ensemble for another 100 ps at 1.0 bar using the Berendsen pressure coupling method [98,99]. These steps were crucial for ensuring that the systems reached a stable state prior to the molecular dynamics (MD) simulations. The MD simulations were subsequently run for 200 ns under constant conditions of 310 K and 1 bar. A 1.0 nm cut-off was applied for short-range, non-bonded interactions, while long-range, electrostatic interactions were handled using the Particle Mesh Ewald method [100]. To preserve structural integrity, the LINCS algorithm was used to constrain bonds involving hydrogen atoms [101]. The simulations employed a timestep of 2 fs with coordinate data saved every 5000 steps (10 ps), a strategy that ensured accurate and reliable simulation results [71,72].

To assess the fidelity and stability of the ligand–enzyme interactions, a comprehensive analysis of the MD trajectory data was conducted, utilizing metrics such as the root mean square deviation (RMSD), root mean square fluctuation (RMSF), radius of gyration (Rg), and hydrogen bond profiles. This methodological approach reflects a rigorous commitment to obtaining precise and meaningful insights into the molecular dynamics of the systems under investigation.

4.3. MM-PBSA Calculation

In this study, the Molecular Mechanics Poisson–Boltzmann Surface Area (MM-PBSA) method was applied to estimate binding free energies based on MD trajectory snapshots [76]. This method, known for its effectiveness in accurately assessing binding energies, enabled a comprehensive analysis of the energetic interactions between the radio-iodinated compounds ([^{125}I]anastrozole and [^{125}I]epirubicin) and the co-crystallized ligand within the AKT1 enzyme's active site. MM-PBSA calculations were performed during the production phase of the MD simulations, with snapshots taken at 100 ps intervals over the final 20 ns (from 180 to 200 ns). These calculations were conducted using the g_mmpbsa tool within the GROMACS software package [102,103], allowing for a detailed evaluation of interaction strengths.

This method offered valuable insights into the binding efficiencies of [^{125}I]anastrozole and [^{125}I]epirubicin relative to the co-crystallized ligand, elucidating the molecular features that could influence their therapeutic potential. The equations used for calculating the binding free energy of the ligand–enzyme complex in solvent, which form the foundation of the MM-PBSA method, have been extensively documented in the existing literature [71,72].

5. Conclusions

This study represents the first comprehensive in silico exploration of the binding potential of radioiodinated anastrozole ([^{125}I]anastrozole) and epirubicin ([^{125}I]epirubicin) against the AKT1 enzyme, a critical target in breast cancer therapy. Utilizing a range of computational techniques, including molecular docking, molecular dynamics simulations (200 ns), and MM-PBSA calculations, the therapeutic potential of these compounds was thoroughly evaluated. Molecular docking results indicate that [^{125}I]epirubicin displays the highest binding affinity to AKT1 with a ΔG_{bind} of -11.84 kcal/mol, followed by [^{125}I]anastrozole at -10.68 kcal/mol, and the co-crystallized ligand at -9.53 kcal/mol. These findings suggest that [^{125}I]epirubicin may provide superior inhibitory effects against AKT1 due to its stronger molecular interactions within the active site. Molecular dynamics simulations further corroborated these results, showing that both [^{125}I]anastrozole and [^{125}I]epirubicin formed stable complexes with AKT1, as reflected by RMSD values consistently around 2.2 Å, while the co-crystallized ligand stabilized at approximately 2.69 Å. RMSF and radius of gyration analyses revealed minimal disruption to the enzyme's overall structure and dynamics, which is crucial for preserving its biological function. Specifically, the consistent RMSF patterns and average Rg values (~16.8 $\pm$ 0.1 Å) suggest that these ligands do not significantly alter the enzyme's compactness or flexibility. Additionally, MM-PBSA calculations supported the findings from docking and dynamics simulations, confirming that [^{125}I]epirubicin exhibits the most favorable binding energy (-23.57 ± 0.14 kcal/mol), driven by significant electrostatic and van der Waals interactions. The binding free energies of [^{125}I]anastrozole and the co-crystallized ligand also indicated their potential as AKT1 inhibitors, although to a slightly lesser degree compared to [^{125}I]epirubicin. This study highlights the promising inhibitory potential of [^{125}I]anastrozole and [^{125}I]epirubicin, particularly [^{125}I]epirubicin, against AKT1, suggesting that these compounds could serve as valuable leads in the development of new therapeutic agents for breast cancer. Future research should focus on empirical validation through in vitro and in vivo testing to confirm these computational predictions and fully assess the therapeutic efficacy and safety profiles of these compounds. Planned studies include cell-based assays to evaluate cytotoxicity and enzyme inhibition, as well as animal models to assess pharmacokinetics, biodistribution, and therapeutic outcomes. The broader impact of these findings lies in their contribution to the field of radiopharmaceuticals and personalized cancer therapy. By demonstrating the potential of [^{125}I]anastrozole and [^{125}I]epirubicin as AKT1-targeted radiopharmaceuticals, this study advances the development of novel diagnostic and therapeutic agents that can be tailored to the molecular profile of individual tumors, thereby enhancing treatment specificity and efficacy. Such advancements could pave the way for more effective and personalized treatment strategies in oncology, particularly in managing breast cancer.

Supplementary Materials: The following supporting information can be downloaded at: https://www.mdpi.com/article/10.3390/molecules29174203/s1, Figure S1. Radio-iodinated anastrozole (a) and epirubicin (b). Figure S2. Superimposition (a) and 2D interactions analysis of the co-crystallized ligand (green C, red O, and blue N) (b) and redocked ligand (orange C, red O, and blue N) (c). The crystal structure of human Akt1 kinase domain in complex with pyrrolopyrimidine inhibitor (3OCB.pdb) (RMSD is ~0.98 Å).

Funding: This research received no specific grant from any funding agency in the public, commercial, or not-for-profit sectors.

Institutional Review Board Statement: Not applicable.

Informed Consent Statement: Not applicable.

Data Availability Statement: Data are contained in manuscript and Supplementary Materials.

Acknowledgments: The author wishes to express his gratitude to Imam Mohammad Ibn Saud Islamic University (IMSIU) for its unending support of this research.

Conflicts of Interest: The author declares no conflict of interest.

References

1. Torre, L.A.; Islami, F.; Siegel, R.L.; Ward, E.M.; Jemal, A. Global cancer in women: Burden and trends. *Cancer Epidemiol. Biomark. Prev.* **2017**, *26*, 444–457. [CrossRef] [PubMed]
2. Arnold, M.; Morgan, E.; Rumgay, H.; Mafra, A.; Singh, D.; Laversanne, M.; Vignat, J.; Gralow, J.R.; Cardoso, F.; Siesling, S. Current and future burden of breast cancer: Global statistics for 2020 and 2040. *Breast* **2022**, *66*, 15–23. [CrossRef]
3. Ferlay, J.; Colombet, M.; Soerjomataram, I.; Parkin, D.M.; Piñeros, M.; Znaor, A.; Bray, F. Cancer statistics for the year 2020: An overview. *Int. J. Cancer* **2021**, *149*, 778–789. [CrossRef] [PubMed]
4. Azubuike, S.O.; Muirhead, C.; Hayes, L.; McNally, R. Rising global burden of breast cancer: The case of sub-Saharan Africa (with emphasis on Nigeria) and implications for regional development: A review. *World J. Surg. Oncol.* **2018**, *16*, 1–13. [CrossRef]
5. Bhushan, A.; Gonsalves, A.; Menon, J.U. Current state of breast cancer diagnosis, treatment, and theranostics. *Pharmaceutics* **2021**, *13*, 723. [CrossRef] [PubMed]
6. Joshi, A.; GK, A.V.; Sakorikar, T.; Kamal, A.M.; Vaidya, J.S.; Pandya, H.J. Recent advances in biosensing approaches for point-of-care breast cancer diagnostics: Challenges and future prospects. *Nanoscale Adv.* **2021**, *3*, 5542–5564. [CrossRef]
7. López-Portugués, C.; Montes-Bayón, M.; Díez, P. Biomarkers in Ovarian Cancer: Towards Personalized Medicine. *Proteomes* **2024**, *12*, 8. [CrossRef]
8. Abeelh, E.A.; AbuAbeileh, Z. Comparative Effectiveness of Mammography, Ultrasound, and MRI in the Detection of Breast Carcinoma in Dense Breast Tissue: A Systematic Review. *Cureus* **2024**, *16*, e59054. [CrossRef]
9. Rowe, S.P.; Pomper, M.G. Molecular imaging in oncology: Current impact and future directions. *CA Cancer J. Clin.* **2022**, *72*, 333–352. [CrossRef]
10. Manafi-Farid, R.; Ranjbar, S.; Jamshidi Araghi, Z.; Pilz, J.; Schweighofer-Zwink, G.; Pirich, C.; Beheshti, M. Molecular imaging in primary staging of prostate cancer patients: Current aspects and future trends. *Cancers* **2021**, *13*, 5360. [CrossRef] [PubMed]
11. Sgouros, G.; Bodei, L.; McDevitt, M.R.; Nedrow, J.R. Radiopharmaceutical therapy in cancer: Clinical advances and challenges. *Nat. Rev. Drug Discov.* **2020**, *19*, 589–608. [CrossRef] [PubMed]
12. Barca, C.; Griessinger, C.M.; Faust, A.; Depke, D.; Essler, M.; Windhorst, A.D.; Devoogdt, N.; Brindle, K.M.; Schäfers, M.; Zinnhardt, B. Expanding theranostic radiopharmaceuticals for tumor diagnosis and therapy. *Pharmaceuticals* **2021**, *15*, 13. [CrossRef] [PubMed]
13. Signore, A.; Lauri, C.; Auletta, S.; Varani, M.; Onofrio, L.; Glaudemans, A.W.; Panzuto, F.; Marchetti, P. Radiopharmaceuticals for Breast Cancer and Neuroendocrine Tumors: Two Examples of How Tissue Characterization May Influence the Choice of Therapy. *Cancers* **2020**, *12*, 781. [CrossRef] [PubMed]
14. Crișan, G.; Moldovean-Cioroianu, N.S.; Timaru, D.-G.; Andrieș, G.; Căinap, C.; Chiș, V. Radiopharmaceuticals for PET and SPECT imaging: A literature review over the last decade. *Int. J. Mol. Sci.* **2022**, *23*, 5023. [CrossRef]
15. Theodoropoulos, A.S.; Gkiozos, I.; Kontopyrgias, G.; Charpidou, A.; Kotteas, E.; Kyrgias, G.; Tolia, M. Modern radiopharmaceuticals for lung cancer imaging with positron emission tomography/computed tomography scan: A systematic review. *SAGE Open Med.* **2020**, *8*, 2050312120961594. [CrossRef]
16. Alqahtani, F.F. SPECT/CT and PET/CT, related radiopharmaceuticals, and areas of application and comparison. *Saudi Pharm. J.* **2023**, *31*, 312–328. [CrossRef] [PubMed]
17. Linden, H.M.; Stekhova, S.A.; Link, J.M.; Gralow, J.R.; Livingston, R.B.; Ellis, G.K.; Petra, P.H.; Peterson, L.M.; Schubert, E.K.; Dunnwald, L.K. Quantitative fluoroestradiol positron emission tomography imaging predicts response to endocrine treatment in breast cancer. *J. Clin. Oncol.* **2006**, *24*, 2793–2799. [CrossRef]
18. Venema, C.M.; Apollonio, G.; Hospers, G.A.; Schröder, C.P.; Dierckx, R.A.; de Vries, E.F.; Glaudemans, A.W. Recommendations and technical aspects of 16α-[18F] fluoro-17β-estradiol PET to image the estrogen receptor in vivo: The Groningen experience. *Clin. Nucl. Med.* **2016**, *41*, 844–851. [CrossRef]
19. Beauregard, J.-M.; Croteau, É.; Ahmed, N.; van Lier, J.E.; Bénard, F. Assessment of human biodistribution and dosimetry of 4-fluoro-11β-methoxy-16α-^{18}F-fluoroestradiol using serial whole-body PET/CT. *J. Nucl. Med.* **2009**, *50*, 100–107. [CrossRef]
20. McLarty, K.; Cornelissen, B.; Scollard, D.A.; Done, S.J.; Chun, K.; Reilly, R.M. Associations between the uptake of 111 In-DTPA-trastuzumab, HER2 density and response to trastuzumab (Herceptin) in athymic mice bearing subcutaneous human tumour xenografts. *Eur. J. Nucl. Med. Mol. Imaging* **2009**, *36*, 81–93. [CrossRef]
21. Lin, M.; Paolillo, V.; Le, D.B.; Macapinlac, H.; Ravizzini, G.C. Monoclonal antibody based radiopharmaceuticals for imaging and therapy. *Curr. Probl. Cancer* **2021**, *45*, 100796. [CrossRef]
22. Mankoff, D.A.; Peterson, L.M.; Tewson, T.J.; Link, J.M.; Gralow, J.R.; Graham, M.M.; Krohn, K.A. [^{18}F] fluoroestradiol radiation dosimetry in human PET studies. *J. Nucl. Med.* **2001**, *42*, 679–684. [PubMed]
23. Peterson, L.M.; Mankoff, D.A.; Lawton, T.; Yagle, K.; Schubert, E.K.; Stekhova, S.; Gown, A.; Link, J.M.; Tewson, T.; Krohn, K.A. Quantitative imaging of estrogen receptor expression in breast cancer with PET and ^{18}F-fluoroestradiol. *J. Nucl. Med.* **2008**, *49*, 367–374. [CrossRef] [PubMed]
24. Hospers, G.; Helmond, F.; De Vries, E.; Dierckx, R.; De Vries, E. PET imaging of steroid receptor expression in breast and prostate cancer. *Curr. Pharm. Des.* **2008**, *14*, 3020–3032. [CrossRef]
25. Dehdashti, F.; Mortimer, J.E.; Trinkaus, K.; Naughton, M.J.; Ellis, M.; Katzenellenbogen, J.A.; Welch, M.J.; Siegel, B.A. PET-based estradiol challenge as a predictive biomarker of response to endocrine therapy in women with estrogen-receptor-positive breast cancer. *Breast Cancer Res. Treat.* **2009**, *113*, 509–517. [CrossRef] [PubMed]

26. Dehdashti, F.; Flanagan, F.L.; Mortimer, J.E.; Katzenellenbogen, J.A.; Welch, M.J.; Siegel, B.A. Positron emission tomographic assessment of "metabolic flare" to predict response of metastatic breast cancer to antiestrogen therapy. *Eur. J. Nucl. Med.* **1999**, *26*, 51–56. [CrossRef]
27. Kurland, B.F.; Peterson, L.M.; Lee, J.H.; Schubert, E.K.; Currin, E.R.; Link, J.M.; Krohn, K.A.; Mankoff, D.A.; Linden, H.M. Estrogen receptor binding (^{18}F-FES PET) and glycolytic activity (^{18}F-FDG PET) predict progression-free survival on endocrine therapy in patients with ER+ breast cancer. *Clin. Cancer Res.* **2017**, *23*, 407–415. [CrossRef]
28. van Kruchten, M.; Glaudemans, A.W.; de Vries, E.F.; Beets-Tan, R.G.; Schröder, C.P.; Dierckx, R.A.; de Vries, E.G.; Hospers, G.A. PET imaging of estrogen receptors as a diagnostic tool for breast cancer patients presenting with a clinical dilemma. *J. Nucl. Med.* **2012**, *53*, 182–190. [CrossRef]
29. Venema, C.; de Vries, E.; Glaudemans, A.; Poppema, B.; Hospers, G.; Schröder, C. 18F-FES PET has added value in staging and therapy decision making in patients with disseminated lobular breast cancer. *Clin. Nucl. Med.* **2017**, *42*, 612–614. [CrossRef]
30. van Kruchten, M.; de Vries, E.G.; Glaudemans, A.W.; van Lanschot, M.C.; van Faassen, M.; Kema, I.P.; Brown, M.; Schröder, C.P.; de Vries, E.F.; Hospers, G.A. Measuring residual estrogen receptor availability during fulvestrant therapy in patients with metastatic breast cancer. *Cancer Discov.* **2015**, *5*, 72–81. [CrossRef]
31. Yang, Z.; Sun, Y.; Xue, J.; Yao, Z.; Xu, J.; Cheng, J.; Shi, W.; Zhu, B.; Zhang, Y.; Zhang, Y. Can Positron Emission Tomography/Computed Tomography with the Dual Tracers Fluorine-18 Fluoroestradiol and Fluorodeoxyglucose Predict Neoadjuvant Chemotherapy Response of Breast Cancer?—A Pilot Study. *PLoS ONE* **2013**, *8*, e78192. [CrossRef] [PubMed]
32. de Vries, E.; Venema, C.; Glaudemans, A.; Jager, A.; Garner, F.; O'Neill, A.; Patki, A.; van Menke-van der Houven, C. A phase 1 study of RAD1901, an oral selective estrogen receptor degrader, to determine changes in the F-18-FES uptake and tumor responses in ER-positive, HER2-negative, advanced breast cancer patients. *Cancer Res.* **2017**, *77*, P2-08-08. [CrossRef]
33. van Kruchten, M.; de Vries, E.F.; Arts, H.J.; Jager, N.M.; Bongaerts, A.H.; Glaudemans, A.W.; Hollema, H.; de Vries, E.G.; Hospers, G.A.; Reyners, A.K. Assessment of estrogen receptor expression in epithelial ovarian cancer patients using 16α-^{18}F-fluoro-17β-estradiol PET/CT. *J. Nucl. Med.* **2015**, *56*, 50–55. [CrossRef]
34. Antunes, I.F.; Willemsen, A.T.; Sijbesma, J.W.; Boerema, A.S.; van Waarde, A.; Glaudemans, A.W.; Dierckx, R.A.; de Vries, E.G.; Hospers, G.A.; de Vries, E.F. In vivo quantification of Erβ expression by pharmacokinetic modeling: Studies with ^{18}F-FHNP PET. *J. Nucl. Med.* **2017**, *58*, 1743–1748. [CrossRef] [PubMed]
35. Güleç, B.A.; Yurt, F. Treatment with radiopharmaceuticals and radionuclides in breast cancer: Current options. *Eur. J. Breast Health* **2021**, *17*, 214. [CrossRef] [PubMed]
36. Filippi, L.; Urso, L.; Ferrari, C.; Guglielmo, P.; Evangelista, L. The impact of PET imaging on triple negative breast cancer: An updated evidence-based perspective. *Eur. J. Nucl. Med. Mol. Imaging* **2024**, 1–17. [CrossRef]
37. Vito, A.; Rathmann, S.; Mercanti, N.; El-Sayes, N.; Mossman, K.; Valliant, J. Combined radionuclide therapy and immunotherapy for treatment of triple negative breast cancer. *Int. J. Mol. Sci.* **2021**, *22*, 4843. [CrossRef]
38. Raguraman, R.; Srivastava, A.; Munshi, A.; Ramesh, R. Therapeutic approaches targeting molecular signaling pathways common to diabetes, lung diseases and cancer. *Adv. Drug Deliv. Rev.* **2021**, *178*, 113918. [CrossRef] [PubMed]
39. Miricescu, D.; Totan, A.; Stanescu-Spinu, I.-I.; Badoiu, S.C.; Stefani, C.; Greabu, M. PI3K/AKT/mTOR signaling pathway in breast cancer: From molecular landscape to clinical aspects. *Int. J. Mol. Sci.* **2020**, *22*, 173. [CrossRef]
40. Ortega, M.A.; Fraile-Martínez, O.; Asúnsolo, Á.; Buján, J.; García-Honduvilla, N.; Coca, S. Signal transduction pathways in breast cancer: The important role of PI3K/Akt/mTOR. *J. Oncol.* **2020**, *2020*, 9258396. [CrossRef]
41. Mitsiades, C.S.; Mitsiades, N.; Koutsilieris, M. The Akt pathway: Molecular targets for anti-cancer drug development. *Curr. Cancer Drug Targets* **2004**, *4*, 235–256. [CrossRef]
42. Kumar, A.; Rajendran, V.; Sethumadhavan, R.; Purohit, R. AKT kinase pathway: A leading target in cancer research. *Sci. World J.* **2013**, *2013*, 756134. [CrossRef]
43. Nitulescu, G.M.; Margina, D.; Juzenas, P.; Peng, Q.; Olaru, O.T.; Saloustros, E.; Fenga, C.; Spandidos, D.A.; Libra, M.; Tsatsakis, A.M. Akt inhibitors in cancer treatment: The long journey from drug discovery to clinical use. *Int. J. Oncol.* **2016**, *48*, 869–885. [CrossRef] [PubMed]
44. Lin, J.; Sampath, D.; Nannini, M.A.; Lee, B.B.; Degtyarev, M.; Oeh, J.; Savage, H.; Guan, Z.; Hong, R.; Kassees, R. Targeting activated Akt with GDC-0068, a novel selective Akt inhibitor that is efficacious in multiple tumor models. *Clin. Cancer Res.* **2013**, *19*, 1760–1772. [CrossRef] [PubMed]
45. He, Y.; Sun, M.M.; Zhang, G.G.; Yang, J.; Chen, K.S.; Xu, W.W.; Li, B. Targeting PI3K/Akt signal transduction for cancer therapy. *Signal Transduct. Target. Ther.* **2021**, *6*, 425. [CrossRef]
46. Nandan, D.; Zhang, N.; Yu, Y.; Schwartz, B.; Chen, S.; Kima, P.E.; Reiner, N.E. Miransertib (ARQ 092), an orally-available, selective Akt inhibitor is effective against Leishmania. *PLoS ONE* **2018**, *13*, e0206920. [CrossRef]
47. Ibrahim, A.; Sakr, T.; Khoweysa, O.; Motaleb, M.; Abd El-Bary, A.; El-Kolaly, M. Radioiodinated anastrozole and epirubicin as potential targeting radiopharmaceuticals for solid tumor imaging. *J. Radioanal. Nucl. Chem.* **2015**, *303*, 967–975. [CrossRef]
48. Binmujlli, M.A. Radiological and Molecular Analysis of Radioiodinated Anastrozole and Epirubicin as Innovative Radiopharmaceuticals Targeting Methylenetetrahydrofolate Dehydrogenase 2 in Solid Tumors. *Pharmaceutics* **2024**, *16*, 616. [CrossRef] [PubMed]
49. Tenenbaum, F.; Ricard, M. Peroperative detection probes. Evaluation and perspectives in endocrinology. In *Annales d'Endocrinologie*; Elsevier: Masson, The Netherlands, 1997; pp. 2213–3941.

50. Mross, K.; Mayer, U.; Langenbuch, T.; Hamm, K.; Burk, K.; Hossfeld, D. Toxicity, pharmacokinetics and metabolism of iododoxorubicin in cancer patients. *Eur. J. Cancer Clin. Oncol.* **1990**, *26*, 1156–1162. [CrossRef]
51. Twelves, C.; Dobbs, N.; Lawrence, M.; Ramirez, A.; Summerhayes, M.; Richards, M.; Towlson, K.; Rubens, R. Iododoxorubicin in advanced breast cancer: A phase II evaluation of clinical activity, pharmacology and quality of life. *Br. J. Cancer* **1994**, *69*, 726–731. [CrossRef]
52. Formelli, F.; Carsana, R.; Pollini, C. Pharmacokinetics of 4′-deoxy-4′-iodo-doxorubicin in plasma and tissues of tumor-bearing mice compared with doxorubicin. *Cancer Res.* **1987**, *47*, 5401–5406. [PubMed]
53. Amine, A.; Mohammadi, H.; Bourais, I.; Palleschi, G. Enzyme inhibition-based biosensors for food safety and environmental monitoring. *Biosens. Bioelectron.* **2006**, *21*, 1405–1423. [CrossRef]
54. Copeland, R.A. *Evaluation of Enzyme Inhibitors in Drug Discovery: A Guide for Medicinal Chemists and Pharmacologists*; John Wiley & Sons: New York, NY, USA, 2013.
55. Alonso, H.; Bliznyuk, A.A.; Gready, J.E. Combining docking and molecular dynamic simulations in drug design. *Med. Res. Rev.* **2006**, *26*, 531–568. [CrossRef]
56. Sanabria-Chanaga, E.E.; Betancourt-Conde, I.; Hernández-Campos, A.; Téllez-Valencia, A.; Castillo, R. In silico hit optimization toward AKT inhibition: Fragment-based approach, molecular docking and molecular dynamics study. *J. Biomol. Struct. Dyn.* **2019**, *37*, 4301–4311. [CrossRef]
57. Kaur, T.; Madgulkar, A.; Bhalekar, M.; Asgaonkar, K. Molecular docking in formulation and development. *Curr. Drug Discov. Technol.* **2019**, *16*, 30–39. [CrossRef] [PubMed]
58. Alhawarri, M.B. Exploring the Anticancer Potential of Furanpydone A: A Computational Study on its Inhibition of MTHFD2 Across Diverse Cancer Cell Lines. *Cell Biochem. Biophys.* **2024**, 1–18. [CrossRef]
59. Ibrahim, M.M.; Azmi, M.N.; Alhawarri, M.B.; Kamal, N.N.S.N.M.; AbuMahmoud, H. Synthesis, characterization and bioactivity of new pyridine-2(H)-one, nicotinonitrile, and furo[2,3-b]pyridine derivatives. *Mol. Divers.* **2024**, 1–19. [CrossRef]
60. Alidmat, M.M.; Alhawarri, M.B.; Al-Refai, M.; Mansi, I.A.; Al-Balas, Q.; Ibrahim, M.M. Synthesis, Characterization and Glyoxalase inhibitory activity of 4,6-Diheteroarylpyrimidine-2-amine derivatives: In vitro and in silico studies. *Egypt. J. Chem.* **2024**. [CrossRef]
61. Ferreira, L.G.; Dos Santos, R.N.; Oliva, G.; Andricopulo, A.D. Molecular docking and structure-based drug design strategies. *Molecules* **2015**, *20*, 13384–13421. [CrossRef]
62. De Ruyck, J.; Brysbaert, G.; Blossey, R.; Lensink, M.F. Molecular docking as a popular tool in drug design, an in silico travel. *Adv. Appl. Bioinform. Chem.* **2016**, *9*, 1–11. [CrossRef]
63. Mukherjee, S. Docking Platform and Validation Resources for Structure-based Drug Design. Ph.D. Thesis, State University of New York, Stony Brook, NY, USA, 2012.
64. Alhawarri, M.B.; Al-Thiabat, M.G.; Dubey, A.; Tufail, A.; Fouad, D.; Alrimawi, B.H.; Dayoob, M. ADME profiling, molecular docking, DFT, and MEP analysis reveal cissamaline, cissamanine, and cissamdine from Cissampelos capensis Lf as potential anti-Alzheimer's agents. *RSC Adv.* **2024**, *14*, 9878–9891. [CrossRef] [PubMed]
65. Alhawarri, M.B.; Olimat, S. Potential Serotonin 5-HT2A Receptor Agonist of Psychoactive Components of Silene undulata Aiton: LC-MS/MS, ADMET, and Molecular Docking Studies. *Curr. Pharm. Biotechnol.* **2024**. [CrossRef]
66. Yunos, N.M.; Al-Thiabat, M.G.; Sallehudin, N.J. Quassinoids from Eurycoma longifolia as Potential Dihydrofolate Reductase Inhibitors: A Computational Study. *Curr. Pharm. Biotechnol.* **2024**, *25*, 2154–2165. [CrossRef] [PubMed]
67. Yunos, N.M.; Wahab, H.A.; Al-Thiabat, M.G.; Sallehudin, N.J.; Jauri, M.H. In Vitro and In Silico Analysis of the Anticancer Effects of Eurycomanone and Eurycomalactone from Eurycoma longifolia. *Plants* **2023**, *12*, 2827. [CrossRef]
68. Alhawarri, M.B.; Dianita, R.; Rawa, M.S.A.; Nogawa, T.; Wahab, H.A. Potential Anti-Cholinesterase Activity of Bioactive Compounds Extracted from *Cassia grandis* Lf and *Cassia timoriensis* DC. *Plants* **2023**, *12*, 344. [CrossRef] [PubMed]
69. Amir Rawa, M.S.; Al-Thiabat, M.G.; Nogawa, T.; Futamura, Y.; Okano, A.; Wahab, H.A. Naturally Occurring 8ß, 13ß-kaur-15-en-17-al and Anti-Malarial Activity from *Podocarpus polystachyus* Leaves. *Pharmaceuticals* **2022**, *15*, 902. [CrossRef]
70. Alidmat, M.M.; Khairuddean, M.; Kamal, N.N.S.N.M.; Muhammad, M.; Wahab, H.A.; Althiabat, M.G.; Alhawarri, M.B. Synthesis, Characterization, Molecular Docking and Cytotoxicity Evaluation of New Thienyl Chalcone Derivatives against Breast Cancer Cells. *Syst. Rev. Pharm.* **2022**, *13*, 1.
71. Shalayel, M.H.F.; Al-Mazaideh, G.M.; Alanezi, A.A.; Almuqati, A.F.; Alotaibi, M. Diosgenin and Monohydroxy Spirostanol from *Prunus amygdalus* var *amara* Seeds as Potential Suppressors of EGFR and HER2 Tyrosine Kinases: A Computational Approach. *Pharmaceuticals* **2023**, *16*, 704. [CrossRef]
72. Shalayel, M.H.F.; Al-Mazaideh, G.M.; Alanezi, A.A.; Almuqati, A.F.; Alotaibi, M. The Potential Anti-Cancerous Activity of *Prunus amygdalus* var. *amara* Extract. *Processes* **2023**, *11*, 1277. [CrossRef]
73. Li, D.-D.; Wu, T.-T.; Yu, P.; Wang, Z.-Z.; Xiao, W.; Jiang, Y.; Zhao, L.-G. Molecular dynamics analysis of binding sites of epidermal growth factor receptor kinase inhibitors. *ACS Omega* **2020**, *5*, 16307–16314. [CrossRef]
74. Janati-Fard, F.; Housaindokht, M.R.; Monhemi, H. Investigation of structural stability and enzymatic activity of glucose oxidase and its subunits. *J. Mol. Catal. B Enzym.* **2016**, *134*, 16–24. [CrossRef]

75. da Fonseca, A.M.; Caluaco, B.J.; Madureira, J.M.C.; Cabongo, S.Q.; Gaieta, E.M.; Djata, F.; Colares, R.P.; Neto, M.M.; Fernandes, C.F.C.; Marinho, G.S. Screening of potential inhibitors targeting the main protease structure of SARS-CoV-2 via molecular docking, and approach with molecular dynamics, RMSD, RMSF, H-bond, SASA and MMGBSA. *Mol. Biotechnol.* **2023**, *66*, 1919–1933. [CrossRef]
76. Ben-Shalom, I.Y.; Pfeiffer-Marek, S.; Baringhaus, K.-H.; Gohlke, H. Efficient approximation of ligand rotational and translational entropy changes upon binding for use in MM-PBSA calculations. *J. Chem. Inf. Model.* **2017**, *57*, 170–189. [CrossRef]
77. Robertson, J.F.; Coleman, R.E.; Cheung, K.-L.; Evans, A.; Holcombe, C.; Skene, A.; Rea, D.; Ahmed, S.; Jahan, A.; Horgan, K. Proliferation and AKT activity biomarker analyses after capivasertib (AZD5363) treatment of patients with ER+ invasive breast cancer (STAKT). *Clin. Cancer Res.* **2020**, *26*, 1574–1585. [CrossRef]
78. Luboff, A.J.; DeRemer, D.L. Capivasertib: A novel AKT inhibitor approved for hormone-receptor-positive, HER-2-negative metastatic breast cancer. *Ann. Pharmacother.* **2024**, 10600280241241531. [CrossRef] [PubMed]
79. Voorhees, P.M.; Spencer, A.; Sutherland, H.J.; O'Dwyer, M.E.; Huang, S.-Y.; Stewart, K.; Chari, A.; Rosenzwieg, M.; Nooka, A.K.; Rosenbaum, C.A. Novel AKT inhibitor afuresertib in combination with bortezomib and dexamethasone demonstrates favorable safety profile and significant clinical activity in patients with relapsed/refractory multiple myeloma. *Blood* **2013**, *122*, 283.
80. Shariati, M.; Meric-Bernstam, F. Targeting AKT for cancer therapy. *Expert Opin. Investig. Drugs* **2019**, *28*, 977–988. [CrossRef]
81. Arceci, R.J.; Allen, C.E.; Dunkel, I.; Jacobsen, E.D.; Whitlock, J.; Vassallo, R.; Borrello, I.M.; Oliff, A.; Morris, S.R.; Reedy, B.A.M. Evaluation of Afuresertib, an oral pan-AKT inhibitor, in patients with Langerhans cell histiocytosis. *Blood* **2013**, *122*, 2907. [CrossRef]
82. Sharif Siam, M.K.; Sarker, A.; Sayeem, M.M.S. In silico drug design and molecular docking studies targeting Akt1 (RAC-alpha serine/threonine-protein kinase) and Akt2 (RAC-beta serine/threonine-protein kinase) proteins and investigation of CYP (cytochrome P450) inhibitors against MAOB (monoamine oxidase B) for OSCC (oral squamous cell carcinoma) treatment. *J. Biomol. Struct. Dyn.* **2021**, *39*, 6467–6479.
83. Zhong, S.; Hou, Y.; Zhang, Z.; Guo, Z.; Yang, W.; Dou, G.; Lv, X.; Wang, X.; Ge, J.; Wu, B. Identification of novel natural inhibitors targeting AKT Serine/Threonine Kinase 1 (AKT1) by computational study. *Bioengineered* **2022**, *13*, 12003–12020. [CrossRef]
84. Kong, W.; Zhu, L.; Li, T.; Chen, J.; Fan, B.; Ji, W.; Zhang, C.; Cai, X.; Hu, C.; Sun, X. Azeliragon inhibits PAK1 and enhances the therapeutic efficacy of AKT inhibitors in pancreatic cancer. *Eur. J. Pharmacol.* **2023**, *948*, 175703. [CrossRef] [PubMed]
85. Halder, A.K.; Cordeiro, M.N.D. AKT inhibitors: The road ahead to computational modeling-guided discovery. *Int. J. Mol. Sci.* **2021**, *22*, 3944. [CrossRef] [PubMed]
86. Westbrook, J.; Feng, Z.; Chen, L.; Yang, H.; Berman, H.M. The protein data bank and structural genomics. *Nucleic Acids Res.* **2003**, *31*, 489–491. [CrossRef]
87. Blake, J.F.; Kallan, N.C.; Xiao, D.; Xu, R.; Bencsik, J.R.; Skelton, N.J.; Spencer, K.L.; Mitchell, I.S.; Woessner, R.D.; Gloor, S.L. Discovery of pyrrolopyrimidine inhibitors of Akt. *Bioorganic Med. Chem. Lett.* **2010**, *20*, 5607–5612. [CrossRef] [PubMed]
88. Biovia, D.S. *Dassault Systèmes*; Discovery Studio Visualizer: San Diego, CA, USA, 2017; Volume 936, Available online: https://discover.3ds.com/discovery-studio-visualizer-download (accessed on 29 April 2024).
89. Land, H.; Humble, M.S. YASARA: A tool to obtain structural guidance in biocatalytic investigations. In *Protein Engineering*; Springer: Berlin/Heidelberg, Germany, 2018; pp. 43–67.
90. Abdelbagi, M.E.; Al-Mazaideh, G.M.; Ahmed, A.E.; Al-Rimawi, F.; Ayyal Salman, H.; Almutairi, A.; Abuilaiwi, F.A.; Wedian, F. Exploring Securigera securidaca Seeds as a Source of Potential CDK1 Inhibitors: Identification of Hippeastrine and Naringenin as Promising Hit Candidates. *Processes* **2023**, *11*, 1478. [CrossRef]
91. Gordon, J.C.; Myers, J.B.; Folta, T.; Shoja, V.; Heath, L.S.; Onufriev, A. H++: A server for estimating p K as and adding missing hydrogens to macromolecules. *Nucleic Acids Res.* **2005**, *33*, W368–W371. [CrossRef] [PubMed]
92. Norgan, A.P.; Coffman, P.K.; Kocher, J.-P.A.; Katzmann, D.J.; Sosa, C.P. Multilevel parallelization of AutoDock 4.2. *J. Cheminform.* **2011**, *3*, 1–9. [CrossRef]
93. Hou, X.; Du, J.; Zhang, J.; Du, L.; Fang, H.; Li, M. How to improve docking accuracy of AutoDock 4.2: A case study using different electrostatic potentials. *J. Chem. Inf. Model.* **2013**, *53*, 188–200. [CrossRef]
94. Jaramillo-Botero, A.; Naserifar, S.; Goddard, W.A., III. General multiobjective force field optimization framework, with application to reactive force fields for silicon carbide. *J. Chem. Theory Comput.* **2014**, *10*, 1426–1439. [CrossRef]
95. O'Boyle, N.M.; Banck, M.; James, C.A.; Morley, C.; Vandermeersch, T.; Hutchison, G.R. Open Babel: An open chemical toolbox. *J. Cheminform.* **2011**, *3*, 1–14. [CrossRef]
96. Fuhrmann, J.; Rurainski, A.; Lenhof, H.P.; Neumann, D. A new Lamarckian genetic algorithm for flexible ligand-receptor docking. *J. Comput. Chem.* **2010**, *31*, 1911–1918. [CrossRef]
97. Schmid, N.; Eichenberger, A.P.; Choutko, A.; Riniker, S.; Winger, M.; Mark, A.E.; Van Gunsteren, W.F. Definition and testing of the GROMOS force-field versions 54A7 and 54B7. *Eur. Biophys. J.* **2011**, *40*, 843–856. [CrossRef] [PubMed]
98. Rühle, V. Pressure coupling/barostats. *J. Club.* **2008**, 1–5. Available online: https://www.mpip-mainz.mpg.de/en/home (accessed on 25 April 2024).
99. Berendsen, H.J.C.; Postma, J.P.M.; van Gunsteren, W.F.; DiNola, A.; Haak, J.R. Molecular dynamics with coupling to an external bath. *J. Chem. Phys.* **1984**, *81*, 3684–3690. [CrossRef]
100. Petersen, H.G. Accuracy and efficiency of the particle mesh Ewald method. *J. Chem. Phys.* **1995**, *103*, 3668–3679. [CrossRef]

101. Hess, B.; Bekker, H.; Berendsen, H.J.; Fraaije, J.G. LINCS: A linear constraint solver for molecular simulations. *J. Comput. Chem.* **1997**, *18*, 1463–1472. [CrossRef]
102. Kumari, R.; Kumar, R.; Consortium, O.S.D.D.; Lynn, A. g_mmpbsa—A GROMACS tool for high-throughput MM-PBSA calculations. *J. Chem. Inf. Model.* **2014**, *54*, 1951–1962. [CrossRef]
103. Verma, S.; Grover, S.; Tyagi, C.; Goyal, S.; Jamal, S.; Singh, A.; Grover, A. Hydrophobic interactions are a key to MDM2 inhibition by polyphenols as revealed by molecular dynamics simulations and MM/PBSA free energy calculations. *PLoS ONE* **2016**, *11*, e0149014. [CrossRef] [PubMed]

molecules

MDPI

Article

Ensemble-Based Virtual Screening Led to the Discovery of Novel Lead Molecules as Potential NMBAs

Yi Zhang [1,2], Gonghui Ge [3], Xiangyang Xu [2,*] and Jinhui Wu [1,*]

1 School of Medicine, Nanjing University, Nanjing 210093, China; cruckzhang0304@163.com
2 Jiangsu Key Laboratory of Central Nervous System Drug Research and Development, Jiangsu Nhwa Pharmaceutical Co., Ltd., Xuzhou 221116, China
3 School of Pharmacy, China Medical University, Shenyang 110122, China
* Correspondence: xuxiangyang@nhwa-group.com (X.X.); wuj@nju.edu.cn (J.W.)

Abstract: Neuromuscular blocking agents (NMBAs) are routinely used during anesthesia to relax skeletal muscle. Nicotinic acetylcholine receptors (nAChRs) are ligand-gated ion channels; NMBAs can induce muscle paralysis by preventing the neurotransmitter acetylcholine (ACh) from binding to nAChRs situated on the postsynaptic membranes. Despite widespread efforts, it is still a great challenge to find new NMBAs since the introduction of cisatracurium in 1995. In this work, an effective ensemble-based virtual screening method, including molecular property filters, 3D pharmacophore model, and molecular docking, was applied to discover potential NMBAs from the ZINC15 database. The results showed that screened hit compounds had better docking scores than the reference compound *d*-tubocurarine. In order to further investigate the binding modes between the hit compounds and nAChRs at simulated physiological conditions, the molecular dynamics simulation was performed. Deep analysis of the simulation results revealed that ZINC257459695 can stably bind to nAChRs' active sites and interact with the key residue Asp165. The binding free energies were also calculated for the obtained hits using the MM/GBSA method. In silico ADMET calculations were performed to assess the pharmacokinetic properties of hit compounds in the human body. Overall, the identified ZINC257459695 may be a promising lead compound for developing new NMBAs as an adjunct to general anesthesia, necessitating further investigations.

Keywords: NMBAs; pharmacophore model; molecular docking; molecular dynamics; virtual screening

1. Introduction

Neuromuscular blocking agents (NMBAs), commonly referred to as muscle relaxants, are frequently used to facilitate tracheal intubation and provide skeletal muscle relaxation during surgery or mechanical ventilation [1]. There are two primary categories of NMBAs, defined according to their blocking mechanisms: depolarizing and nondepolarizing agents [2]. Succinylcholine is the only available depolarizing NMBAs still in clinical use. Due to its rapid onset and short duration, the utility of succinylcholine is limited by mechanism-related side effects, such as myalgia, hyperkalemia, and malignant hyperthermia [3,4]. The majority of currently used NMBAs are nondepolarizing blockers (Figure 1), classified structurally into benzylisoquinolines (e.g., cisatracurium and mivacurium) and aminosteroids (e.g., rocuronium, vecuronium, pipecuronium, and pancuronium). Nondepolarizing blockers are characterized by two quaternary ammonium groups, and the distance between the two protonated nitrogen atoms is approximately 14 Å [5], corresponding to a 10-atom separation ("14-Å rule" or "10-atom rule") [6]. The nondepolarizing NMBAs inhibit nicotinic acetylcholine receptors (nAChRs) located postsynaptically on skeletal muscle membranes, thereby inducing skeletal muscle relaxation [5].

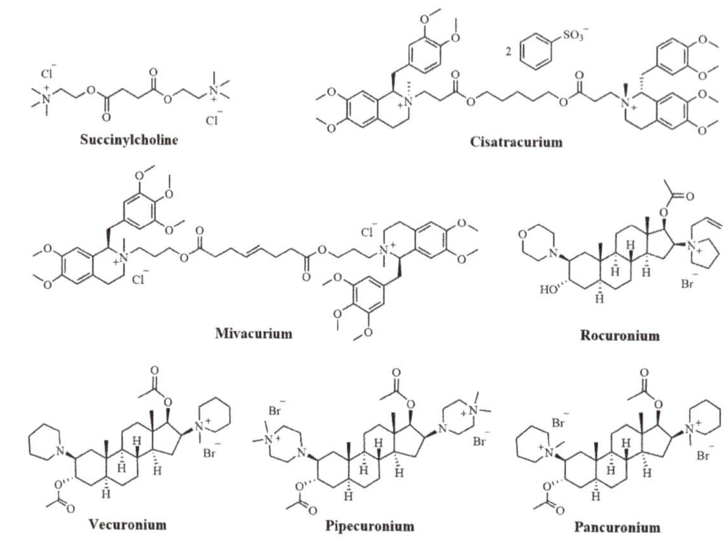

Figure 1. The chemical structures of marketed NMBAs.

Muscular nAChRs are ligand-gated pentameric ion channels physiologically activated by acetylcholine (ACh), and the activation of nAChRs initiates the electrical signal that triggers action potentials, leading to muscle contraction [7]. As shown in Figure 2A (PDB ID: 7SMS), muscular nAChRs structurally consist of five protein subunits organized around a central pore: with two alpha (2α) subunits, one beta (β) subunit, and one delta (δ) subunit, accompanied by either an epsilon (ε) or a gamma (γ) subunit, depending on the developmental stage of the muscle [8,9]. The adult muscle receptors have an epsilon subunit, whereas in infants the gamma type is present [10]. Each muscular nAChR has two binding pockets for ACh in the extracellular domain, which are located at the interfaces of the α-δ and α-ε(γ) subunits [11]. In these pockets, electron-rich amino acid residues (e.g., tyrosine, tryptophan) can interact electrostatically with the positively charged ammonium group of ACh [12,13]. The nondepolarizing NMBAs act as competitive antagonists and impede the interaction between ACh and nAChRs through occupying the ACh binding sites, thereby preventing the opening of the channel (Figure 2B). In fact, positive charges at the quaternary ammonium sites of NMBAs mimic the quaternized nitrogen atom of ACh, which is the structural reason for the attraction of these compounds to muscular nAChRs [14,15].

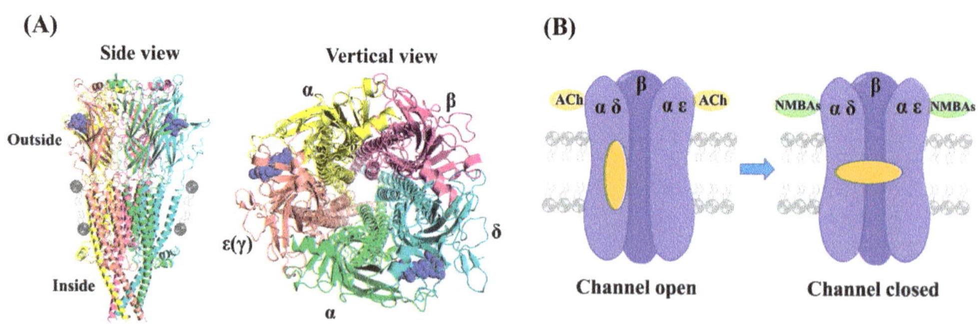

Figure 2. (**A**) The structure of muscular nAChRs (PDB ID: 7SMS). (**B**) The schematic diagram of adult nAChRs inhibited by NMBAs binding both ACh sites.

Currently, muscle relaxation remains a mainstay of modern anesthesia and intensive care [16,17]. NMBAs are used to ease tracheal intubation and to decrease the doses of the general anesthetics [18,19]. Other uses include in acute respiratory distress syndrome (ARDS) [20], elevated intracranial pressure (ICP) [21], and therapeutic hypothermia (TTM) [22]. To meet different pharmacological needs, a variety of structurally diverse compounds have been reported as potential NMBAs [14,23–25]. However, the progress in NMBAs development has been impeded since the discovery of cisatracurium over 25 years ago. High-throughput screening techniques, especially computer-aided virtual screening, have been widely employed for the discovery of lead molecules with new scaffolds of specified targets from large chemical databases [26]. These computer-assisted design methods can be classified as ligand- and structure-based virtual screening approaches [27,28]. The ligand-based virtual screening (LBVS) methods, including pharmacophore modeling and qualitative structure–activity relationships (QSAR), focus on comparative molecular similarity analysis of compounds with unknown and known activity [29,30]. The structure-based virtual screening (SBVS) method, such as molecular docking, is an effective tool to discover putative targets for a particular ligand [31]. Up to now, hundreds of compounds with neuromuscular blocking activities have been reported by many research groups (Table S1). Furthermore, high-resolution cryo-EM structures of muscle-type nicotinic receptor with *d*-tubocurarine have recently been solved (PDB ID: 7SMS) [32]. Therefore, in this study, molecular property filtering and 3D pharmacophore modelling were applied as ligand-based screening techniques to identify compounds with similar characteristics to known neuromuscular blockers. Subsequently, structure-based docking studies were conducted to assess their binding affinities and poses. As shown in Figure 3, the three virtual screening methods were combined to discover potential NMBAs from ZINC15 database. Moreover, molecular dynamics simulation was applied to investigate the stability of the complexes and the mechanism of interaction between nAChRs and hit compounds at physiological conditions.

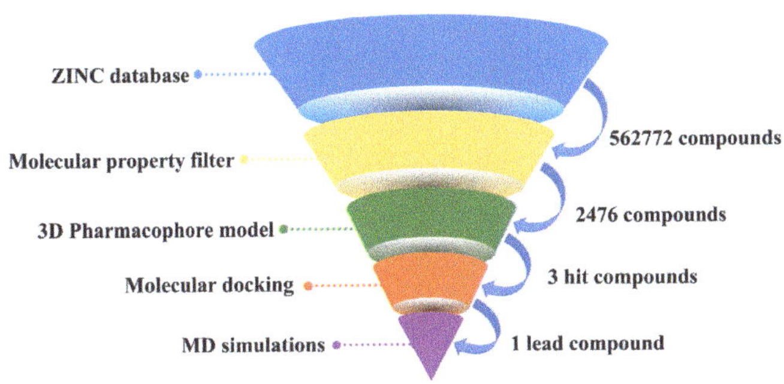

Figure 3. The workflow for discovery of potential NMBAs in this research.

2. Results

2.1. Molecular Property Filters

Muscle relaxants are mainly quaternary ammonium compounds with a large molecular size, making them unique among the drugs associated with disease [33]. For this reason, classical screening strategies, such as the Lipinski's rule of five, may not be appropriate for the discovery of promising lead molecules. Therefore, a dataset of 294 reported compounds with neuromuscular blocking activities was initially compiled from the earlier literature (Table S1). Then, three key properties (molecular weight, charge, and logP) of compounds in the NMBAs database were calculated using the ChemDraw software (Version 18.2). The 997 million compounds in the ZINC15 database were filtered as preliminary screening based on the 3 properties. As we can see in Figure 4A, compounds in the database have large molecular weights, with only a few less than 400 Da, so a filter was applied to exclude compounds with a molecular weight below this threshold. Given that NMBAs are primarily composed of bis-quaternary ammonium salts, as shown in Figure 4B, 93.9% of compounds in the database are double-charged cationic molecules. Consequently, all double positively charged compounds were selected to ensure that the molecules had desirable physicochemical properties as drug molecules. Figure 4C shows that the majority of compounds with calculated logP values fall within the range of −4 to 4; a third filter was, therefore, employed to exclude the compounds with logP values outside of this window. The filtration resulted in obtaining nearly 562 thousand compounds from the ZINC15 database in SDF format. In addition, the compounds were minimized using MOE software (Version 2019.01) and then selected for virtual screening through the pharmacophore model.

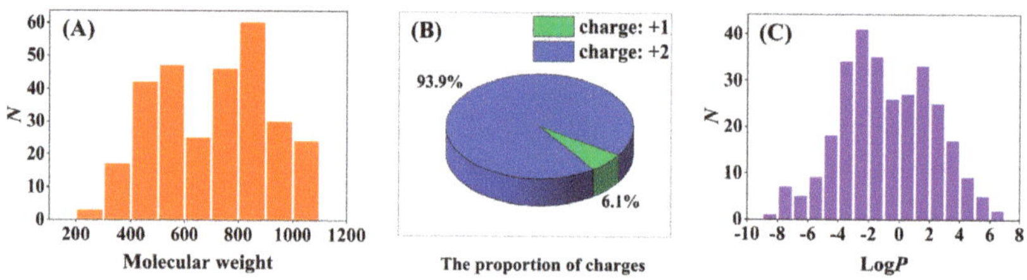

Figure 4. Molecular properties of ligands in the NMBAs database. (**A**) Frequency distribution of molecular weights. (**B**) The proportion of ligands with single-charged versus double-charged. (**C**) Frequency distribution of molecular calculated logP values.

2.2. Pharmacophore Modelling

A pharmacophore is a representation of steric and electronic features required for interaction with a macromolecular target, which results in a pharmacological response [34]. We developed a ligand-based pharmacophore model in MOE based on the six marketed nondepolarizing NMBAs mentioned in Figure 1. The selection of these compounds to build the pharmacophore model was based on their high potency and structural diversity. The 3D pharmacophore was generated by the flexible alignment of all six structures, and the 3D features that they shared were identified through the pharmacophore consensus module [35]. The common 3D pharmacophore model is composed of four hydrophobic groups (F1, F2, F3, F4), one aromatic ring (F5), and two cations (F6, F7). In addition, previous study has indicated that the distance of the two quaternized N atoms falling in between 11 and 14 Å is desirable for maximizing bioactivity [6,12]. As shown in Figure 5, the distance between F6 and F7 falls between 11.7 and 15.3 Å, which was in compliance with the "14-Å rule". The model offered a molecular framework for the virtual screening of databases utilizing suitable pharmacophore features (Figure 5). A total of 562 thousand chemical compounds selected from the ZINC15 database were screened through the generated 3D pharmacophore model. As a result of pharmacophore-based virtual screening, 2476 hit molecules were identified to fit the pharmacophore features.

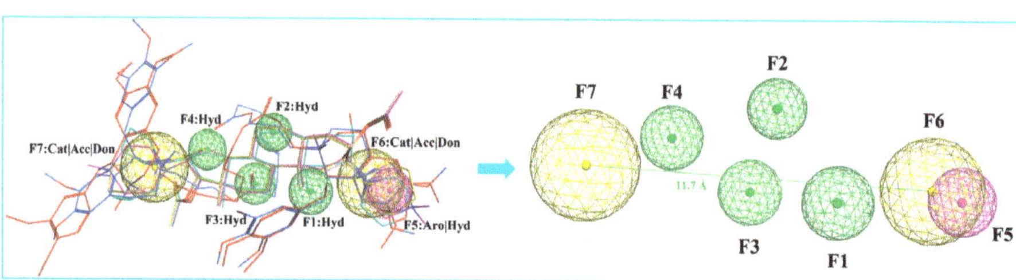

Figure 5. Pharmacophore model based on the aligned NMBAs. Color code: cisatracurium, red; mivacurium, blue; rocuronium, cyan; vecuronium, pink; pipecuronium, magenta; pancuronium, yellow. Pharmacophore features: Hyd, hydrophobic groups; Aro, aromatic rings; Cat, Cation; Acc, H-bond Acceptor; Don, H-bond donor.

2.3. Molecular Docking

Molecular docking provides important data in rational drug design, which can predict the binding affinities, spatial orientation, and predominant binding modes of the small molecule drug candidates to the active site of target proteins [36]. In this investigation, we employed structure-based virtual screening approach that involves a docking simulation of 2476 molecules identified through the pharmacophore model into the active pocket of nAChRs (PDB ID: 7SMS) [32]. Muscular nAChRs contain two binding sites in the extracellular domain (ECD), which are located at the α-γ and α-δ (infant) or α-δ and α-ε (adult) subunit interfaces, respectively [8]. As shown in Figure 6, the transmembrane domain (TMD) of each subunit comprises four helices, M1-M4, with M2 lining the ion channel and M4 being most peripheral [32]. The intracellular domain (ICD) of each subunit is formed by a partially ordered loop between the M3 and M4 helices: an amphipathic MX helix following M3, and a long helix called MA that leads into and is continuous with M4. The MA helices form a bundle at their N-termini and frame lateral portals for ion flux [37]. We performed molecular docking simulations of the compounds binding to the nAChRs α-δ site because it is present in both these nAChR subtypes [38]. The computationally derived potential hits, along with the binding energy values, were shown in Figure 6. As lower binding energy corresponds to a higher binding affinity, four compounds, namely ZINC257357801, ZINC257459695, ZINC8926303, and ZINC1293069436, emerged as the best compounds with better scoring energies (−8.60 kcal/mol, −8.52 kcal/mol, −8.36 kcal/mol, and −8.25 kcal/mol, respectively) than the standard compound d-tubocurarine (−8.14 kcal/mol).

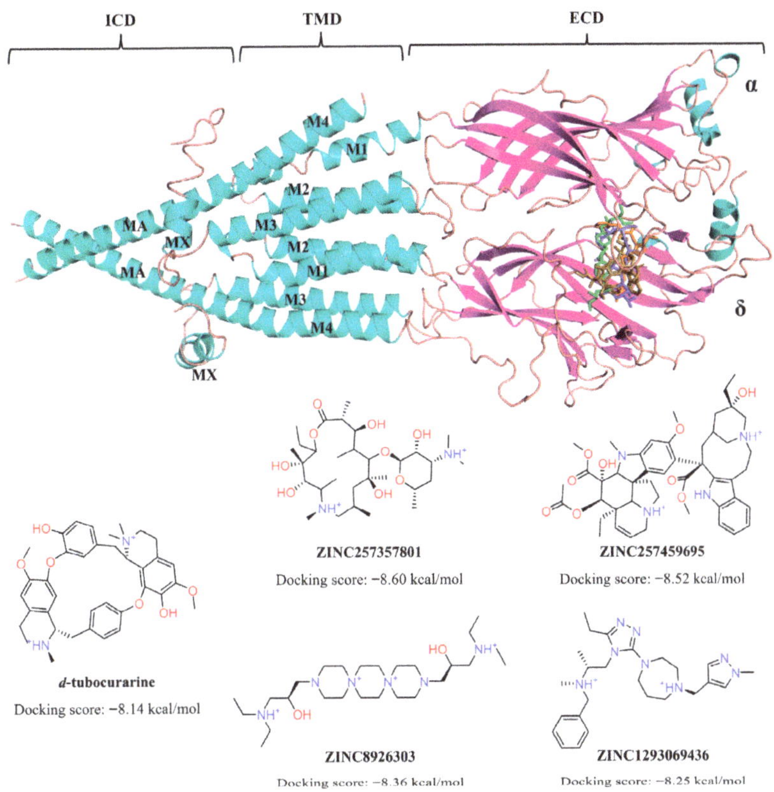

Figure 6. The chemical structures and docking scores (kcal/mol) of screened hit compounds (PDB ID: 7SMS).

The four compounds with desired properties were selected as potential NMBAs by the ensemble-based virtual screening strategy. The molecular interactions between the hits and nAChRs produced by MOE are displayed in Figure 7. Through the analysis of the cryo-EM structure, we found that the quaternized NH group of d-tubocurarine can form hydrogen bonds with the key residue Asp165 (Figure S1), which played important roles in binding. As shown in Figure 7A, residue Asp165 formed a hydrogen bond with the dimethylamino group of ZINC257357801, and residues of Cys192 and Trp149 formed hydrogen bonds with the hydrogen atom on the hydroxyl group in the middle and right side, respectively. The residue Tyr198 formed a Pi-H bond with the ethyl group, and residue Trp57 formed a Pi-cation with the quaternized nitrogen atom of the ring. For ZINC257459695, as shown in Figure 7B, residue Asp165 formed a hydrogen bond with the hydrogen atom of the pyrrolidine ring, residue Cys192 formed a hydrogen bond with the oxygen atom on the carbonyl group, and residue Asp180 formed a hydrogen bond with the methyl ester at the right end. For ZINC8926303, residues of Asp165 and Ile148 formed hydrogen bonds with hydrogen atoms on two terminal quaternary ammonium groups, residue Tyr190 formed a Pi-H bond with the piperazine ring, and residue Tyr198 formed a Pi-H bond with the ethyl group in the right side. For ZINC1293069436, residue Asp59 formed a hydrogen bond with the hydrogen atom on the methylene group, residue Tyr198 formed a hydrophobic interaction with the benzene ring, and residue Met163 formed a Pi-H bond with the pyrazole ring. Obviously, it was evident from Figure 7 that ZINC257357801, ZINC257459695, and ZINC8926303 could interact with the key residue Asp165 in the active site of nAChRs. Therefore, the stability and molecular interaction pattern of these three hits and d-tubocurarine at simulated physiological condition were further probed by applying molecular dynamic (MD) simulation.

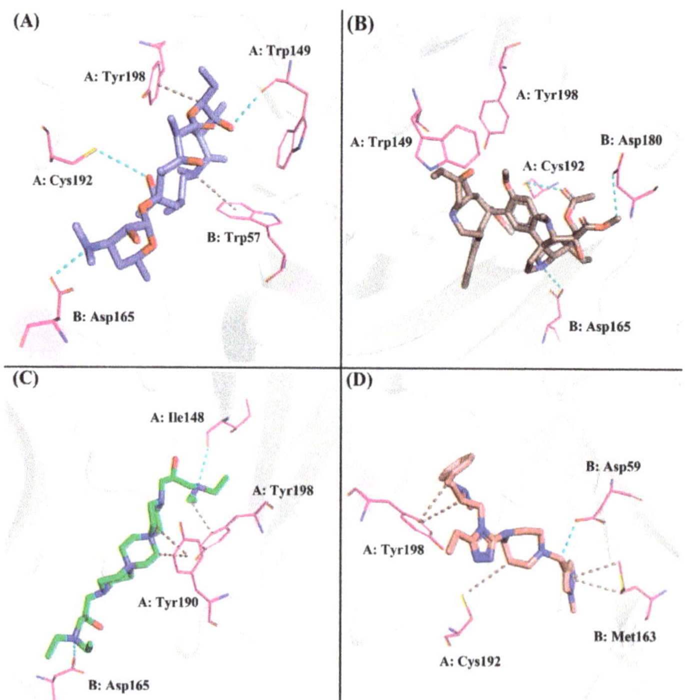

Figure 7. Preferred binding poses of ZINC257357801 (**A**), slate sticks, ZINC257459695 (**B**), sand sticks, ZINC8926303 (**C**), green sticks, and ZINC1293069436 (**D**), salmon sticks, bound to the nAChRs α-δ site (PDB ID: 7SMS).

2.4. Molecular Dynamic Simulation

In this section, we conducted a molecular dynamics simulation to investigate the flexibility and solvation effect of biomolecular systems. To this end, we ran 25,000 ps molecular dynamics production runs on the 3 hit candidates (ZINC257357801, ZINC257459695, and ZINC8926303) and the reference compound (d-tubocurarine). The MD trajectories were analyzed by the flowing parameters: root mean square deviation (RMSD), root mean square fluctuation (RMSF), radius of gyration (Rg), solvent accessible surface area (SASA), hydrogen bonds (H-bond), dynamic cross-correlation matrix (DCCM), and binding pattern analysis.

2.4.1. Structural Deviation and Compactness Analysis

The structural deviation and stability were measured by the RMSD and SASA, and the fluctuation was assessed by RMSF, and the compactness was measured by Rg. From the observation of the RMSD graph presented in Figure 8A, all the protein–ligand entities showed steady and stable behavior throughout the simulation time; the RMSD fluctuated around by 0.15–0.35 nm. However, in the same simulation time, the ZINC257459695 graph line showed much stable behavior as compared to other complexes, which remained steady and stable at an RMSD value of around 0.18 nm. The RMSD graph lines of ZINC257357801, ZINC8926303, and *d*-tubocurarine displayed an increasing trend, with RMSD values ranging from 0 to 0.34 nm from 0 to 5000 ps. The overall RMSD graphs results showed stable behavior in the backbone of all docked complexes.

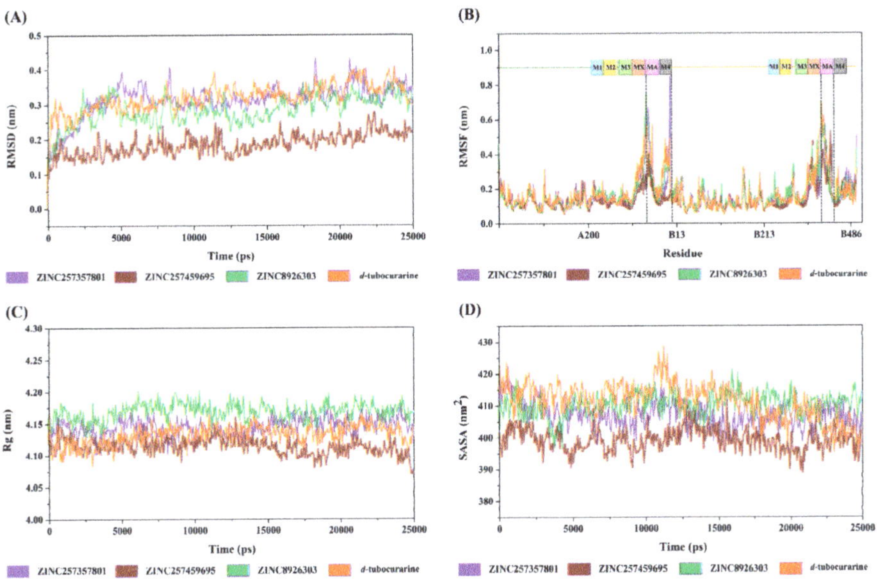

Figure 8. The RMSD (**A**), RMSF (**B**), Rg (**C**), and SASA (**D**) curves of the ZINC257357801, ZINC257459695, ZINC8926303, and *d*-tubocurarine systems.

The root mean square fluctuation calculates the fluctuation in the individual amino acid residues throughout the simulation process in the presence of different ligand molecules. As represented in Figure 8B, all the promising lead molecules show a very similar RMSF pattern with the reference compound. Some residues in the MA and M4 domains (chain A), and MA domain (chain B) showed greater flexibility, and RMSF fluctuated between 0.11 and 0.76 nm. The α-helixes of MA and M4 are located around the frame lateral portals for ion flux, which may be the reason for the flexibility.

The Rg parameter was used to evaluate structural compactness of a protein–ligand in a biological system. A stable folded structure is described by a relatively constant Rg value, whereas an unfolded structure will cause the Rg value to fluctuate with time. As shown in Figure 8C, the values of Rg for the four complexes fluctuated between 4.09 and 4.17 nm during the simulation. The calculated average Rg values for ZINC257357801, ZINC257459695, ZINC8926303, and *d*-tubocurarine are 4.15 nm, 4.11 nm, 4.17 nm, and 4.13 nm, respectively. With lower Rg values compared to all other compounds, ZINC257459695-nAChRs may be regarded as the most compact biomolecular system.

The solvent accessible surface area of protein has always been considered as a decisive factor in protein folding and stability studies. As we can see in Figure 8D, the SASA for the four systems showed little variation from 390 to 425 nm^2, which suggested that the four systems were relatively stable. The binding of ZINC257459695, compared to other compounds, resulted in reduced SASA values, as the surface of the protein becomes unexposed to the solvent after ligand binding.

2.4.2. Hydrogen Bond Analysis

The stability of the protein–ligand complex is facilitated by the formation of hydrogen bonds between the receptor and ligand. Therefore, the total number of hydrogen bonds were investigated in the complexes after the 25,000 ps simulation time. As exhibited in Figure 9A,D, for the ZINC257357801 and *d*-tubocurarine complexes, two to three hydrogen bonds were identified. ZINC8926303, on the other hand, was shown to form three to four hydrogen bonds. Interestingly, for the ZINC257459695–nAChRs system, a maximum of five hydrogen bonds were observed, which could form more hydrogen bonds than that of the reference compound *d*-tubocurarine during the entire simulation period. Obviously, the three hits could form more hydrogen bonds than the reference compound. Furthermore, through the above-detailed H-bond analysis, we can conclude that the compound ZINC257459695 was bound to the nAChRs protein more effectively and tightly when compared to the other three compounds. The results of the H-bond analysis were consistent with the earlier analysis of RMSD and SASA metrics.

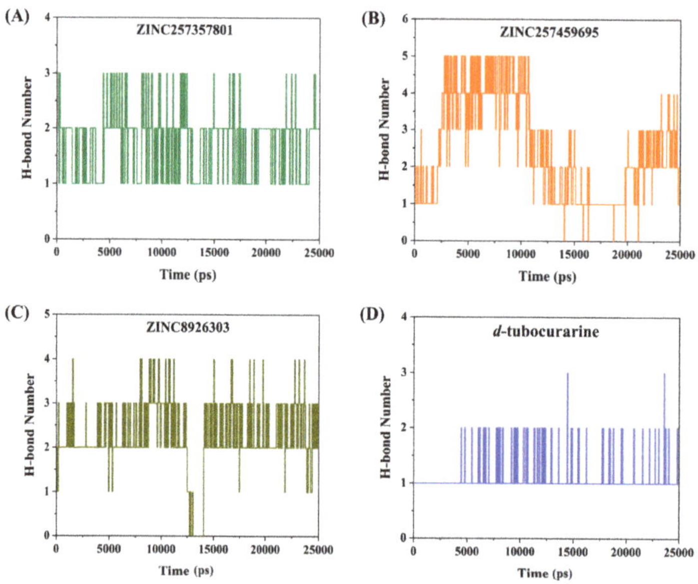

Figure 9. Number of hydrogen bonds for systems: ZINC257357801 (**A**), ZINC257459695 (**B**), ZINC8926303 (**C**), and *d*-tubocurarine (**D**) systems during the MD simulations.

2.4.3. Dynamic Cross-Correlation Map Analysis

The analysis of DCCM was to check the correlated motion of structural domains to achieve a stable conformation of the receptor following the binding of ligands. Highly positive sections (cyan) indicate strongly positive correlated movement of residues in the same direction, while the negative regions (pink) represent strongly anti-correlated motions. As illustrated in Figure 10, the binding of ZINC257357801, ZINC257459695, ZINC8926303, and *d*-tubocurarine generated obvious influences on the internal dynamic behavior of nAChRs. For the ZINC257459695 complex, both chain A and B produced strongly positive correlated motions independently (Figure 10B). This phenomenon suggested that the binding of ZINC257459695 may have resulted in conformational changes in the protein. The ZINC257357801 complex has maximum residues in a positive correlation, while ZINC8926303 complex residues have a slightly weaker but still positive correlation (Figure 10A,C). From the results, it is concluded that the presence of ZINC257357801 and ZINC8926303 induced significant correlated motions in protein, whereas slightly anti-correlated motions are observed. The *d*-tubocurarine complex has the most residues in a negative correlation, but the overall correlation between the binding site and the amino acids has increased (Figure 10D).

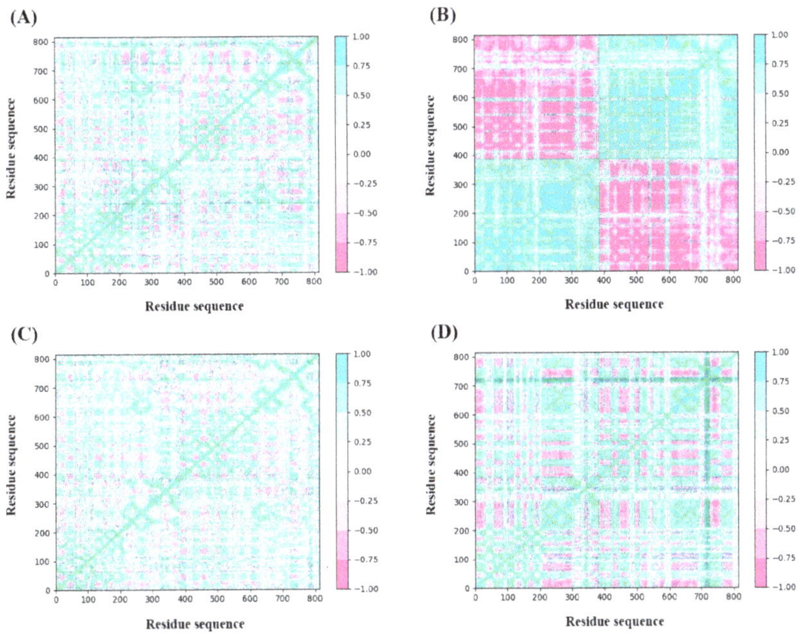

Figure 10. DCCM comparative analysis of ZINC257357801 (**A**), ZINC257459695 (**B**), ZINC8926303 (**C**), and *d*-tubocurarine (**D**).

2.4.4. The Binding Modes Refined through the MD Simulations

The Gibbs free energy landscape for the four systems was displayed in Figure 11A–D. A very weak or unstable receptor–ligand interaction can result in many minimal energy clusters, whereas a strong and stable interaction can generate one conformation cluster in the potential energy map [39]. As depicted in Figure 11, only a single energy minima was found in the case of the ZINC257357801, ZINC257459695, and ZINC8926303 complexes, whereas there are two energy minima found for the *d*-tubocurarine complex. In addition, the narrow and shallow energy basin denotes limited structural conformation stability. It can be observed from Figure 11 that the binding of the nAChRs protein with ZINC257357801 and ZINC257459695 has resulted in a noticeable single narrow energy minima basin related to its conformational state, suggesting a stable and strong receptor–ligand conformation. Although the *d*-tubocurarine complex has two energy minima, they are completely separated from one another by an energy barrier making it less stable. This is also in agreement with earlier RMSD, H-bond, and DCCM analysis results.

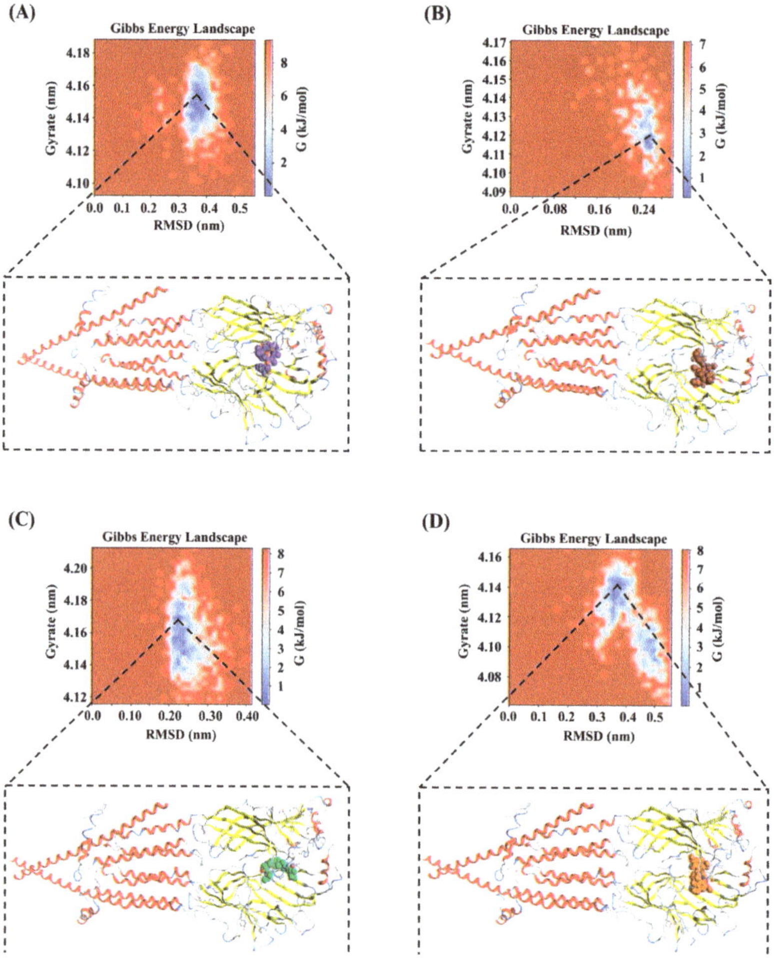

Figure 11. The Gibbs free energy landscapes of ZINC257357801 (**A**), ZINC257459695 (**B**), ZINC8926303 (**C**), and *d*-tubocurarine (**D**) systems during the MD simulations.

Based on the calculation of the Gibbs free energy, the most stable conformations of each system were extracted from the lowest energy field (dark blue) to explore the key residues and interactions between these compounds and nAChRs. For the ZINC257357801 system, as shown in Figure 12A, residue Asp165 formed two different hydrogen bonds with the dimethylamino group. ZINC257357801 and Trp57 formed a Pi-H interaction, and residue Trp149 formed a hydrogen bond with the hydroxyl group at the right end. As exhibited in Figure 12B, based on the most energetically favorable binding mode, residues of Asp165 and Cys193 formed two hydrogen bonds with ZINC257459695, while residues of Tyr198 and Trp57 were observed to form hydrophobic interactions. For the ZINC8926303 system, as shown in Figure 12C, ZINC8926303 formed hydrophobic interactions with residues of Trp149, Tyr151, and Cys193. Notably, the hydrogen bond between the terminal quaternary ammonium group and residue Asp165 disappeared, which could be the reason for the decreased stability after the simulation. For the *d*-tubocurarine system, as displayed in Figure 12D, the quaternized NH group acted as a hydrogen bond donor with residue Asp165, and Tyr190 interacted with *d*-tubocurarine as a hydrophobic interaction.

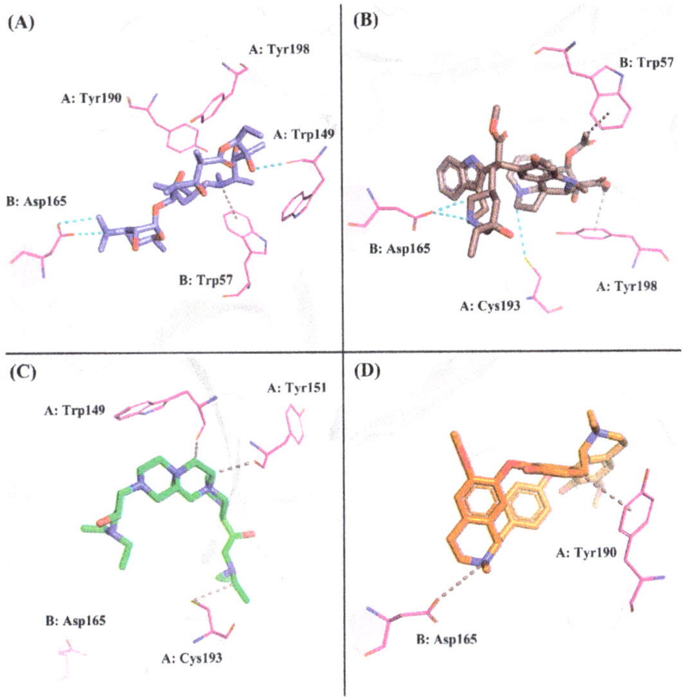

Figure 12. The binding modes of ZINC257357801 (**A**), ZINC257459695 (**B**), ZINC8926303 (**C**), and *d*-tubocurarine (**D**) given by the molecular dynamics simulation.

2.4.5. Binding Free Energy Calculation by MM/GBSA Method

To better assess the reliability of the different nAChRs–ligand complexes, the corresponding protein–ligand binding free energies were evaluated from the MD coordinates extracted from the last 10 ns of simulation. The molecular mechanics generalized Born surface area (MM-GBSA) method was used for the calculation. The binding free energy ΔG_{bind} was comprised of the Van der Waals interaction (ΔE_{vdw}), electrostatic interaction (ΔE_{ele}), polar solubility energy (ΔG_{ps}), and non-polar solubility energy (ΔG_{nps}). Among these interactions, ΔE_{vdw}, ΔE_{ele}, and ΔG_{nps} were considered to be beneficial for the ΔG_{bind}, but ΔG_{ps} was considered to be unfavorable to the ΔG_{bind}. The analysis of the MM-GBSA results showed that the electrostatic force of interaction is majorly contributing in the protein–ligand binding compared to the Van der Waals force of interaction. As shown in Table 1, ZINC257459695 had a maximum affinity with nAChRs, followed by ZINC257357801 and ZINC8926303, showing -50.40 ± 3.61, -40.52 ± 5.49, and -31.01 ± 6.22 kcal/mol, respectively. The estimated total binding energy suggested that the ZINC257459695 has the most energetically favored binding mode as compared to the reference, *d*-tubocurarine. However, compared with the cisatracurium and rocuronium, it can be seen that the ΔE_{vdw} and ΔG_{nps} of ZINC257459695 were much lower, indicating that the two marketed NMBAs possess superior binding energy. In addition, the binding free energy of the ZINC8926303 is the weakest. The ΔE_{ele} of ZINC8926303 complex was the lowest for all three hit compounds, mainly because the ΔG_{bind} of the ZINC8926303 was weaker than that of the standard *d*-tubocurarine.

Table 1. Binding free energy calculated by the MM-GBSA method (kcal/mol).

System	ΔG_{bind}	ΔE_{vdw}	ΔE_{ele}	ΔG_{ps}	ΔG_{nps}
ZINC257357801	−40.52 ± 5.49	−42.87 ± 3.28	−726.05 ± 4.22	735.10 ± 1.23	−6.70 ± 0.05
ZINC257459695	−50.40 ± 3.61	−33.82 ± 0.84	−1453.27 ± 0.09	1442.28 ± 3.51	−5.59 ± 0.00
ZINC8926303	−31.01 ± 6.22	−37.21 ± 2.20	−586.86 ± 3.01	598.35 ± 4.97	−5.28 ± 0.02
d-tubocurarine	−39.57 ± 3.11	−55.84 ± 0.11	−657.70 ± 2.99	680.80 ± 0.83	−6.82 ± 0.16
Cisatracurium	−66.86 ± 3.07	−92.50 ± 2.01	−669.38 ± 0.33	706.26 ± 2.30	−11.24 ± 0.08
Rocuronium	−57.65 ± 2.50	−62.59 ± 1.72	−340.73 ± 1.67	353.45 ± 0.70	−7.78 ± 0.01

2.5. *In Silico Pharmacokinetic Profile (ADMET)*

In the pursuit of efficient drug discovery, data on absorption, distribution, metabolism, excretion, and toxicity (ADMET) are crucial for identifying and developing novel drug candidates [40,41]. Thus, the pharmacokinetic profile of the top three hits, along with *d*-tubocurarine, cisatracurium, and rocuronium, were assessed using the PreADMET method [42] and summarized in Table 2. The aqueous solubility of a drug is a vital factor that can significantly affect its bioavailability, and the three hit compounds exhibited moderate to high water solubility levels. Blood–brain barrier penetration was used to evaluate the distribution of the compounds, and the BBB values for ZINC257357801, ZINC257459695, and ZINC8926303 were 0.04, 0.11, and 0.05, respectively. During the metabolic phase, the ZINC257357801, ZINC8926303, and rocuronium were discovered to be inhibitors of cytochrome P450 2D6. In addition, the Ames test result of ZINC257459695 was negative, implying that it was probably unable to induce gene mutation. However, the usage of ZINC257459695 could be limited due to its blockage of hERG channels, warranting further optimization.

Table 2. The ADMET prediction for the investigated compounds.

Compound	Buffer Solubility [1]	BBB [2]	PPB [3]	CYP2D6 Inhibition	Ames Test	hERG Inhibition
ZINC257357801	1132	0.04	3.83	Inhibitor	Non-mutagen	Ambiguous
ZINC257459695	0.30	0.11	13.97	Non	Non-mutagen	High
ZINC8926303	88,380	0.05	19.16	Inhibitor	Mutagen	Ambiguous
d-tubocurarine	1.18	1.22	67.68	Non	Non-mutagen	High risk
Cisatracurium	0.0081	0.80	72.49	Non	Non-mutagen	Medium risk
Rocuronium	121.69	0.26	18.97	Inhibitor	Mutagen	Low risk

[1] Buffer solubility: water solubility in buffer system (SK atomic types, mg/L). [2] BBB: blood–brain barrier penetration (C.brain/C.blood). [3] PPB: plasma protein binding (%).

3. Materials and Methods

3.1. *Ligand-Based Pharmacophore Generation*

The alignments of six marketed NMBAs (i.e., cisatracurium, mivacurium, rocuronium, vecuronium, pipecuronium, and pancuronium) were generated using the Flexible Alignment module of the MOE (Molecular Operating Environment software, Version 2019.01) [43]. The 3D chemical structures of the NMBAs were built and energy-minimized using the MMFF94 force field [44]. Then, small molecules were aligned by maximizing the structural overlap of steric features while exploring alternative conformations with low ligand strain. Finally, a collection of alignments, along with a score for each alignment, was generated, and a final alignment with the lowest S score was obtained for the following pharmacophore generation.

The identification of common structural features among six aligned NMBAs was performed using a ligand-based pharmacophore approach. To generate a pharmacophore model with good quality, a set of pharmacophore features, namely hydrogen-bond donor (Don), hydrogen-bond acceptor (Acc), aromatic center (Aro), hydrophobic atom (HydA), anionic atom (Ani), and cationic atom (Cat) were mapped. The Pharmacophore Consensus module was used to generate suggested features for a pharmacophore query from the aligned structures, and a pharmacophore model containing seven common chemical features was developed. Based on the pharmacophore model generated, virtual screening was conducted using a Pharmacophore Search protocol in MOE through the EHT scheme [45]. A pharmacophore query consisting of identified pharmacophoric features was used to filter the database of molecular conformations.

3.2. Molecular Docking

The cryo-EM structures of muscle-type nicotinic receptor (PDB ID: 7SMS, Resolution 3.18 Å) [32] were obtained from the RCSB Protein Data Bank (PDB, https://www.rcsb.org/, accessed on 3 January 2024). As a part of preparing the target protein, the water molecules were removed from the protein and hydrogen atoms were added to optimize the structure by using energy minimization. For the docking parameters, we set the force field to Amber10 and used the triangle matcher placement algorithm, which returned thirty poses; we further used the rigid receptor refinement method, which returned five poses [32]. The GBVI/WSA dG method was applied to score the poses in both steps [46]. The scores were generated by accounting for the individual contributions of energy terms, including hydrogen bonds, electrostatics, and hydrophobicity. The scoring of ligands determines which ligand pose is the most energetically favorable and ranks the library of screened molecules to indicate which compounds are most likely to be active and suitable for further analysis. The selected complexes were illustrated using the PyMOL software (Version 2.5.0) [47].

3.3. Molecular Dynamics Simulation

All simulations were performed with the GROMACS 2019.6 package [48] (https://www.gromacs.org, accessed on 2 February 2024) and were carried out using the amber99sb-ildn force field at 298 K. General amber force field (GAFF) parameters were assigned to the ligands [49], whereas partial charges were calculated using the AM1-BCC method [50]. All ligand–protein complexes were placed in the cubic water-box and set to be 15 Å away from the box edge, using a TIP3P explicit solvent model [51]. Either Na^+ or Cl^- ions were added as counterions for the neutralization of the systems. A 2 fs time step of integration was chosen for all MD simulations, and the systems were equilibrated in the NVE ensemble for 50,000 steps, followed by equilibration in the NVT ensemble for an additional 50,000 steps. The minimized complexes were used as starting conformations for the MD simulations. Periodic boundary conditions and particle mesh Ewald (PME) electrostatics were employed in the simulations [52]. Finally, 25,000 ps molecular dynamics simulations were performed at 298 K with a trajectory recording interval of 50 ps. In addition, various dynamic analysis parameters, like RMSD, RMSF, Rg, SASA, H-bond, DCCM, and others, were carried out using the GROMACS tool.

3.4. Binding Free Energy Calculation

The binding free energy of protein–ligand complexes of all three hits and reference *d*-tubocurarine with nAChRs were calculated using the molecular mechanics generalized Born surface area (MM/GBSA) approach [53]. The following equation was used to calculate the binding free energy from the MD simulation:

$$\Delta G_{\text{bind}} = \Delta E_{\text{MM}} + \Delta G_{\text{solv}} \tag{1}$$

where ΔG_{bind} denotes the binding free energy. The changes in the gas phase molecular mechanics (ΔE_{MM}) and solvation Gibbs energy (ΔG_{solv}) are determined as follows:

$$\Delta E_{\text{MM}} = \Delta E_{\text{vdw}} + \Delta E_{\text{ele}} \tag{2}$$

$$\Delta G_{solv} = \Delta G_{ps} + \Delta G_{nps} \tag{3}$$

where ΔE_{MM} is the sum of the changes in the Van der Waals energies ΔE_{vdw} and the electrostatic energies ΔE_{ele}. The polar solvation ΔG_{ps} was calculated using the generalized Born model, while the non-polar solvation ΔG_{nps} was estimated by the solvent-accessible surface area. The solvent dielectric constant of 78.5 and the non-polar surface tension constant of 0.0072 kcal/mol·Å^2 were used for the MM/GBSA calculations.

3.5. ADMET Property Prediction

In this section, the Pre-ADMET server application (https://preadmet.qsarhub.com, accessed on 12 April 2024) was used to calculate ADMET parameters, such as buffer solubility, blood–brain barrier penetration (BBB), plasma protein binding (PPB), cytochrome P450 2D6 inhibition (CYP2D6 inhibition), Ames test, and hERG inhibition. The Pre-ADMET approach is based on different classes of molecular parameters, which are considered for generating quantitative structure properties.

4. Conclusions

Through the effective ensemble-based virtual screening approach, including molecular property filters, a 3D pharmacophore model, and molecular docking, three hit compounds were identified as potential NMBAs from the ZINC15 database. In order to further investigate the binding modes between three hits and nAChRs at simulated physiological conditions, the molecular dynamics simulation was performed. Based on the common results of the RMSD, RMSF, Rg, SASA, H-bond, and DCCM, ZINC257459695 displayed stable binding patterns and was considered as a promising lead compound. Furthermore, deeper analysis of the MD results revealed that ZINC257459695 could stably bind to nAChRs' active site and interact with the key residue Asp165. From the MM-GBSA analysis, the identified ZINC257459695 was the most reliable binding mode for nAChRs, with the binding free energy of −50.40 kcal/mol. Additionally, the ADMET properties revealed that ZINC257459695 can be further developed as potential drug candidates. Overall, ZINC257459695, which possesses huge potential to serve as a promising lead compound in developing novel NMBAs as an adjunct to general anesthesia, warrants further optimization.

Supplementary Materials: The following supporting information can be downloaded at: https://www.mdpi.com/article/10.3390/molecules29091955/s1, Table S1: The NMBAs database collected from the literature and used in the molecular property filters; Figure S1: The binding modes of *d*-tubocurarine observed in the cryo-EM structure (PDB ID: 7SMS).

Author Contributions: Conceptualization, Y.Z.; investigation, X.X. and J.W.; methodology, Y.Z. and J.W.; software, Y.Z. and G.G.; supervision, J.W.; validation, Y.Z., G.G. and X.X.; writing—original draft, Y.Z.; project administration, X.X. and J.W. All authors have read and agreed to the published version of the manuscript.

Funding: This research was funded by Jiangsu Funding Program for Excellent Postdoctoral Talent, grant number: 2023ZB462.

Institutional Review Board Statement: Not applicable.

Informed Consent Statement: Not applicable.

Data Availability Statement: The raw data supporting the conclusions of this article will be made available by the authors, without undue reservation.

Acknowledgments: The authors would like to extend their sincere thanks to the National Post-Doctor Regulatory Commission of China and the Department of Human Resources and Social Security of Jiangsu Province.

Conflicts of Interest: Authors Yi Zhang and Xiangyang Xu were employed by the company Jiangsu Nhwa Pharmaceuti cal Co., Ltd. The remaining authors declare that the research was conducted in the absence of any commercial or financial relationships that could be construed as a potential conflict of interest.

References

1. Stäuble, C.G.; Blobner, M. The future of neuromuscular blocking agents. *Curr. Opin. Anaesthesiol.* **2020**, *33*, 490–498. [CrossRef] [PubMed]
2. Tuba, Z.; Maho, S.; Vizi, E.S. Synthesis and structure-activity relationships of neuromuscular blocking agents. *Curr. Med. Chem.* **2002**, *9*, 1507–1536. [CrossRef] [PubMed]
3. Kim, Y.B.; Sung, T.Y.; Yang, H.S. Factors that affect the onset of action of non-depolarizing neuromuscular blocking agents. *Korean J. Anesthesiol.* **2017**, *70*, 500–510. [CrossRef]
4. Tran, D.T.; Newton, E.K.; Mount, V.A.; Lee, J.S.; Wells, G.A.; Perry, J.J. Rocuronium versus succinylcholine for rapid sequence induction intubation. *Cochrane Database Syst. Rev.* **2015**, Cd002788. [CrossRef] [PubMed]
5. Selinger, A.J.; Cavallin, N.A.; Yanai, A.; Birol, I.; Hof, F. Template-Directed Synthesis of Bivalent, Broad-Spectrum Hosts for Neuromuscular Blocking Agents. *Angew. Chem. Int. Ed. Engl.* **2022**, *61*, e202113235. [CrossRef] [PubMed]
6. Lee, C. Conformation, action, and mechanism of action of neuromuscular blocking muscle relaxants. *Pharmacol. Ther.* **2003**, *98*, 143–169. [CrossRef] [PubMed]
7. Akkol, E.K.; Karatoprak, G.; Carpar, E.; Hussain, Y.; Khan, H.; Aschner, M. Effects of Natural Products on Neuromuscular Junction. *Curr. Neuropharmacol.* **2022**, *20*, 594–610. [CrossRef]
8. Karlin, A. Emerging structure of the nicotinic acetylcholine receptors. *Nat. Rev. Neurosci.* **2002**, *3*, 102–114. [CrossRef]
9. Unwin, N. Nicotinic acetylcholine receptor and the structural basis of neuromuscular transmission: Insights from Torpedo postsynaptic membranes. *Q. Rev. Biophys.* **2013**, *46*, 283–322. [CrossRef]
10. Liu, Y.; Sugiura, Y.; Padgett, D.; Lin, W. Postsynaptic development of the neuromuscular junction in mice lacking the gamma-subunit of muscle nicotinic acetylcholine receptor. *J. Mol. Neurosci.* **2010**, *40*, 21–26. [CrossRef]
11. Unwin, N. Refined structure of the nicotinic acetylcholine receptor at 4A resolution. *J. Mol. Biol.* **2005**, *346*, 967–989. [CrossRef] [PubMed]
12. Gyermek, L. Development of ultra short-acting muscle relaxant agents: History, research strategies, and challenges. *Med. Res. Rev.* **2005**, *25*, 610–654. [CrossRef] [PubMed]
13. Cecchini, M.; Changeux, J.P. The nicotinic acetylcholine receptor and its prokaryotic homologues: Structure, conformational transitions & allosteric modulation. *Neuropharmacology* **2015**, *96*, 137–149. [CrossRef] [PubMed]
14. El-Subbagh, H.I.; El-Azab, A.S.; Hassan, G.S.; El-Messery, S.M.; Abdel-Aziz, A.A.; El-Taher, K.E.H. Thiadiazolodiazepine analogues as a new class of neuromuscular blocking agents: Synthesis, biological evaluation and molecular modeling study. *Eur. J. Med. Chem.* **2017**, *126*, 15–23. [CrossRef] [PubMed]
15. Hu, H.; Rao, Z.; Xu, J.; Zhu, Q.; Altenbach, H.J.; Chen, H.; Zhou, D.; Xiao, Y.; Ke, X.; Guo, H.; et al. 16-morpholino quaternary ammonium steroidal derivatives as neuromuscular blocking agents: Synthesis, biological evaluation and in silico probe of ligand-receptor interaction. *Eur. J. Med. Chem.* **2012**, *56*, 332–347. [CrossRef] [PubMed]
16. Murphy, G.S.; Vender, J.S. Neuromuscular-blocking drugs. Use and misuse in the intensive care unit. *Crit. Care Clin.* **2001**, *17*, 925–942. [CrossRef]
17. Plaud, B.; Baillard, C.; Bourgain, J.L.; Bouroche, G.; Desplanque, L.; Devys, J.M.; Fletcher, D.; Fuchs-Buder, T.; Lebuffe, G.; Meistelman, C.; et al. Guidelines on muscle relaxants and reversal in anaesthesia. *Anaesth. Crit. Care Pain Med.* **2020**, *39*, 125–142. [CrossRef] [PubMed]
18. Klucka, J.; Kosinova, M.; Zacharowski, K.; De Hert, S.; Kratochvil, M.; Toukalkova, M.; Stoudek, R.; Zelinkova, H.; Stourac, P. Rapid sequence induction: An international survey. *Eur. J. Anaesthesiol.* **2020**, *37*, 435–442. [CrossRef]
19. Schlaich, N.; Mertzlufft, F.; Soltész, S.; Fuchs-Buder, T. Remifentanil and propofol without muscle relaxants or with different doses of rocuronium for tracheal intubation in outpatient anaesthesia. *Acta Anaesthesiol. Scand.* **2000**, *44*, 720–726. [CrossRef]
20. Hraiech, S.; Yoshida, T.; Annane, D.; Duggal, A.; Fanelli, V.; Gacouin, A.; Heunks, L.; Jaber, S.; Sottile, P.D.; Papazian, L. Myorelaxants in ARDS patients. *Intensive Care Med.* **2020**, *46*, 2357–2372. [CrossRef]
21. McCall, M.; Jeejeebhoy, K.; Pencharz, P.; Moulton, R. Effect of neuromuscular blockade on energy expenditure in patients with severe head injury. *JPEN J. Parenter Enter. Nutr.* **2003**, *27*, 27–35. [CrossRef] [PubMed]
22. Lascarrou, J.B.; Le Gouge, A.; Dimet, J.; Lacherade, J.C.; Martin-Lefèvre, L.; Fiancette, M.; Vinatier, I.; Lebert, C.; Bachoumas, K.; Yehia, A.; et al. Neuromuscular blockade during therapeutic hypothermia after cardiac arrest: Observational study of neurological and infectious outcomes. *Resuscitation* **2014**, *85*, 1257–1262. [CrossRef] [PubMed]
23. Savarese, J.J.; Sunaga, H.; McGilvra, J.D.; Belmont, M.R.; Murrell, M.T.; Jeannotte, E.; Cooke, F.E.; Wastila, W.B.; Heerdt, P.M. Preclinical Pharmacology in the Rhesus Monkey of CW 1759-50, a New Ultra-short Acting Nondepolarizing Neuromuscular Blocking Agent, Degraded and Antagonized by L-Cysteine. *Anesthesiology* **2018**, *129*, 970–988. [CrossRef]

24. Herrera-Arozamena, C.; Estrada-Valencia, M.; Martí-Marí, O.; Pérez, C.; de la Fuente Revenga, M.; Villalba-Galea, C.A.; Rodríguez-Franco, M.I. Optical control of muscular nicotinic channels with azocuroniums, photoswitchable azobenzenes bearing two N-methyl-N-carbocyclic quaternary ammonium groups. *Eur. J. Med. Chem.* **2020**, *200*, 112403. [CrossRef] [PubMed]
25. Goswami, L.N.; Olds, T.J.; Monk, T.G.; Johnson, Q.L.; Dilger, J.P.; Shanawaz, M.A.; Jalisatgi, S.S.; Hawthorne, M.F.; Kracke, G.R. Isomeric Carborane Neuromuscular Blocking Agents. *ChemMedChem* **2019**, *14*, 1108–1114. [CrossRef]
26. Blanes-Mira, C.; Fernández-Aguado, P.; de Andrés-López, J.; Fernández-Carvajal, A.; Ferrer-Montiel, A.; Fernández-Ballester, G. Comprehensive Survey of Consensus Docking for High-Throughput Virtual Screening. *Molecules* **2022**, *28*, 175. [CrossRef] [PubMed]
27. Lyne, P.D. Structure-based virtual screening: An overview. *Drug Discov. Today* **2002**, *7*, 1047–1055. [CrossRef]
28. Ripphausen, P.; Nisius, B.; Bajorath, J. State-of-the-art in ligand-based virtual screening. *Drug Discov. Today* **2011**, *16*, 372–376. [CrossRef]
29. Yang, S.Y. Pharmacophore modeling and applications in drug discovery: Challenges and recent advances. *Drug Discov. Today* **2010**, *15*, 444–450. [CrossRef]
30. Tropsha, A.; Golbraikh, A. Predictive QSAR modeling workflow, model applicability domains, and virtual screening. *Curr. Pharm. Des.* **2007**, *13*, 3494–3504. [CrossRef]
31. Ferreira, L.G.; Dos Santos, R.N.; Oliva, G.; Andricopulo, A.D. Molecular docking and structure-based drug design strategies. *Molecules* **2015**, *20*, 13384–13421. [CrossRef] [PubMed]
32. Rahman, M.M.; Basta, T.; Teng, J.; Lee, M.; Worrell, B.T.; Stowell, M.H.B.; Hibbs, R.E. Structural mechanism of muscle nicotinic receptor desensitization and block by curare. *Nat. Struct. Mol. Biol.* **2022**, *29*, 386–394. [CrossRef] [PubMed]
33. Erkan, S. Activity of the rocuronium molecule and its derivatives: A theoretical calculation. *J. Mol. Struct.* **2019**, *1189*, 257–264. [CrossRef]
34. Udayakumar, M.; Kumar, P.S.; Hemavathi, K.; Shanmugapriya, P.; Seenivasagam, R. Receptor-based pharmacophore tool for design and development of next-generation drugs. *Int. J. Bioinform. Res. Appl.* **2013**, *9*, 487–516. [CrossRef] [PubMed]
35. Almahmoud, S.; Jin, W.; Geng, L.; Wang, J.; Wang, X.; Vennerstrom, J.L.; Zhong, H.A. Ligand-based design of GLUT inhibitors as potential antitumor agents. *Bioorganic Med. Chem.* **2020**, *28*, 115395. [CrossRef]
36. Guedes, I.A.; de Magalhães, C.S.; Dardenne, L.E. Receptor-ligand molecular docking. *Biophys. Rev.* **2014**, *6*, 75–87. [CrossRef] [PubMed]
37. Miyazawa, A.; Fujiyoshi, Y.; Stowell, M.; Unwin, N. Nicotinic acetylcholine receptor at 4.6 A resolution: Transverse tunnels in the channel wall. *J. Mol. Biol.* **1999**, *288*, 765–786. [CrossRef] [PubMed]
38. Benoit, E.; Couesnon, A.; Lindovsky, J.; Iorga, B.I.; Aráoz, R.; Servent, D.; Zakarian, A.; Molgó, J. Synthetic Pinnatoxins A and G Reversibly Block Mouse Skeletal Neuromuscular Transmission In Vivo and In Vitro. *Mar. Drugs* **2019**, *17*, 306. [CrossRef] [PubMed]
39. Majumder, R.; Mandal, M. Screening of plant-based natural compounds as a potential COVID-19 main protease inhibitor: An in silico docking and molecular dynamics simulation approach. *J. Biomol. Struct. Dyn.* **2022**, *40*, 696–711. [CrossRef]
40. Wang, Y.; Liu, H.; Fan, Y.; Chen, X.; Yang, Y.; Zhu, L.; Zhao, J.; Chen, Y.; Zhang, Y. In Silico Prediction of Human Intravenous Pharmacokinetic Parameters with Improved Accuracy. *J. Chem. Inf. Model.* **2019**, *59*, 3968–3980. [CrossRef]
41. Ali, M.I.; Thirukovela, N.S.; Kumar, G.B.; Dasari, G.; Badithapuram, V.; Manchal, R.; Bandari, S. Design, synthesis, in silico molecular docking, and ADMET studies of quinoxaline-isoxazole-piperazine conjugates as EGFR-targeting agents. *Chem. Biol. Drug Des.* **2024**, *103*, e14499. [CrossRef] [PubMed]
42. Viana Nunes, A.M.; das Chagas Pereira de Andrade, F.; Filgueiras, L.A.; de Carvalho Maia, O.A.; Cunha, R.; Rodezno, S.V.A.; Maia Filho, A.L.M.; de Amorim Carvalho, F.A.; Braz, D.C.; Mendes, A.N. preADMET analysis and clinical aspects of dogs treated with the Organotellurium compound RF07: A possible control for canine visceral leishmaniasis? *Environ. Toxicol. Pharmacol.* **2020**, *80*, 103470. [CrossRef] [PubMed]
43. Vilar, S.; Cozza, G.; Moro, S. Medicinal chemistry and the molecular operating environment (MOE): Application of QSAR and molecular docking to drug discovery. *Curr. Top. Med. Chem.* **2008**, *8*, 1555–1572. [CrossRef] [PubMed]
44. Halgren, T.A. MMFF VI. MMFF94s option for energy minimization studies. *J. Comput. Chem.* **1999**, *20*, 720–729. [CrossRef]
45. Yin, K.; Zhao, G.; Xu, C.; Qiu, X.; Wen, B.; Sun, H.; Liu, G.; Liu, Y.; Zhao, Q.; Wei, Q.; et al. Prediction of Toxoplasma gondii virulence factor ROP18 competitive inhibitors by virtual screening. *Parasites Vectors* **2019**, *12*, 98. [CrossRef] [PubMed]
46. Corbeil, C.R.; Williams, C.I.; Labute, P. Variability in docking success rates due to dataset preparation. *J. Comput. Aided Mol. Des.* **2012**, *26*, 775–786. [CrossRef] [PubMed]
47. Mooers, B.H.M.; Brown, M.E. Templates for writing PyMOL scripts. *Protein Sci.* **2021**, *30*, 262–269. [CrossRef] [PubMed]
48. Van Der Spoel, D.; Lindahl, E.; Hess, B.; Groenhof, G.; Mark, A.E.; Berendsen, H.J. GROMACS: Fast, flexible, and free. *J. Comput. Chem.* **2005**, *26*, 1701–1718. [CrossRef] [PubMed]
49. Wang, J.; Wolf, R.M.; Caldwell, J.W.; Kollman, P.A.; Case, D.A. Development and testing of a general amber force field. *J. Comput. Chem.* **2004**, *25*, 1157–1174. [CrossRef]
50. Jakalian, A.; Jack, D.B.; Bayly, C.I. Fast, efficient generation of high-quality atomic charges. AM1-BCC model: II. Parameterization and validation. *J. Comput. Chem.* **2002**, *23*, 1623–1641. [CrossRef]
51. Price, D.J.; Brooks, C.L., 3rd. A modified TIP3P water potential for simulation with Ewald summation. *J. Chem. Phys.* **2004**, *121*, 10096–10103. [CrossRef] [PubMed]

52. Simmonett, A.C.; Brooks, B.R. A compression strategy for particle mesh Ewald theory. *J. Chem. Phys.* **2021**, *154*, 054112. [CrossRef] [PubMed]
53. Wang, E.; Sun, H.; Wang, J.; Wang, Z.; Liu, H.; Zhang, J.Z.H.; Hou, T. End-Point Binding Free Energy Calculation with MM/PBSA and MM/GBSA: Strategies and Applications in Drug Design. *Chem. Rev.* **2019**, *119*, 9478–9508. [CrossRef] [PubMed]

molecules

MDPI

Article

Photophysical Exploration of Alectinib and Rilpivirine: Insights from Theory and Experiment

Chun Zhang [1], Yuting Yang [1], Suya Gan [1], Aimin Ren [2], Yu-Bo Zhou [3,4], Jia Li [3,4], Da-Jun Xiang [5,*] and Wen-Long Wang [1,*]

1 School of Life Sciences and Health Engineering, Jiangnan University, Wuxi 214122, China
2 Institute of Theoretical Chemistry, College of Chemistry, Jilin University, Liutiao Road 2#, Changchun 130061, China
3 National Center for Drug Screening, State key Laboratory of Drug Research, Shanghai Institute of Materia Medica, Chinese Academy of Sciences, Shanghai 201203, China
4 Zhongshan Institute for Drug Discovery, Shanghai Institute of Materia Medica, Chinese Academy of Sciences, SSIP Healthcare and Medicine Demonstration Zone, Zhongshan Tsuihang New District, Zhongshan 528400, China
5 Xishan People's Hospital of Wuxi City, Wuxi 214105, China
* Correspondence: xiangdjxshospital@yeah.net (D.-J.X.); wenlongwang@jiangnan.edu.cn (W.-L.W.)

Abstract: Due to the excellent characteristics of fluorescence-based imaging, such as non-invasive detection of biomarkers in vitro and in vivo with high sensitivity, good spatio-temporal resolution and fast response times, it has shown significant prospects in various applications. Compounds with both biological activities and fluorescent properties have the potential for integrated diagnosis and treatment application. Alectinib and Rilpivirine are two excellent drugs on sale that represent a clinically approved targeted therapy for ALK-rearranged NSCLC and have exhibited more favorable safety and tolerance profiles in Phase III clinical trials, ECHO and THRIVE, respectively. The optical properties of these two drugs, Alectinib and Rilpivirine, were deeply explored, firstly through the simulation of molecular structures, electrostatic potential, OPA/TPA and emission spectral properties and experiments on UV-vis spectra, fluorescence and cell imaging. It was found that Alectinib exhibited 7.8% of fluorescence quantum yield at the 450 nm excited wavelength, due to a larger electronic transition dipole moment (8.41 Debye), bigger charge transition quantity (0.682 e) and smaller reorganization energy (2821.6 cm^{-1}). The stronger UV-vis spectra of Rilpivirine were due to a larger electron–hole overlap index (Sr: 0.733) and were also seen in CDD plots. Furthermore, Alectinib possessed obvious active two-photon absorption properties ($\delta_{max}^{TPA} * \phi$ = 201.75 GM), which have potential TPA imaging applications in bio-systems. Lastly, Alectinib and Rilpivirine displayed green fluorescence in HeLa cells, suggesting the potential ability for biological imaging. Investigation using theoretical and experimental methods is certainly encouraged, given the particular significance of developing integrated diagnosis and treatment.

Keywords: fluorescence imaging; Alectinib; Rilpivirine; quantum chemistry

1. Introduction

In recent years, optical imaging modalities have shown promise in various health-related applications, such as disease diagnosis and optical-guided surgery [1,2]. Compared to traditional methods, fluorescence-based imaging offers several advantages, including non-invasive detection of biomarkers in vitro and in vivo with high sensitivity, fast response times and excellent spatio-temporal resolution [3,4]. Fluorescence-based imaging often relies on the use of chemical tools, which are fluorescent molecules. Small fluorescent molecules are an important kind of fluorescent compound and have the potential to achieve the real-time monitoring of organ function and the visualization of organ-related processes at the cellular level [5]. An excellent small fluorescent molecule should possess the following characteristics: high fluorescence quantum yield, adequate water solubility, fluorophore with good photo-stability and acceptable bio-compatibility [1]. In recent years, more and more fluorescent molecules combined with drugs, aptamers, peptide sequences and other easily modified ligands have been used to develop novel integrated diagnosis and treatment [6–9]. Along with the increase in reports about the potential UV-vis/fluorescence spectral properties of some (chemotherapy) drugs, research on integrated diagnosis and treatment has attracted widespread attention in the international fields of biology, materials and medicine [10–12]. The technology of integrated diagnosis and treatment is able to track the development, occurrence and treatment process of a lesion site (such as cancer) in real time, implement effective and precise treatment, improve the cure effect, reduce side effects, detect differentiation and metastasis for the lesion site and take measures to maximize patient survival and timely recovery rates. However, the strategy of combining fluorophores and suitable ligands faces some limitations, for example, complicated synthesis, poor permeability, low bioavailability and more. Therefore, none are currently used in clinical applications. Following the significant demand of quality of human life, a small organic drug molecule combining near-infrared (NIR) fluorescence imaging has significance for promoting clinical applications of integrated diagnosis and treatment [13,14].

Alectinib and Rilpivirine (shown in Table 1) are two excellent drugs on sale with USD 1.292 billion and USD 964 million of retail sales volume in 2020, respectively [15]. Alectinib is a highly selective, second-generation inhibitor of the tyrosine kinase anaplastic lymphoma kinase (ALK), and is a clinically approved targeted therapy for ALK-rearranged non-small lung cancers (NSCLCs) [16–18]. Importantly, Alectinib is also effective for treating brain metastasis of ALK-rearranged NSCLCs, suggesting its high brain penetrance. In 2020, it was reported that Alectinib could provide a personalized maximum benefit for patients with high-grade serous ovarian cancer who are positive for EML4-ALK [19]. On the other hand, Rilpivirine is a new-generation NNRTI and is considered as a recommended or alternative key drug in the current ART guidelines. It exhibited more favorable safety and tolerance profiles compared with Efavirenz in Phase III clinical trials, ECHO and THRIVE [20,21]. The most commonly observed mutation in patients with Rilpivirine-containing treatment failure is E138K [22]. I135T/L, escape mutations from HLA-B*51/52-restricted cytotoxic T lymphocytes, may predispose HIV-1 to harbor E138K upon failure of Rilpivirine-containing ART and the mutation patterns of drug resistance may vary due to baseline polymorphic mutations [22]. Developing small organic drug molecules with NIR spectra could bring about new opportunities for improving disease diagnosis and effective therapeutics. But it is extremely difficult, so it has not yet been reported. In this study, we thoroughly studied the molecular properties and explored the photo-physical luminous mechanism for these two drugs on sale (Alectinib and Rilpivirine) in order to offer more theoretical foundations for traceable drugs and promote the development of integrated diagnosis and treatment.

Table 1. The chemical structure, stable geometries of the ground state (S_0) and first excited state (S_1) for compounds Alectinib and Rilpivirine.

Molecules	S_0	S_1
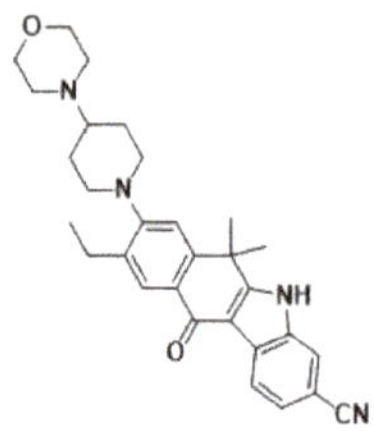 Alectinib	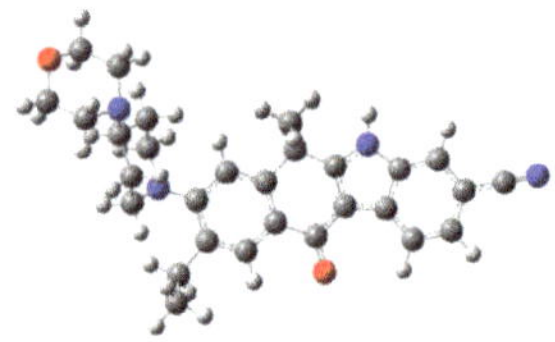 Top view 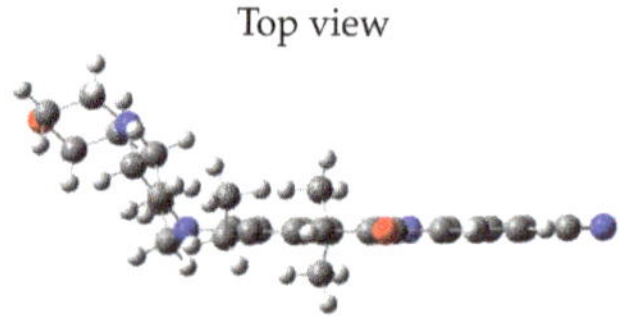 Side view	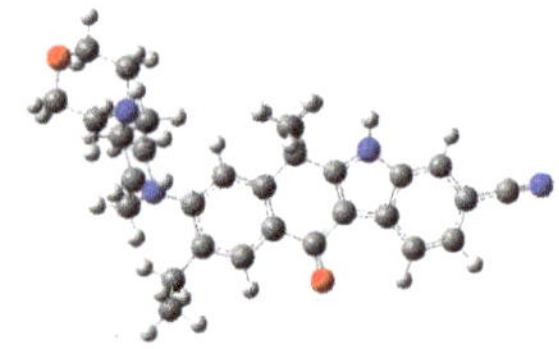 Top view 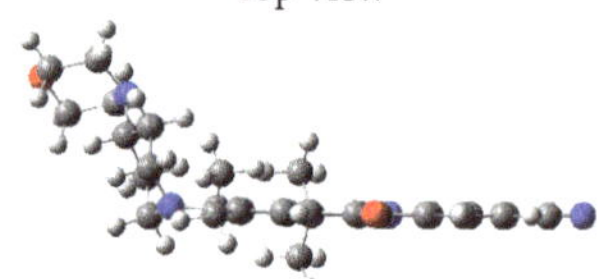 Side view
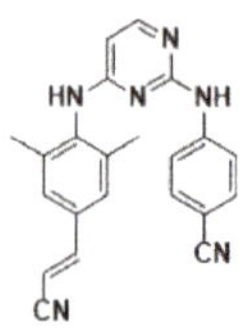 Rilpivirine	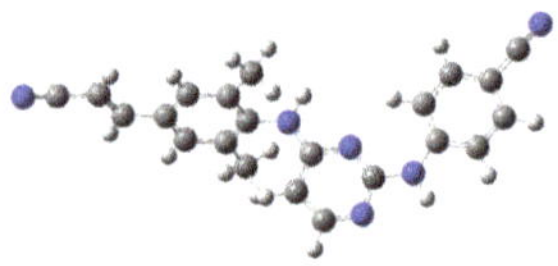 Top view 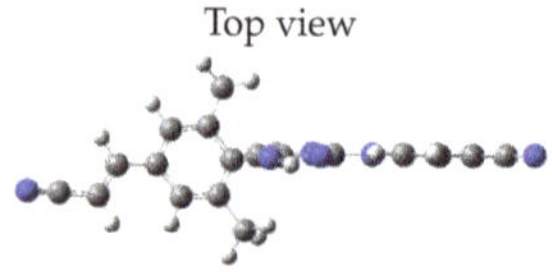 Side view	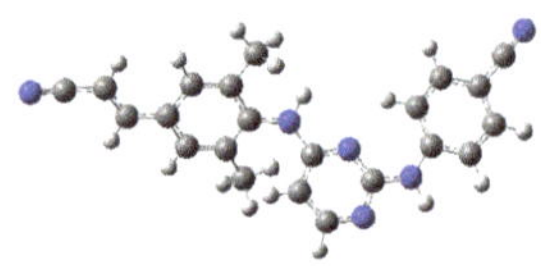 Top view 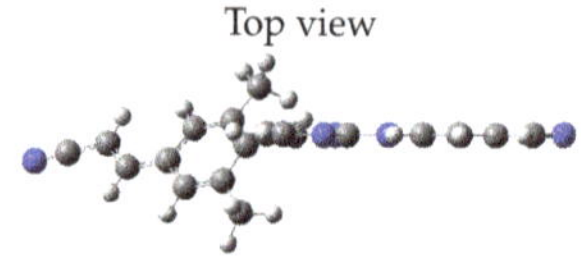 Side view

2. Results and Discussion

2.1. Molecular Structural Characteristics

Firstly, the geometric characteristic is very important to study the electronic and photophysical properties of a compound. In order to learn the geometric characteristics, the structural parameters of stable ground state (S_0) and first excited state (S_1) including the main bond lengths (Å) and dihedral angles (°) were calculated for drugs Alectinib and Rilpivirine, and the resulting data were listed in Tables 1 and S1. From Table 1 about the geometries from molecular top and side view, it can be seen that in not only S_0 but also S_1, molecule Alectinib had a π-conjugation planar molecular skeleton besides the terminal 4-(piperidin-4-yl) morpholine group. And the dihedral angle (see Table S1) between parent molecular plane and 4-(piperidin-4-yl) morpholine group for Alectinib was decreased from the ground state (at about 61.2°) to the excited state (at about 46.6°), suggesting the larger planarity in the excited state. To make the bond length change more intuitive from S_0 to S_1, the data about bond length in Table S1 were drawn in Figure 1. As expected, the single bond in S_1 was shorter (such as C36-C48: 1.48 > 1.43 Å) and the double bond was longer (such as C48-O49: 1.24 < 1.30 Å) than those in S_0, implying the bond length alternation was smaller in S_1 (seen Figure 1A), and further implying the enhanced π-conjugation effect in S_1. In addition, Alectinib possessed strong electron-withdrawing group (–CN) connected with π-conjugation indole group directly. It may lead to marked intramolecular charge transfer (ICT) and benefit for electron transition. The drug molecule Rilpivirine had two conjugated planes, which were distorted by about 145.8° in S_0 and decreased to about 111.0° in S_1 (see Table S1), likely due to the steric effect from two methyl substituents on the benzene ring. The two terminal electron acceptors (–CN) in Rilpivirine also were connected with a π-conjugation benzene ring, which maybe benefit electron transition. The same for the molecule Rilpivirine, the important bond length change (in Figure 1B) was smaller in S_1, which also increased the conjugated electronic structures. To conclude, drugs Alectinib and Rilpivirine exhibited the structural characteristics: planar parent skeleton, stronger π-conjugation effect in S_1 and electron acceptors connecting conjugated structures, which would be a benefit for superior photo-physical performance.

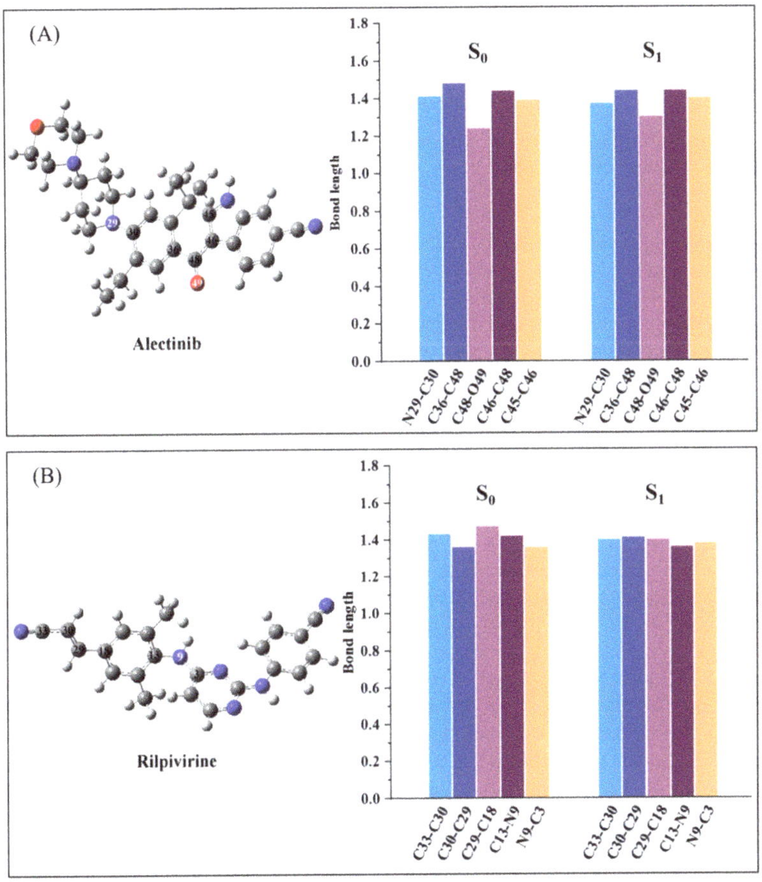

Figure 1. The important bond length (Å) of molecules Alectinib (**A**) and Rilpivirine (**B**) in S_0 and S_1.

2.2. Molecular Electrostatic Potential

The electrostatic potential is a representation of an electric charge distribution for a single molecule, which is an important property for binding with protein. The blue and red regions represent positive and negative electronic potential regions, respectively. The darker color is a "more positive" or "more negative" potential. A negative electric potential means that a positive charge group will be attracted easily. For Alectinib (shown in Figure 2), the red color indicates a higher electron density around the oxygen and cyan group representing that it will benefit the forming interaction with a positive charge group (such as amide group) in protein. On the other hand, the amino group of compounds Alectinib and Rilpivirine (shown in Figure 2) exhibit obvious positive electric potential, which will attract the negative charge group (such as carbonyl group) in protein.

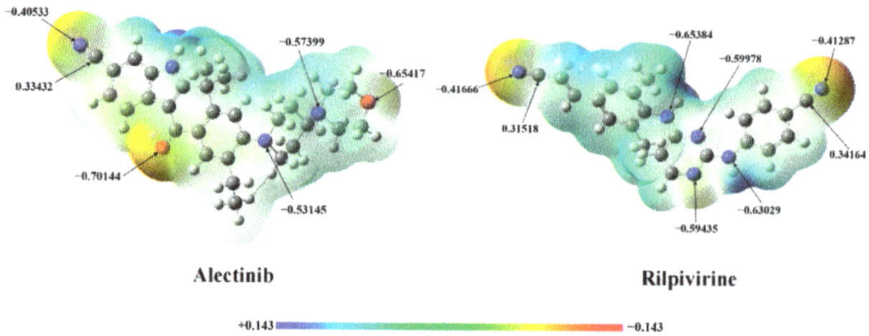

Figure 2. Molecular electrostatic potential surfaces plotted for the Alectinib and Rilpivirine (The blue and red regions represent electronic potential regions of positive and negative potential, respectively.).

2.3. *UV-vis Experiments and One-Photon Absorption (OPA) Spectral Properties*

To evaluate the UV-vis spectra of molecules Alectinib and Rilpivirine, we first tested the solvent-dependent excited spectra in organic solvent 100%DMSO (dimethyl sulfoxide), 50%DMSO-50%PBS (phosphate-buffered saline) and 10%DMSO-90%PBS (shown in Figure S1, Supplementary Materials). As displayed in Figure S1, the UV-vis spectra of Alectinib and Rilpivirine at the three solvent exhibited one maximum characteristic peak and were centered at about 345 nm and 310 nm, respectively. Upon adding PBS, the intensity of excited characteristic peak for Alectinib was decreased gradually, accompanied with a red-shift wavelength from 338 nm → 348 nm → 349 nm in Figure S1A. As well as for Rilpivirine, following with increased PBS, the intensity of absorption peak also was reduced from Figure S1B. It was worth noted that the solution became some turbid for Alectinib and Rilpivirine upon adding 25%PBS solvent in UV-vis experiments, which was in accordance with the reported poor/medium soluble for Alectinib and Rilpivirine [23]. The poor/medium soluble may lead to the reduced intensity of UV-vis spectra along with increased PBS proportion. To further explore the photo-physical properties of UV-vis spectra, theoretical calculation for OPA was performed in the next, which would be discussed in detail.

Spectra is closely related with optical properties and electronic characteristics. So, the simulated OPA spectral properties of Alectinib and Rilpivirine in water were obtained and listed in Table 2, concluding OPA wavelength (λ_{max}^{OPA}), oscillator strengths ($f^{\,O}$), vertical excitation energies (E^{0f}), transition dipole moment (μ^{0f}) and transition characteristics. TDDFT methods were employed to obtain the OPA spectra according quantum chemical calculation. The solvation model density (SMD) [24,25] with default parameters of H_2O was used to implicitly consider the homogeneous dielectric solvation effects. Additionally, compared with the results from Figure S1 and Table 2, we found that the calculated wavelengths of Alectinib and Rilpivirine were in reasonable agreement with the experimental results generally.

The main absorption peak ($349^{Exp.}$ nm) in 10%DMSO-90%PBS of Alectinib was connected with the first excited state S_1, which was derived from HOMO → LUMO transition (88.1%) and had localized excitation (LE) and charge transfer (CT) characteristics (see Figure S2). The CT characteristic mainly came from the 1-(2-ethylphenyl) piperidine group to the indole group due to the electron-accepting group (–CN) connecting with π-conjugated indole ring. For Rilpivirine, the maximum excited wavelength ($311^{Exp.}$ nm) in 10%DMSO-90%PBS was also derived from $S_0 \rightarrow S_1$ electron transition with a major contribution of HOMO-1 → LUMO (56.7%). Furthermore, the transition density matrix (TDM) and charge density difference (CDD) were adopted to describe the electron transfer process directly from the ground and excited state in the whole molecule [26–28]. Here, we mainly studied the electronic transition process of $S_0 \rightarrow S_1$ for Alectinib and Rilpivirine in the OPA spectrum, and the electronic transition properties were analyzed qualitatively by TDM and CDD. In the meantime, more transition indexes were also listed in Table S2, including the centroid distance of the electrons and holes (D), electron-hole overlap index (Sr), average distribution breadth of the electrons and holes (H), hole delocalization index (HDI) and electron delocalization index (EDI) [29]. Firstly, the two-dimensional diagram showed the charge density difference, in which the green isosurface represented electron distribution while the blue isosurface represented hole distribution. In the transition process from the S_0 to S_1, the molecule Alectinib showed the characteristic of local excitation and weak charge transfer, which was mainly concentrated on the 7-ethyl-4-methylnaphthalen-1(4*H*)-one group in Figure 3 (CDD). The smaller average distribution breadth (H: 2.038 Å) of the electrons and holes for Alectinib was due to the local excitation from 7-ethyl-4-methylnaphthalen-1(4*H*)-one group mainly. The centroid distance (D: 1.084 Å) of the electrons and holes for Alectinib was smaller, likely because of the weak charge transfer from the carbanyl group to 6-ethyl-1-methyl-1,4-dihydronaphthalene. For molecule Rilpivirine, owning to the existence of two cyano groups in the left terminal, larger local excitation occurred in the left half of the molecule and the electronic transfer happened from right to left. Combining with TDM plots in Figure 3, it could be seen that stronger contribution of electron transfer for Rilpivirine resulted from the 8th nitrogen atom in linker to the 10th carbon atom in benzene ring (atomic numbers in Figure S3) [30]. Lastly, the larger electron-hole overlap index (Sr: 0.733) in Rilpivirine implied the more electron-hole overlap, which was also seen in CDD plots.

Table 2. Calculated OPA spectra properties of Alectinib and Rilpivirine in water concluding OPA wavelength (λ^{OPA}), oscillator strengths (f^{O}), vertical excitation energies (E^{0f}), transition dipole moment (μ^{0f}) and transition characteristics.

Molecule	λ_{max}^{OPA}/nm	f^{O}	E^{0f}/eV	μ^{0f}/Debye	Transition Characteristics
Alectinib	318.0 [a] (349 [Exp.])	0.7456 [a]	3.90 [a]	1.1 [a]	$S_0 \rightarrow S_1$ (HOMO → LUMO 88.1%) [a]
Rilpivirine	298.6 [a] (311 [Exp.])	1.4836 [a]	4.15 [a]	1.5 [a]	$S_0 \rightarrow S_1$ (HOMO-1 → LUMO 56.7%) [a]

[a]—was the data from the calculated OPA properties. [Exp.]—represented the data from experiments.

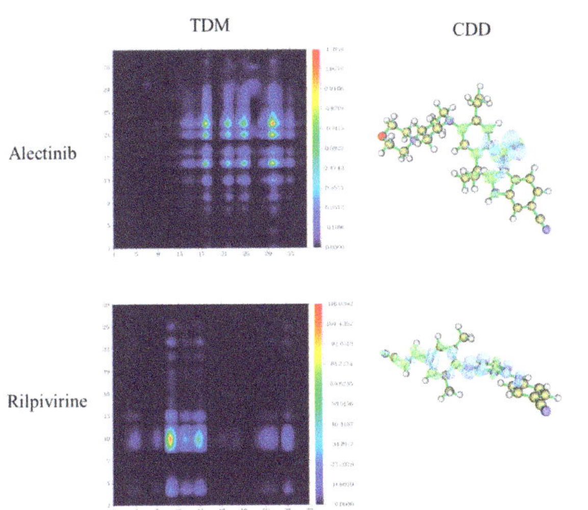

Figure 3. The transition density matrix (TDM) and charge density difference (CDD, green represented electron distribution and blue was hole distribution.) plots of Alectinib and Rilpivirine.

2.4. TPA Spectral Properties

The traditional fluorescence imaging technique applyies one-photon microscopy (OPM), which involves the application of UV-vis light as an excitation source and achieves the lower tissue penetrations depth (generally 100 mm), limiting the application of OPM in living systems [31]. Over the past decades, the fluorescence imaging technique basing on multiphoton absorption uses near-infrared (NIR) as excitation sources, which has been proved to be one of the most effective tools in biomedical imaging applications [5,32]. For two-photon microscopy (TPM), the electron of fluorophore is excited to excited states after absorbing two photons simultaneously, using half the energy of photons compared to OPM [2]. Maria Goppert Mayer, as a Nobel laureate, first envisioned the concept of using two light quanta to excite a fluorophore. In her honor, the unit 'GM' was used to represent the TPA cross section values of a molecule [33]. Subsequently, the first cellular images was obtained by Webb et al. according to simultaneous excitation of fluorophore with two photons of NIR wavelength (700 nm–1100 nm) using femtosecond (*fs*) pulsed laser, which was in biological optical window to excite the fluorophore. The TPM technology has exclusive advantages in fluorescence imaging, including deeper imaging depth (down to 1 mm), less photobleaching, weaker background fluorescence, and higher spatiotemporal resolution [34]. These promising features have inspired more and more scientists to develop novel fluorescent molecules with enhanced TPA properties by designing molecules with an appropriate donor–acceptor systems, suitable dipolars, π-bridges, quadrupolars, octupolar characteristics and more [35–37]. Thus, the potential TPA properties of Alectinib and Rilpivirine were also predicted in this section, hoping to improve their bio-imaging applications.

As we all known, effective TPA imaging is influenced by the TPA cross-section (δ^{TPA}) and fluorescent quantum yield (Φ) simultaneously. The δ^{TPA} denotes the TPA probability of a molecule. The larger the TPA cross-section, the larger the probability is for reaching the excited state after absorbing two photons simultaneously. In this work, we used the response function theory method to obtain the TPA properties [38,39]. We performed the calculation for the TPA spectral properties of both Alectinib and Rilpivirine, including the maximum TPA cross sections (δ^{TPA}_{max}) and corresponding TPA wavelengths (λ^{TPA}_{max}) by DALTON software (Dalton2021.alpha, http://daltonprogram.org) in the 550 nm–1000 nm region [40]. Firstly, in order to decrease the deviation of simulated TPA spectra, two common TD-DFT functionals (Cam-B3LYP and B3LYP) for predicting the TPA properties were adopted here. The TPA spectra in gas and water (with PCM solvent) were obtained and listed in Tables 3 and S3, using Cam-B3LYP and B3LYP functional, respectively. We could draw the following: (i) not only by B3LYP but also by Cam-B3LYP, the TPA wavelength and cross section of Alectinib in water was longer and larger than that in gas, such as 647.4 nm/159 GM (water) > 623.00 nm/1.0 GM (gas), suggesting the potential application in biological systems, as well as for compound Rilpivirine. (ii) Molecules Alectinib and Rilpivirine generally exhibited shorter TPA wavelength and smaller TPA cross section under Cam-B3LYP functional compared with those under B3LYP functional, which were in agreement with the reported investigation [12]. But the transition characteristic for the TPA spectra of Alectinib and Rilpivirine by the two functionals was consistent (from $S_0 \rightarrow S_1$). Considering the reported better results from B3LYP compared with the experimental data, we adopted the calculated TPA properties by B3LYP functional for the later discussion [41,42]. (iii) In a water environment, the compound Alectinib exhibited a larger TPA cross-section (δ^{TPA}_{max}: 269.0 GM) at 772.5 nm, which was in the NIR wavelength region. From the next fluorescence experiment, the fluorescence quantum yield of Alectinib was 7.5%, so its action TPA cross-section ($\delta^{TPA}_{max}*\Phi$) was 201.75 GM, which was larger than 50 GM and was suitable for applications in biological samples with reasonable incident laser power [31]. (iv) For Rilpivirine, the TPA cross section in water was medium (δ^{TPA}_{max}: 159.0 GM) at 744.6 nm excited wavelength, but the fluorescence quantum yield (Φ: 1.1%) was lower. Thus, its smaller action cross-section ($\delta^{TPA}_{max}*\Phi$: 17.49 GM) might restrict the potential application in TPA bio-imaging. The latter sections were devoted to discussing important aspects that affected their TPA properties.

Table 3. The calculated TPA properties of Alectinib and Rilpivirine including the maximum TPA cross-section (δ_{max}^{TPA}), corresponding TPA wavelength (λ_{max}^{TPA}), transition nature and charge transfer amount in gas (a) and water (b) by B3LYP functional.

Molecules	δ_{max}^{TPA}/GM	λ_{max}^{TPA}/nm	Transition Nature	q_{TPA}^{CT}/e
Alectinib	44.8 [a] 269.0 [b]	700.5 [a] 772.5 [b]	$S_0 \rightarrow S_1$ [a] (HOMO → LUMO) $S_0 \rightarrow S_1$ [b] (HOMO → LUMO)	0.723 [b]
Rilpivirine	21.8 [a] 159.0 [b]	756.0 [a] 744.6 [b]	$S_0 \rightarrow S_1$ [a] (HOMO → LUMO) $S_0 \rightarrow S_1$ [b] (HOMO → LUMO)	0.546 [b]

[a] were the calculated TPA properties in gas. [b] represented the calculated TPA properties in water.

In order to further clarify the origin of the TPA activity of Alectinib and Rilpivirine and explain the calculated TPA spectra, the two-state approximation expression (X) related to the TPA cross section was adopted here [43,44]:

$$\delta_{\text{max}}^{\text{TPA}} \propto \frac{\left(M^{01}\right)^2 \left(\left|\Delta\mu^{01}\right|\right)^2}{\left(E^{01}\right)^2} \tag{1}$$

The values of transition/state dipole moment vectors and transition energy involved in two-state approximation model were listed in Table 4. As shown in Table 4, it was clear that the larger TPA cross-section of Alectinib resulted from the smaller transition energy (E^{01} = 3.83 eV), larger transition state dipole moment (2.99 Debye), and bigger difference of state dipole moment ($\left|\Delta\mu^{01}\right|$ = 2.48 Debye) mainly. Additionally, the simulated TPA tensor elements basing on the quadratic response theory were also listed in Table 5 to reveal the structural characteristics for TPA properties of Alectinib and Rilpivirine. By using the TPA tensor elements in forma (4), the maximum TPA cross-sections in atomic units (a.u.) were obtained and displayed in Table 5. From Table 5, it could be seen that the Sxx component had a significant contribution in promoting the TPA process for Alectinib and Rilpivirine, which happened in the direction of ICT from Figure S2. Thus, the larger TPA cross-section for Alectinib might result from the bigger charge transfer amount (Alectinib (0.723 e) > Rilpivirine (0.546 e), in Table 3) during the TPA transition process.

Table 4. Parameters related to TPA transition process of Alectinib and Rilpivirine.

Molecules	Excited State	μ^{00}/Debye	μ^{11}/Debye	$\left|\Delta\mu^{01}\right|$/Debye	M^{01}/Debye	E^{01}/eV
Alectinib	S_1	10.96	13.44	2.48	2.99	3.83
Rilpivirine	S_1	9.32	10.54	1.22	2.00	3.99

Table 5. TPA tensor elements (S_{ab}) and TPA cross sections (σ^{TPA}) (in au) for Alectinib and Rilpivirine molecules calculated in water solvent by DALTON software with B3LYP functional.

Molecules	Excited State	S_{ab}/a.u.						σ^{TPA}/a.u.
		S_{xx}	S_{yy}	S_{zz}	S_{xy}	S_{xz}	S_{yz}	
Alectinib	S_1	422.7	−31.7	7.5	6.6	63.2	−1.0	1,068,860.6
Rilpivirine	S_1	286.6	18.5	−2.9	2.5	4.7	95.6	590,613.0

2.5. Fluorescence Spectral Properties

For optical imaging and detection, it is essential to have high fluorescence efficiency. Basing on the systematical analysis of absorption spectral properties, how about the fluorescence properties of Alectinib and Rilpivirine? Thus, the fluorescence spectra of the molecules Alectinib and Rilpivirine in different solvents (100%DMSO, 75%DMSO-25%PBS, 50%DMSO-50%PBS and 25%DMSO-75%PBS) were measured by fluorescence experiments, respectively, and were drawn in Figure 4. As shown in Figure 4, we noticed that the emission wavelength of Alectinib and Rilpivirine were about 450 nm and 500 nm, respectively, and the fluorescence efficiency of Alectinib was higher than that of Rilpivirine (Φ: 7.5% > 1.1%). Furthermore, the fluorescence intensity in 100%DMSO was strongest for the two compounds, and it exhibited an obvious decreasing tendency along with the increase in PBS content. What is the reason? Of particular note was that in the fluorescent experiments, the solution became more turbid for the compounds Alectinib and Rilpivirine upon adding 25%PBS, which was in agreement with the reported poor/medium soluble [23]. Thus, the fluorescence spectra were further obtained in different concentrations for Alectinib and Rilpivirine, respectively, which were displayed in Figure S4. It could be clearly seen that, following the increased sample concentration (50 μM → 100 μM → 150 μM), the emission intensity was increased gradually. Those demonstrated that the decreased fluorescent intensities of Alectinib and Rilpivirine in more PBS proportion were due to their lower solubility. Additionally, there was a red shift of fluorescence peak following with the increased sample concentration in Figure S4. Most of the chemo drugs benefited the florescence properties due to their aromatic rings. The red shift of fluorescence peak took place at a dense solution due to the photon reabsorption effects when the stokes shift between absorption/fluorescence spectra was sufficiently small as it happened for the chemo drugs of interest here [45,46]. According to Figure S4, the small Stokes shift lucidly appeared that was the origin of photon reabsorption effects and subsequent red shift. To explore the origin of these emission spectra, the theoretical study was analyzed next.

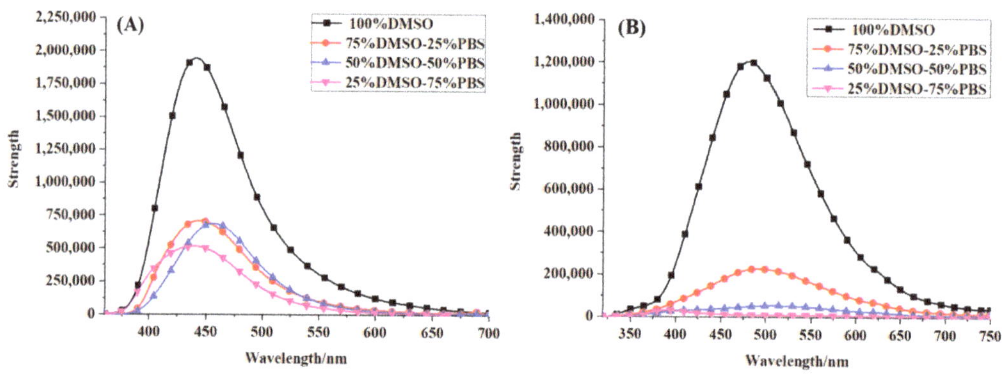

Figure 4. Fluorescence spectra of Alectinib (**A**) and Rilpivirine (**B**) with 100 μM by excitation at 340 nm and 310 nm, respectively.

To deeply explore the origins of the fluorescence properties about these two molecules, we adapted the following TD-DFT//B3LYP/6-31+G(d) calculations and obtained the detailed excited properties. As we all known, the fluorescence quantum yield (Φ) is an important index for measuring fluorescence efficiency, which is determined by the radiative decay rate (K_r) and the nonradiative decay rate (K_{nr}) theoretically. According to Kasha's rule, the electron can transfer from the first singlet excited state (S_1) to the ground state (S_0) through the radiative and non-radiative decay processes. Thus, we thoroughly explored the S_1 excited-state properties for the compounds Alectinib and Rilpivirine, and the results including emission wavelength (λ^{EMI}), oscillator strengths (f^E), transition dipole moment (μ^{f0}_{EMI}), charge transfer quantity (q^{CT}_{EMI}), charge transfer distance (d^{CT}_{EMI}) and transition characteristics, and were listed in Table 6. It could be found that the maximum emission wavelength of Alectinib was at 456.7 nm (450 nm $^{Exp.}$), which was originated from the $S_1 \rightarrow S_0$ electron transition with a major contribution of LUMO→HOMO (98.3%). For Rilpivirine, its maximum emission wavelength was also derived from the electron transition $S_1 \rightarrow S_0$, which was mainly constructed by LUMO→HOMO (97.6%).

Table 6. Fluorescence quantum yield (Φ) from experiments and calculated emission spectra properties of Alectinib and Rilpivirine concluding emission wavelength (λ^{EMI}), oscillator strengths (f^E), transition dipole moment (μ^{f0}_{EMI}), charge transfer quantity (q^{CT}_{EMI}), charge transfer distance (d^{CT}_{EMI}) and transition characteristics using B3LYP functional and 6-31+G(d) basis set.

Molecule	λ^{EMI}/nm	f^E	μ^{f0}_{EMI}/Debye	q^{CT}_{EMI}/e	d^{CT}_{EMI}/Å	Transition Characteristics	Φ/%^{Exp}
Alectinib	456.7 $450^{Exp.}$	0.7469	8.41	0.682	2.568	$S_1 \rightarrow S_0$ (LUMO → HOMO 98.3%)	7.5
Rilpivirine	435.4 $500^{Exp.}$	1.3784	6.33	0.502	2.627	$S_1 \rightarrow S_0$ (LUMO → HOMO 97.6%)	1.1

$^{Exp.}$ referred to the results from fluorescence experiments.

From above fluorescence experiment, Alectinib had a larger fluorescence quantum yield than Rilpivirine (Φ: 7.5% > 1.1%) which might come from the larger electronic transition dipole moment and charge transition quantity during the emission process in Table 6 (μ^{f0}_{EMI}: 8.41 Debye > 6.33 Debye, q^{CT}_{EMI}: 0.682 e > 0.502 e). Additionally, we also analyzed this phenomenon by means of reorganization energy, which was an important parameter for evaluating geometric relaxation and energy component in internal conversion process. From Figure 5, it could be summarized that (i) the total reorganization energy of Alectinib was smaller than that of Rilpivirine (Reorg. Energy: 2821.6 cm^{-1} < 5176.9 cm^{-1}), implying the smaller geometric relaxation for Alectinib. (ii) In Alectinib, the four main vibration models (> 200 cm^{-1}), including molecular skeleton scissoring vibration mode at 13.62 cm^{-1}, benzene ring stretching vibration mode at 1486.36 cm^{-1}, 1676.46 cm^{-1} and 1681.77 cm^{-1}, made the contributions to the total reorganization energy. The C-O bond of benzene ring stretching vibration mode at 1486.36 cm^{-1} had a higher reorganization energy up to 330.34 cm^{-1}, and occupied 11.71% of the total reorganization energy. (iii) For the molecule Rilpivirine, in the high-frequency regions, the C-C bond of the benzene ring attached to a double bond stretching vibration mode at 1708.87 cm^{-1} exhibited particular large reorganization energy up to 940.95 cm^{-1}, which contributed 55.06% to its total reorganization energy. Concluding, compared to the compound Rilpivirine, molecule Alectinib possessed larger electronic transition dipole moment and charge transition quantity, which were beneficial for the radiative decay process, and the smaller geometric relaxation, which weakened the nonradiative decay process. On the other hand, the molecular aromatic rings had important influences on the fluorescence properties [39,41,47]. The Alectinib possessed stronger molecular planarity due to the appropriate aromatic ring substitution than that of Rilpivirine (46.6° < 111.0°), which also contributed the higher fluorescence quantum yield. Thus, Alectinib had higher fluorescence quantum yield than Rilpivirine in organic solvent. Lastly, how were their fluorescence properties in cell imaging?

2.6. Cell Imaging Application

To evaluate the bio-imaging ability, the compounds Alectinib and Rilpivirine to stain living cells of HeLa cell line were studied by confocal laser scanning fluorescence microscopy. HeLa cells were cultured onto glass-bottom Petri dishes for 12 h before treatment. Live cells were treated with the compounds Alectinib and Rilpivirine at 20 μM for 16 h, washed with PBS three times, fixed in 4% paraformaldehyde solution for 15 min, and washed with PBS again, respectively. Cell images were obtained by confocal laser scanning fluorescence microscopy (CLSM, Nikon, Ti2-E+A1, Japan) using both 405 nm and 488 nm as the excitation wavelengths. As displayed in Figure 6, HeLa cells stained with the compounds Alectinib and Rilpivirine showed green fluorescence. It suggested that Alectinib and Rilpivirine had better imaging performances in cell imaging, although Rilpivirine had a lower fluorescence quantum yield than Alectinib (1.1% < 7.5%) in an organic solvent. These might come from the complex cellar environments, which formed some interaction with small molecules and enhanced the fluorescence intensity of Rilpivirine. So, Alectinib and Rilpivirine had the potential abilities for biological imaging applications. Compared with commercial dyes or drugs sharing with a single function, drug molecules Alectinib and Rilpivirine have potential practical applications in integrated diagnosis and treatment.

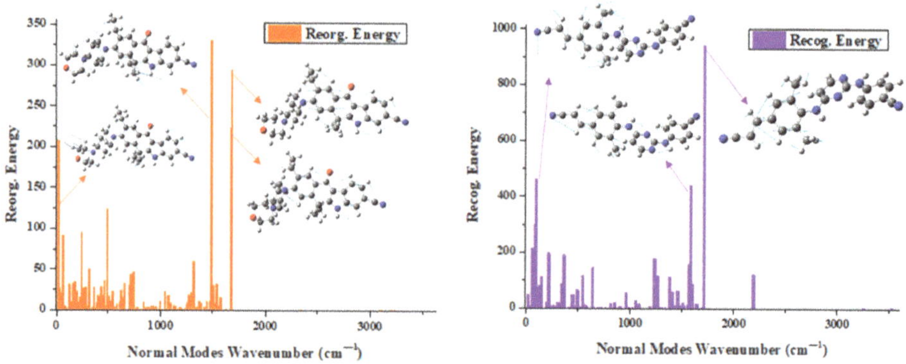

Figure 5. Reorganization energies (cm^{-1}) and crucial displacement vectors for the normal modes with large reorganization energies for compounds Alectinib (origin) and Rilpivirine (purple).

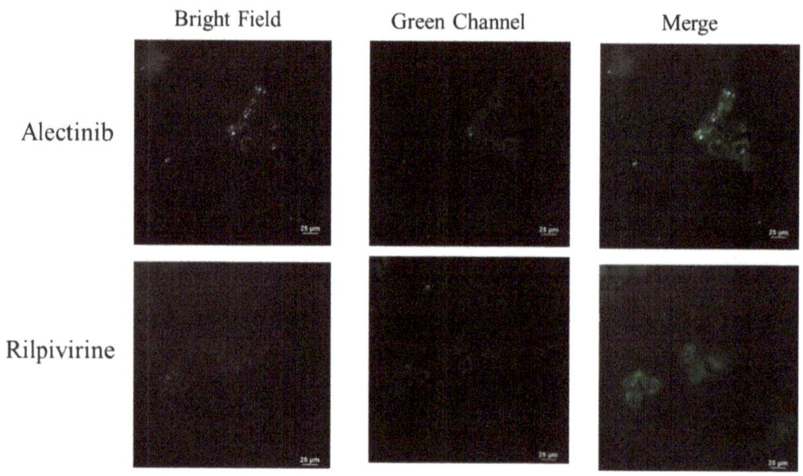

Figure 6. Confocal microscopy colocalization images of the studied drug molecules Alectinib and Rilpivirine in HeLa cells.

3. Material and Methods

In this work, the optimization and frequency calculations of ground-state and geometries structures for Alectinib and Rilpivirine were obtained at the level of M06-2X/6-31+G(d) with the help of Gaussian 16 program package [48,49]. Furthermore, their stable molecular geometries were displayed in Table 1. The three smallest vibrational frequencies of Alectinib and Rilpivirine at the real local minima points were positive values, confirming the stabilities of studied Alectinib and Rilpivirine geometries. The one-photon absorption (OPA) and emission spectral properties were simulated according to the time-dependent density functional theory (TD-DFT) based on stable molecular structures. The M06-2X/6-31+G(d) and B3LYP/6-31+G(d) level were used for OPA and emission spectral simulation, respectively. At the same time, the solvent effect was taken into account within the self-consistent reaction field (SCRF) theory through applying SMD model [24]. On the other hand, the two-photon absorption (TPA) properties were calculated with the help of quadratic response theory in the DALTON program [40]. The reorganization energies with the parameters of normal mode displacements were carried out by the MOMAP program [50,51].

The OPA transition probability (δ^{OPA}) is able to evaluate the OPA intensity. The OPA oscillator strength (f), as an important index for measuring various OPA property of a fluorescent compound, is related to δ^{OPA}. The expression of δ^{OPA} from the ground state S_0 to the excited state Sn is obtained by [52]:

$$\delta^{OPA} = \frac{2\omega_n}{3}\sum_{\alpha}|\langle 0|\hat{\mu_\alpha}|n\rangle|^2 \ (\alpha \in x,\ y,\ z) \tag{2}$$

here, ω_n represents the excited energy. $\langle 0|\hat{\mu_\alpha}|n\rangle$ is the transition dipole moment along with the different coordinate directions, which is related to the wave function integral. These physical parameters about the OPA spectra can be calculated with the help of Gaussian 16 program [49].

The TPA cross-section (δ^{TPA}) is used to evaluate TPA intensity, which is determined by two-photon transition probability (σ^{TPA}) and can be given by [53]:

$$\delta^{TPA} = \frac{4\pi^2\alpha\alpha_0^5\omega^2}{15c\Gamma}\sigma^{TPA} \tag{3}$$

here, α, α_0 and c are on behalf of the fine structure constant, Bohr radius, and the speed of light, respectively; ω represents the photon energy in atomic units, and Γ is the broadening factor, describing spectral broadening of an excitation, which has been assumed to be 0.05 eV to make the theoretical simulation process consistent with the experimental spectra. σ^{TPA} can be calculated as [54]:

$$\sigma^{TPA} = \frac{1}{30}\sum_{ab}\left(FS_{aa}\overline{S}_{bb} + GS_{ab}\overline{S}_{ab} + HS_{ab}\overline{S}_{ba}\right) \tag{4}$$

$a,\ b \in \{x,\ y,\ z\}$, and F, G and H are 2, 2, and 2 for linearly polarized light and were −2, 3 and 2 for the circular case, respectively. Taking electric dipole approximation into consideration, the TPA transition tensor S^{if} between the initial state i and the final state f can be expressed as [53]:

$$S_{ab}^{if} = \sum_{n \neq i}\left\{\frac{\langle i|\mu_a|n\rangle\langle n|\overline{\mu}_b|f\rangle}{\omega_{in} - \omega_1} + \frac{\langle i|\mu_b|n\rangle\langle n|\overline{\mu}_a|f\rangle}{\omega_{in} - \omega_2}\right\} \tag{5}$$

where $\langle i|\mu_\alpha|n\rangle$ is the ath compound of the transition dipole moment between the initial electric state i and the intermediate state n. ω_{in} is excitation energy. In addition, ω_1 and ω_2 presents the energies of two photons. These physical parameters about the TPA spectra can be obtained by quadratic response theory with the help of DALTON software [40].

Fluorescence intensity is determined by the competition of radiative decay and non-radiative decay process [42,55]. The fluorescence efficiency could be evaluated by fluorescence quantum yield (Φ), which was obtained from experiment directly in this investigation. The stronger of radiative decay, the larger of the Φ. Conversely, the smaller the non-radiative decay process, the larger of the Φ. Internal conversion (IC), as a non-radiative decay from the first excited state to ground state, is the most important component during the non-radiative process [56]. The IC rate can be calculated under harmonic oscillator approximation though Fermi's golden rule and is expressed as [57]:

$$K_{ic} = \frac{2\pi}{\hbar}\left|H'_{fi}\right|^2\delta\left(E_{fi} + E_{fv_f} - E_{fv_i}\right) \tag{6}$$

The K_{ic} is closely related to electron–vibration coupling and geometric relaxation of the excited state, which can be evaluated by reorganization energy (λ) or Huang−Rhys factor (HR). These above parameters of fluorescence spectra can be calculated by applying Gaussian 16 and MOMAP program [49,51].

The UV-vis excitation and fluorescence spectroscopy were carried out at room temperature and by the instrument of SHIMADZU 2550 UV–vis spectrophotometer (SHIMADZU, Kyoto, Japan) and SHIMADZU RF-6000 Fluorolog instrument (SHIMADZU, Kyoto, Japan), respectively. The slit bandwidth and sampling interval for Shimadzu RF-600 was 5 nm and 0.5 nm, respectively. Cell images were obtained by confocal laser scanning fluorescence microscopy (CLSM, Nikon, Ti2-E+A1, Tokyo, Japan) using 405 nm and 488 nm as the excitation wavelength. The HeLa cells were human cervical cancer cells in DMEM and EMEM culture medium from Human Fenghui Biotechnology Co. Ltd. (Nanjing, China), and its accession number was CL0134. HeLa cells were cultured onto glass-bottom Petri dishes for 12 h before treatment. Live cells were treated with Alectinib and Rilpivirine at 20 μM for 16 h, washed with PBS for 3 times, fixed in 4% paraformaldehyde solution for 15 min, and washed with PBS again.

4. Conclusions

In conclusion, the optical properties of two drugs Alectinib and Rilpivirine were deeply explored firstly, through the simulation of molecular structures, electrostatic potential, OPA/TPA and emission spectral properties, and experiment of UV-vis spectra, fluorescence and cell imaging. Moreover, the relationships between molecular structures and optical properties for Alectinib and Rilpivirine have been minutely investigated based on molecular modeling. The results suggested that the drugs Alectinib and Rilpivirine with planar parent skeleton, stronger π-conjugation effect in S_1 and electron acceptors connecting conjugated structures could be excited by UV-vis spectra, and subsequently emitted fluorescence at 450 nm and 500 nm, respectively. It was noted that the intensities of UV-vis excited and emission spectra for Alectinib and Rilpivirine were decreased during the increased PBS proportion, which were ascribed to the poor/medium solubility. Meanwhile, we found that the fluorescence quantum yield of Alectinib was higher than that of Rilpivirine (Φ: 7.5% > 1.1%) in organic solvent, which might come from the larger electronic transition dipole moment (μ^{f0}_{EMI}: 8.41 Debye > 6.33 Debye) and charge transition quantity (q^{CT}_{EMI}: 0.682 e > 0.502 e) during the emission process. Additionally, the smaller geometric relaxation of Alectinib due to the lower reorganization energy than Rilpivirine (2821.6 cm^{-1} < 5176.9 cm^{-1}) weakened the non-radiative decay process, and also led to its higher fluorescence intensity. For TPA properties, we found that the action TPA cross-section (δ^{TPA}_{max}*Φ) of Alectinib was 201.75 GM at 772.5 nm excited wavelength, which was larger than 50 GM in NIR region and was suitable for applications in biological samples with reasonable incident laser power [31]. Lastly, Alectinib and Rilpivirine displayed the same green fluorescence in HeLa cells, suggesting their potential bio-imaging applications. We hope that this investigation can provide useful guidance for the design and synthesis of more excellent fluorescent activated molecules.

Supplementary Materials: The following supporting information can be downloaded at: https://www.mdpi.com/article/10.3390/molecules28166172/s1. Tables S1–S4: The main selected bond length (Å) and dihedral angles (°) at the optimized S_0 and S_1 geometries for compounds Alectinib and Rilpivirine. Transition index of Alectinib and Rilpivirine from S_0 to S_1 in OPA spectra, including centroid distance of the electrons and holes (D), electron-hole overlap index (Sr), average distribution breadth of the electrons and holes (H), hole delocalization index (HDI) and electron delocalization index (EDI). Calculated TPA properties including the maximum TPA cross-section (δ^{TPA}_{max}), corresponding TPA wavelength (λ^{TPA}_{max}), and transition nature of Alectinib and Rilpivirine in gas (a) and water (b) by Cam-B3LYP functional. The transition energy gaps (ETEG) of Alectinib in H2O and DMSO (polarity: H_2O > DMSO).; Figure S1–S5: The UV-vis spectra of molecule Alectinib (A) and Rilpivirine (B) with the concentration of 100 μM in different solvent (100%DMSO, 50%DMSO-50%PBS and 10%DMSO-90%PBS). Contour surfaces of eight frontier molecular orbitals of Alectinib and Rilpivirine. Atomic numbers for compounds Alectinib and Rilpivirine. Fluorescence spectra of Alectinib (A) and Rilpivirine (B) in 25%DMSO-75%PBS solvent by excitation at 340 nm and 310 nm, respectively. UV-vis spectra of Alectinib (C) and Rilpivirine (D) in 25%DMSO-75%PBS solvent. The plot of fluorescence intensity (red) and wavelength (black) versus concentration for Alectinib (A) and Rilpivirine (B) in 25%DMSO-75%PBS solvent [58].

Author Contributions: Investigation, C.Z. (Theoretical calculation); Y.Y. (spectral experiments) and S.G. (cell experiments); conceptualization, A.R., Y.-B.Z., J.L., D.-J.X. and W.-L.W.; writing—original draft preparation, C.Z.; writing—review and editing, D.-J.X. and W.-L.W.; supervision, D.-J.X. and W.-L.W.; project administration, D.-J.X. and W.-L.W. All authors have read and agreed to the published version of the manuscript.

Funding: This research was funded by National Natural Science Foundation of China (Nos. 22277043 and 21772068), Fundamental Research Funds for the Central Universities (JUSRP121065) and the High Level Personnel Project of Jiangsu Province (JSSCBS20210848). And The APC was funded by Xishan People's Hospital of Wuxi City.

Institutional Review Board Statement: Not applicable.

Informed Consent Statement: Not applicable.

Data Availability Statement: Data available on request from the authors.

Acknowledgments: This work was supported by National Natural Science Foundation of China (Nos. 22277043 and 21772068), Fundamental Research Funds for the Central Universities (JUSRP121065) and the High Level Personnel Project of Jiangsu Province (JSSCBS20210848). The authors express their gratitude to facility support from the BioDuro-Sundia in Wuxi and high-performance computing cluster platform of the School of Biotechnology in Jiangnan University.

Conflicts of Interest: The authors declare no conflict of interest.

Sample Availability: Samples of the compounds Alectinib and Rilpivirine are available from the authors.

References

1. Han, H.H.; Tian, H.; Zang, Y.; Sedgwick, A.C.; Li, J.; Sessler, J.L.; He, X.P.; James, T.D. Small-molecule fluorescence-based probes for interrogating major organ diseases. *Chem. Soc. Rev.* **2021**, *50*, 9391–9429. [CrossRef] [PubMed]
2. Singh, A.K.; Nair, A.V.; Singh, N.D.P. Small Two-Photon Organic Fluorogenic Probes: Sensing and Bioimaging of Cancer Relevant Biomarkers. *Anal. Chem.* **2022**, *94*, 177–192. [CrossRef] [PubMed]
3. Gao, M.; Yu, F.; Lv, C.; Choo, J.; Chen, L. Fluorescent chemical probes for accurate tumor diagnosis and targeting therapy. *Chem. Soc. Rev.* **2017**, *46*, 2237–2271. [CrossRef] [PubMed]
4. Nguyen, Q.T.; Tsien, R.Y. Fluorescence-guided surgery with live molecular navigation—A new cutting edge. Nature reviews. *Cancer* **2013**, *13*, 653–662. [PubMed]
5. Wu, L.; Liu, J.; Li, P.; Tang, B.; James, T.D. Two-photon small-molecule fluorescence-based agents for sensing, imaging, and therapy within biological systems. *Chem. Soc. Rev.* **2021**, *50*, 702–734. [CrossRef] [PubMed]
6. Wu, X.; Sun, X.; Guo, Z.; Tang, J.; Shen, Y.; James, T.D.; Tian, H.; Zhu, W. In vivo and in situ tracking cancer chemotherapy by highly photostable NIR fluorescent theranostic prodrug. *J. Am. Chem. Soc.* **2014**, *136*, 3579–3588. [CrossRef] [PubMed]
7. Sun, C.-L.; Li, J.; Wang, X.-Z.; Shen, R.; Liu, S.; Jiang, J.-Q.; Li, T.; Song, Q.-W.; Liao, Q.; Fu, H.-B.; et al. Rational Design of Organic Probes for Turn-On Two-Photon Excited Fluorescence Imaging and Photodynamic Therapy. *Chem* **2019**, *5*, 600–616. [CrossRef]
8. Wang, X.; Li, S.; Wang, S.; Zheng, S.; Chen, Z.; Song, H. Protein Binding Nanoparticles as an Integrated Platform for Cancer Diagnosis and Treatment. *Adv. Sci.* **2022**, *9*, e2202453. [CrossRef]
9. Yuan, J.; Zhou, Q.-H.; Xu, S.; Zuo, Q.-P.; Li, W.; Zhang, X.-X.; Ren, T.-B.; Yuan, L.; Zhang, X.-B. Enhancing the Release Efficiency of a Molecular Chemotherapeutic Prodrug by Photodynamic Therapy. *Angew. Chem. Int. Ed.* **2022**, *61*, e202206169. [CrossRef]
10. Hosseini Motlagh, N.S.; Parvin, P.; Ghasemi, F.; Atyabi, F.; Jelvani, S.; Abolhosseini, S. Laser induced fluorescence spectroscopy of chemo-drugs as biocompatible fluorophores: Irinotecan, gemcitabine and navelbine. *Laser Phys. Lett.* **2016**, *13*, 075604. [CrossRef]
11. Motlagh, N.S.; Parvin, P.; Ghasemi, F.; Atyabi, F. Fluorescence properties of several chemotherapy drugs: Doxorubicin, paclitaxel and bleomycin. *Biomed. Opt. Express* **2016**, *7*, 2400–2406. [CrossRef] [PubMed]
12. Zhang, C.; Sun, Y.T.; Gao, L.X.; Feng, B.; Yan, X.; Guo, X.H.; Ren, A.M.; Zhou, Y.B.; Li, J.; Wang, W.L. Theoretical study and application of 2-phenyl-1,3,4-thiadiazole derivatives with optical and inhibitory activity against SHP1. *Phys. Chem. Chem. Phys.* **2022**, *24*, 861–874. [CrossRef] [PubMed]
13. Kim, K.H.; Singha, S.; Jun, Y.W.; Reo, Y.J.; Kim, H.R.; Ryu, H.G.; Bhunia, S.; Ahn, K.H. Far-red/near-infrared emitting, two-photon absorbing, and bio-stable amino-Si-pyronin dyes. *Chem. Sci.* **2019**, *10*, 9028–9037. [CrossRef] [PubMed]
14. He, L.; Xiong, H.; Wang, B.; Zhang, Y.; Wang, J.; Zhang, H.; Li, H.; Yang, Z.; Song, X. Rational Design of a Two-Photon Ratiometric Fluorescent Probe for Hypochlorous Acid with a Large Stokes Shift. *Anal. Chem.* **2020**, *92*, 11029–11034. [CrossRef] [PubMed]

15. Haziq, Q.M. Top 200 Small Molecules Pharmaceuticals by Retail Sales in 2020. Available online: https://njardarson.lab.arizona.edu/sites/njardarson.lab.arizona.edu/files/Top%20200%20Pharmaceuticals%20Small%20Molecule%202020New.pdf (accessed on 3 August 2023).
16. Kinoshita, K.; Asoh, K.; Furuichi, N.; Ito, T.; Kawada, H.; Hara, S.; Ohwada, J.; Miyagi, T.; Kobayashi, T.; Takanashi, K.; et al. Design and synthesis of a highly selective, orally active and potent anaplastic lymphoma kinase inhibitor (CH5424802). *Bioorganic Med. Chem.* **2012**, *20*, 1271–1280. [CrossRef]
17. Peters, S.; Camidge, D.R.; Shaw, A.T.; Gadgeel, S.; Ahn, J.S.; Kim, D.W.; Ou, S.I.; Perol, M.; Dziadziuszko, R.; Rosell, R.; et al. Alectinib versus Crizotinib in Untreated ALK-Positive Non-Small-Cell Lung Cancer. *N. Engl. J. Med.* **2017**, *377*, 829–838. [CrossRef]
18. Hida, T.; Nokihara, H.; Kondo, M.; Kim, Y.H.; Azuma, K.; Seto, T.; Takiguchi, Y.; Nishio, M.; Yoshioka, H.; Imamura, F.; et al. Alectinib versus crizotinib in patients with ALK-positive non-small-cell lung cancer (J-ALEX): An open-label, randomised phase 3 trial. *Lancet* **2017**, *390*, 29–39. [CrossRef] [PubMed]
19. Hui, B.; Zhang, J.; Shi, X.; Xing, F.; Shao, Y.W.; Wang, Y.; Zhang, X.; Wang, S. EML4-ALK, a potential therapeutic target that responds to alectinib in ovarian cancer. *Jpn. J. Clin. Oncol.* **2020**, *50*, 1470–1474. [CrossRef]
20. Cohen, C.J.; Andrade-Villanueva, J.; Clotet, B.; Fourie, J.; Johnson, M.A.; Ruxrungtham, K.; Wu, H.; Zorrilla, C.; Crauwels, H.; Rimsky, L.T.; et al. Rilpivirine versus efavirenz with two background nucleoside or nucleotide reverse transcriptase inhibitors in treatment-naive adults infected with HIV-1 (THRIVE): A phase 3, randomised, non-inferiority trial. *Lancet* **2011**, *378*, 229–237. [CrossRef]
21. Molina, J.M.; Cahn, P.; Grinsztejn, B.; Lazzarin, A.; Mills, A.; Saag, M.; Supparatpinyo, K.; Walmsley, S.; Crauwels, H.; Rimsky, L.T.; et al. Rilpivirine versus efavirenz with tenofovir and emtricitabine in treatment-naive adults infected with HIV-1 (ECHO): A phase 3 randomised double-blind active-controlled trial. *Lancet* **2011**, *378*, 238–246. [CrossRef]
22. Hayashida, T.; Hachiya, A.; Ode, H.; Nishijima, T.; Tsuchiya, K.; Sugiura, W.; Takiguchi, M.; Oka, S.; Gatanaga, H. Rilpivirine resistance mutation E138K in HIV-1 reverse transcriptase predisposed by prevalent polymorphic mutations. *J. Antimicrob. Chemother.* **2016**, *71*, 2760–2766. [CrossRef]
23. Ali, J.; Camilleri, P.; Brown, M.B.; Hutt, A.J.; Kirton, S.B. Revisiting the general solubility equation: In silico prediction of aqueous solubility incorporating the effect of topographical polar surface area. *J. Chem. Inf. Model.* **2012**, *52*, 420–428. [CrossRef] [PubMed]
24. Marenich, A.V.; Cramer, C.J.; Truhlar, D.G. Universal Solvation Model Based on Solute Electron Density and on a Continuum Model of the Solvent Defined by the Bulk Dielectric Constant and Atomic Surface Tensions. *J. Phys. Chem. B* **2009**, *113*, 6378–6396. [CrossRef] [PubMed]
25. Marenich, A.V.; Cramer, C.J.; Truhlar, D.G. Performance of SM6, SM8, and SMD on the SAMPL1 Test Set for the Prediction of Small-Molecule Solvation Free Energies. *J. Phys. Chem. B* **2009**, *113*, 4538–4543. [CrossRef] [PubMed]
26. Lu, T.; Chen, F. Multiwfn: A multifunctional wavefunction analyzer. *J. Comput. Chem.* **2012**, *33*, 580–592. [CrossRef] [PubMed]
27. Mu, X.; Wang, X.; Quan, J.; Sun, M. Photoinduced Charge Transfer in Donor-Bridge-Acceptor in One- and Two-photon Absorption: Sequential and Superexchange Mechanisms. *J. Phys. Chem. C* **2020**, *124*, 4968–4981. [CrossRef]
28. Lu, C.; Li, N.; Jin, Y.; Sun, Y.; Wang, J. Physical Mechanisms of Intermolecular Interactions and Cross-Space Charge Transfer in Two-Photon BDBT-TCNB Co-Crystals. *Nanomaterials* **2022**, *12*, 2757. [CrossRef] [PubMed]
29. Lu, C.; Yu, J.; Sheng, H.; Jiang, Y.; Zhao, F.; Wang, J. Linear and Nonlinear Photon-Induced Cross Bridge/Space Charge Transfer in STC Molecular Crystals. *Nanomaterials* **2022**, *12*, 525. [CrossRef]
30. Wei, J.; Li, Y.; Song, P.; Yang, Y.; Ma, F. Enhancement of one- and two-photon absorption and visualization of intramolecular charge transfer of pyrenyl-contained derivatives. *Spectrochim. Acta. Part A Mol. Biomol. Spectrosc.* **2021**, *245*, 118897. [CrossRef] [PubMed]
31. Juvekar, V.; Lee, H.W.; Lee, D.J.; Kim, H.M. Two-photon fluorescent probes for quantitative bio-imaging analysis in live tissues. *Trends Anal. Chem.* **2022**, *157*, 116787. [CrossRef]
32. Kim, H.M.; Cho, B.R. Small-Molecule Two-Photon Probes for Bioimaging Applications. *Chem. Rev.* **2015**, *115*, 5014–5055. [CrossRef] [PubMed]
33. Göppert-Mayer, M. Über Elementarakte mit zwei Quantensprüngen. *Ann. Der Phys.* **1931**, *401*, 273–294. [CrossRef]
34. Denk, W.; Strickler, J.H.; Webb, W.W. Two-Photon Laser Scanning Fluorescence Microscopy. *Science* **1990**, *248*, 73–76. [CrossRef] [PubMed]
35. Jin, H.; Yang, M.; Sun, Z.; Gui, R. Ratiometric two-photon fluorescence probes for sensing, imaging and biomedicine applications at living cell and small animal levels. *Coord. Chem. Rev.* **2021**, *446*, 214114. [CrossRef]
36. Shaya, J.; Corridon, P.R.; Al-Omari, B.; Aoudi, A.; Shunnar, A.; Mohideen, M.I.H.; Qurashi, A.; Michel, B.Y.; Burger, A. Design, photophysical properties, and applications of fluorene-based fluorophores in two-photon fluorescence bioimaging: A review. *J. Photochem. Photobiol. C Photochem. Rev.* **2022**, *52*, 100529. [CrossRef]
37. Xu, L.; Lin, W.; Huang, B.; Zhang, J.; Long, X.; Zhang, W.; Zhang, Q. The design strategies and applications for organic multi-branched two-photon absorption chromophores with novel cores and branches: A recent review. *J. Mater. Chem. C* **2021**, *9*, 1520–1536. [CrossRef]
38. Wei, X.; Cui, W.B.; Qin, G.Y.; Zhang, X.E.; Sun, F.Y.; Li, H.; Guo, J.F.; Ren, A.M. Theoretical Investigation of Ru(II) Complexes with Long Lifetime and a Large Two-Photon Absorption Cross-Section in Photodynamic Therapy. *J. Med. Chem.* **2023**, *66*, 4167–4178. [CrossRef]

39. Zhang, C.; Ren, A.M.; Guo, J.F.; Wang, D.; Yu, L.Y. Theoretical design and investigation of 1,8-naphthalimide-based two-photon fluorescent probes for detecting cytochrome P450 1A with separated fluorescence signal. *Phys. Chem. Chem. Phys.* **2018**, *20*, 13290–13305. [CrossRef]
40. Aidas, K.; Angeli, C.; Bak, K.L.; Bakken, V.; Bast, R.; Boman, L.; Christiansen, O.; Cimiraglia, R.; Coriani, S.; Dahle, P.; et al. The Dalton quantum chemistry program system. *WIREs Comput. Mol. Sci.* **2014**, *4*, 269–284. [CrossRef]
41. Hao, X.L.; Guo, J.F.; Ren, A.M.; Zhou, L. Persistent and Efficient Multimodal Imaging for Tyrosinase Based on Two-Photon Excited Fluorescent and Room-Temperature Phosphorescent Probes. *J. Phys. Chemistry. A* **2022**, *126*, 7650–7659. [CrossRef]
42. Guo, X.-H.; Yu, L.-B.; Hao, X.-L.; He, T.-F.; Guo, J.-F.; Wei, X.; Cui, W.-B.; Yu, L.-Y.; Qu, Z.-X.; Ren, A.-M. Theoretical Study of a Two-Photon Fluorescent Probe Based on Nile Red Derivatives with Controllable Fluorescence Wavelength and Water Solubility. *J. Chem. Inf. Model.* **2021**, *61*, 5082–5097. [CrossRef] [PubMed]
43. Luo, Y.; Rubio-Pons, O.; Guo, J.D.; Agren, H. Charge-transfer Zn-porphyrin derivatives with very large two-photon absorption cross sections at 1.3-1.5 microm fundamental wavelengths. *J. Chem. Phys.* **2005**, *122*, 096101. [CrossRef] [PubMed]
44. Guo, J.-D.; Wang, C.-K.; Luo, Y.; Ågren, H. Influence of electron-acceptor strength on the resonant two-photon absorption cross sections of diphenylaminofluorene-based chromophores. *Phys. Chem. Chem. Phys.* **2003**, *5*, 3869–3873. [CrossRef]
45. Bavali, A.; Parvin, P.; Mortazavi, S.Z.; Mohammadian, M.; Mousavi Pour, M.R. Red/blue spectral shifts of laser-induced fluorescence emission due to different nanoparticle suspensions in various dye solutions. *Appl. Opt.* **2014**, *53*, 5398. [CrossRef] [PubMed]
46. Mardani, K.; Parvin, P.; Bavali, A.; Ehtesham, A.; Moafi, A. Angular study of laser induced fluorescence emission of hybrid media based on Stern-Volmer formalism. *OSA Contin.* **2021**, *4*, 15. [CrossRef]
47. Chen, T.; Xiao, G.; Wang, Z.; Zou, J.; Wang, J.; Hu, W.; Liu, Y.A.; Yang, H.; Wen, K. s-Tetrazine-Bridged Photochromic Aromatic Framework Material. *ACS Omega* **2022**, *7*, 11276–11284. [CrossRef] [PubMed]
48. Zhao, Y.; Truhlar, D.G. The M06 suite of density functionals for main group thermochemistry, thermochemical kinetics, noncovalent interactions, excited states, and transition elements: Two new functionals and systematic testing of four M06-class functionals and 12 other functionals. *Theor. Chem. Acc.* **2008**, *120*, 215–241.
49. Frisch, M.J.T.; Trucks, G.W.; Schlegel, H.B.; Scuseria, G.E.; Robb, M.A.; Cheeseman, J.R.; Scalmani, G.; Barone, V.; Petersson, G.A.; Nakatsuji, H.; et al. *Gaussian 16, Revision C.01*; Gaussian, Inc.: Wallingford, CT, USA, 2019.
50. Shuai, Z. Thermal Vibration Correlation Function Formalism for Molecular Excited State Decay Rates. *Chin. J. Chem.* **2020**, *38*, 1223–1232. [CrossRef]
51. Shuai, Z.; Peng, Q.; Niu, Y.; Geng, H. MOMAP—Molecular Material Property Prediction Package. Available online: http://www.momap.net.cn (accessed on 3 August 2023).
52. Hilborn, R.C. Einstein coefficients, cross sections, f values, dipole moments, and all that. *Am. J. Phys.* **1982**, *50*, 982–986. [CrossRef]
53. Beerepoot, M.T.; Friese, D.H.; List, N.H.; Kongsted, J.; Ruud, K. Benchmarking two-photon absorption cross sections: Performance of CC2 and CAM-B3LYP. *Phys. Chem. Chem. Phys.* **2015**, *17*, 19306–19314. [CrossRef]
54. Friese, D.H.; Beerepoot, M.T.; Ruud, K. Rotational averaging of multiphoton absorption cross sections. *J. Chem. Phys.* **2014**, *141*, 204103. [CrossRef]
55. Zhang, C.; Guo, J.-F.; Ren, A.-M.; Wang, D. A theoretical investigation of the two-photon absorption and fluorescent properties of coumarin-based derivatives for Pd2+ detection. *RSC Adv.* **2017**, *7*, 49505–49517. [CrossRef]
56. Hao, X.L.; Ren, A.M.; Zhou, L. Theoretical Studies and Design Strategies of Highly Efficient Two-Photon Excited Fluorescent Probes for Hydrogen Sulfide Detection through Simulation of Excited-State Dynamics. *J. Chem. Inf. Model.* **2023**, *63*, 2409–2418. [CrossRef]
57. Englman, R.; Jortner, J. The energy gap law for radiationless transitions in large molecules. *Mol. Phys.* **1970**, *18*, 145–164. [CrossRef]
58. Tarai, A.; Huang, M.; Das, P.; Pan, W.; Zhang, J.; Gu, Z.; Yang, Z. ICT and AIE characteristics two cyano-functionalized probes and their photophysical properties, DFT calculations, cytotoxicity, and cell imaging applications. *Molecules* **2020**, *25*, 585. [CrossRef] [PubMed]

molecules

MDPI

Article

mPGES-1 Inhibitor Discovery Based on Computer-Aided Screening: Pharmacophore Models, Molecular Docking, ADMET, and MD Simulations

Qiqi Huang [1], Tianli Lai [1], Qu Wang [2] and Lianxiang Luo [2,3,*]

[1] The First Clinical College, Guangdong Medical University, Zhanjiang 524023, China; hqq77@gdmu.edu.cn (Q.H.); laitianli@gdmu.edu.cn (T.L.)
[2] The Marine Biomedical Research Institute, Guangdong Medical University, Zhanjiang 524023, China; wangqu0503@gdmu.edu
[3] The Marine Biomedical Research Institute of Guangdong Zhanjiang, Zhanjiang 524023, China
* Correspondence: luolianxiang321@gdmu.edu.cn; Tel.: +86-0759-2388163

Abstract: mPGES-1 is an enzyme, which, when activated by inflammatory factors, can cause prostaglandin E synthesis. Traditional non-steroidal anti-inflammatory drugs are capable of inhibiting prostaglandin production, yet they can also cause gastrointestinal reactions and coagulation disorders. mPGES-1, the enzyme at the conclusion of prostaglandin production, does not cause any adverse reactions when inhibited. Numerous studies have demonstrated that mPGES-1 is more abundant in cancerous cells than in healthy cells, indicating that decreasing the expression of mPGES-1 could be a potential therapeutic strategy for cancer. Consequently, the invention of mPGES-1 inhibitors presents a fresh avenue for the treatment of inflammation and cancer. Incorporating a database of TCM compounds, we collected a batch of compounds that had an inhibitory effect on mPGES-1 and possessed IC50 value. Firstly, a pharmacophore model was constructed, and the TCM database was screened, and the compounds with score cut-off values of more than 1 were retained. Then, the compounds retained after being screened via the pharmacodynamic model were screened for docking at the mPGES-1 binding site, followed by high-throughput virtual screening [HTVS] and standard precision [SP] and super-precision [XP] docking, and the compounds in the top 20% of the XP docking score were selected to calculate the total free binding energy of MM-GBSA. The best ten compounds were chosen by comparing their score against the reference ligand 4U9 and the MM-GBSA_dG_Bind score. ADMET analysis resulted in the selection of ten compounds, three of which had desirable medicinal properties. Finally, the binding energy of the target protein mPGES-1 and the candidate ligand compound was analyzed using a 100 ns molecular dynamics simulation of the reference ligand 4U9 and three selected compounds. After a gradual screening study and analysis, we identified a structure that is superior to the reference ligand 4U9 in all aspects, namely compound 15643. Taken together, the results of this study reveal a structure that can be used to inhibit mPGES-1 compound 15643, thereby providing a new option for anti-inflammatory and anti-tumor drugs.

Keywords: mPGES-1; pharmacophore modeling; molecular docking; ADMET; MD simulation

1. Introduction

Prostaglandin PG plays an important role in mediating cell proliferation, differentiation, and apoptosis after binding with specific receptors. In addition, prostaglandins are also involved in the pathological processes of inflammation, cancer, and various cardiovascular diseases [1]. COX is a key enzyme that mediates the conversion of arachidonic acid into prostaglandin. Traditional non-steroidal anti-inflammatory drugs (NSAIDs) produce anti-inflammatory effects by directly inhibiting COX and blocking the production of prostaglandins. However, direct blocking of COX can cause gastrointestinal mucosal damage and coagulation dysfunction [2]. In order to avoid the side effects of NSAIDs, this study focused on microsomal prostaglandin E synthase-1 (mPGES-1), the terminal enzyme downstream of COX-2 in inducible PGE [3]. mPGES-1 is low in normal tissue and is induced in any inflammation. Studies have shown that it is overexpressed in a variety of human cancers, such as lung cancer [4], stomach cancer [5], colon cancer [6], prostate cancer [7], breast cancer [8], cervical cancer [9], pancreatic cancer [10], melanoma [11], squamous carcinoma of the head and neck [12], papillary carcinoma of the thyroid [13], and glioma [14]. Therefore, mPGES-1 has become an important target for the treatment of acute and chronic inflammatory diseases and cancers as shown in Supplementary Figure S1 [15].

Inhibition of mPGES-1 for playing an anti-tumor and anti-inflammatory role has attracted much attention from researchers. Several mPGES-1 inhibitors have been identified in the past; however, there are currently no mPGES-1 inhibitors on the market. Therefore, the discovery of novel mPGES-1 inhibitors for the treatment of cancer and inflammation provokes our thinking and research [16]. In addition, in recent years, the research of traditional Chinese medicine has been deepening day by day. Traditional Chinese medicine will greatly support its modernization and guide rational modern drug discovery. It has been proven that traditional Chinese medicine plays an important role in inhibiting inflammation and tumors [17,18]. In recent years, with the development and application of Chinese herbal medicine, various Chinese herbal preparations have continuously been discovered and applied in clinical practice. We collected the structural information of small molecules from traditional Chinese medicines that can be found on the internet and integrated it to carry out virtual drug screening for mPGES-1 targets. In this study, we hope to screen out mPGES-1 inhibitors from the traditional Chinese medicine database and provide new choices for the development of anti-inflammatory and anti-tumor drugs.

Computer-aided drug design has been widely used in the discovery of lead compounds [19,20]. This work has greatly accelerated the speed of drug design and has guided researchers to better develop new drugs on the basis of science [21]. In this study, a ligand-based and structure-based approach to drug design was used: pharmacophore models, molecular docking screening, ADMET, and kinetic simulations were used to identify novel compounds as potential mPGES-1 inhibitors. After study and analysis, we determined a compound with better scores and stability than the reference ligand. The specific workflow is shown in Supplementary Figure S2.

2. Results

2.1. Pharmacophore Model Establishment

Pharmacophore comprises the physical and chemical characteristics and spatial arrangement necessary for the molecular recognition of ligands by biomacromolecules. Bioactive compounds with specific targets or similar bioactive compounds can be obtained via a screening compounds database with a pharmacophore model [22]. Pharmacophore models can be structure-based or ligand-based. In this study, a series of three-dimensional pharmacophore models of mPGES-1 were constructed based on the reported inhibitors, and the common characteristics of biological activity were analyzed by superimposing these inhibitory compounds. Supplementary Figure S3 shows the collected compounds. The best pharmacophore model (AAHNR) consists of five features: a hydrogen bond donor, two hydrogen bond acceptors, an aromatic ring, and a negatively charged ion center, as shown in Figure 1.

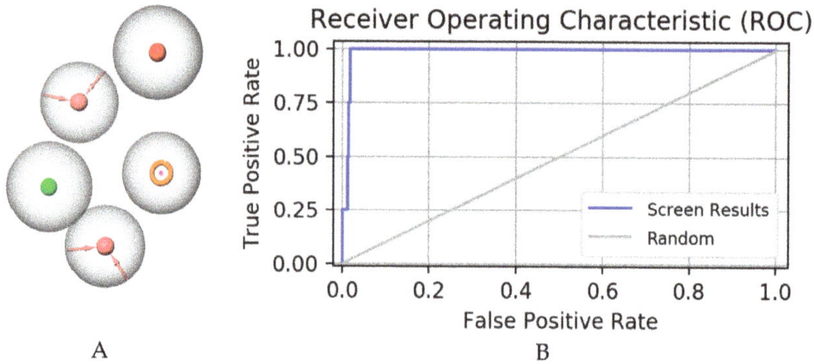

Figure 1. Pharmacophore model constructed as shown in (**A**): a hydrogen bond donor, two hydrogen bond acceptors, an aromatic ring, and a negatively charged ion center; (**B**) represents receiver operating characteristic curve (ROC) of pharmacophore model.

2.2. Pharmacophore Model Verification

A pharmacophore model for screening should have the ability to accurately distinguish between active and inactive compounds [23]. The receiver operating characteristic curve (ROC), which is characterized by a false positive rate (FPR) as the abscissa and a true positive rate (TPR) as the ordinate. The closer the *X*-axis is to zero, the higher the accuracy is. The larger the *Y*-axis, the better the accuracy. The whole figure is divided into two parts according to the position of the curve. The area under the curve is called the AUC (area under the curve) and is used to indicate the accuracy of the forecast [24]. The higher the AUC value, the larger the area under the curve, indicating a higher prediction accuracy. The closer the curve is to the upper left corner (the smaller the X, the larger the Y), the higher the prediction accuracy [25]. In this study, the AUC (area under ROC curve is 0.77, as shown in Figure 2) was verified using this model, which has a good ability to distinguish active compounds from decoy compounds. Figure 3 shows the compounds used to build the verification set.

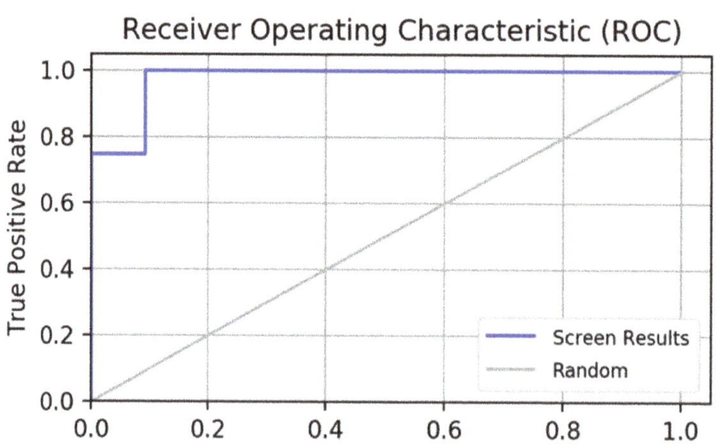

Figure 2. The receiver operating characteristic curve (ROC) of the pharmacophore constructed by this institute validates its results. The full name of ROC is receiver operating characteristic curve, and the abscissa of this curve is the false positive rate (FPR). The ordinate is the true positive rate (TPR).

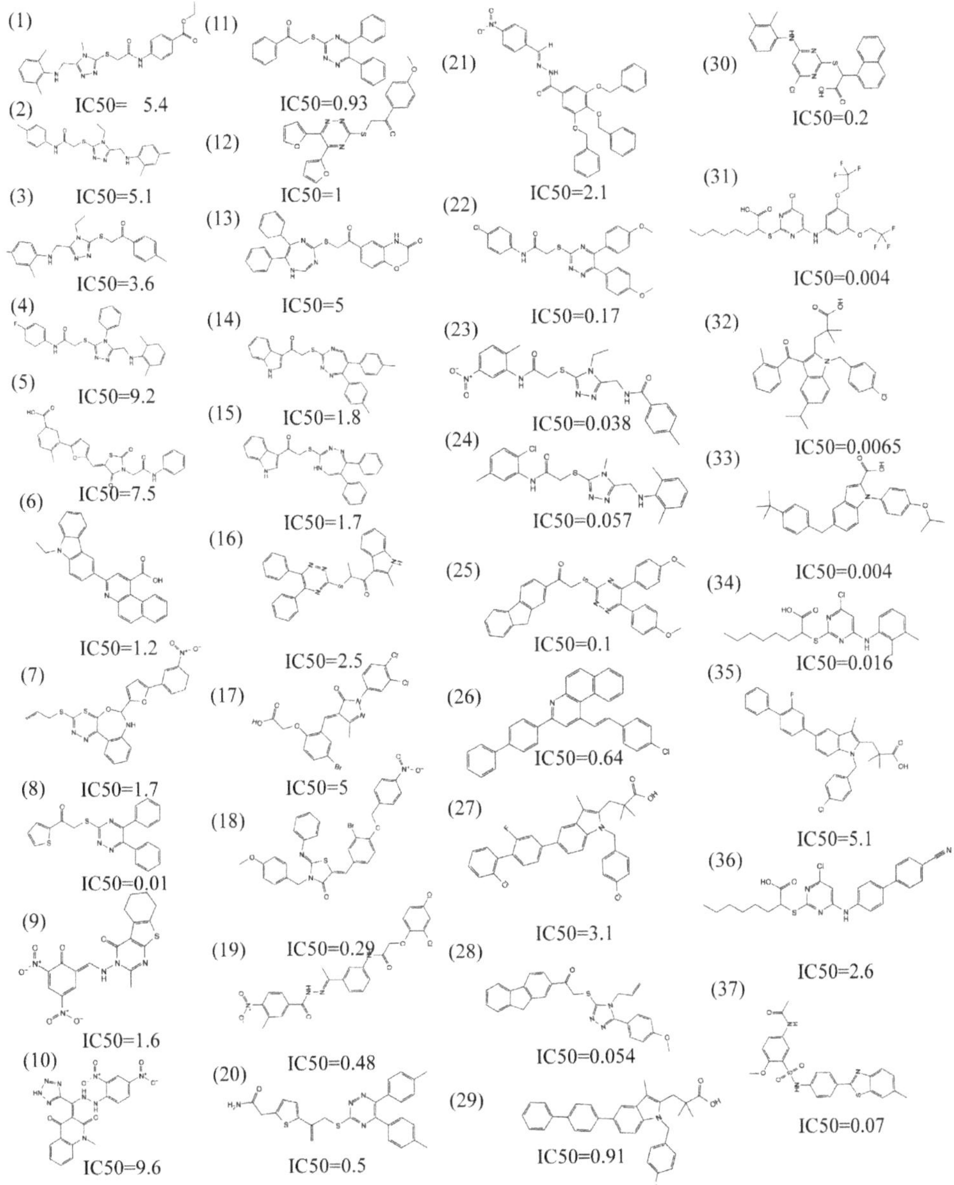

Figure 3. The virtual screening workflow (VSW) used in this study was used to identify targeted mPGES-1 inhibitors.

2.3. Molecular Docking

Molecular docking is an important part of the drug design process, which is used to identify the bioactive conformation of small and medium molecules at protein binding sites and to analyze the interactions between protein ligands [26]. This study evaluated the binding capacity of compounds important to the mPGES-1 protein. Glide is a virtual screening process from HTVS to SP to XP that further eliminates false positives with more extensive sampling and advanced scoring, resulting in higher enrichment. The Glide module processes compounds by searching for the conformation, orientation, and spatial position of the docking ligand. Firstly, the search space is reduced via rough positioning and scoring, and then the energy optimization of the candidate posture is carried out using the OPLS–AA non-bonding potential grid. Finally, the optimal docking posture is selected and evaluated using the function model combining experience and force field [26]. In this study, receptor grids with X = 9.62, Y = 20.16, and Z = 12.64 were prepared using the Glide module based on known binding sites. The Maestro software was used to dock the compounds with mPGES-1 to analyze and evaluate their binding ability. Table 1 shows the comparison of 10 compounds with 4U9 with better butt scores and MMGBSA scores than the reference ligand 4U9. The binding affinities of compound 14294, compound 15643, and compound 14186 were −5.558 kcal/mol, −4.664 kcal/mol, and −5.202 kcal/mol, respectively. Their binding affinity is better than that of the inhibitor (−4.605 kcal/mol) in the protein ligand complex (PDB ID: 4YL3) (https://www.rcsb.org/, accessed on 2 May 2022). The interaction of compound 14294 in the docking complex is shown in Figure 4A,B. The interaction of compound 15643 is shown in Figure 4C,D. The interaction of compound 14186 is shown in Figure 4E,F. Compound 14294 can be observed to form hydrogen bonds with GLN134, GLU77, and THR131. The Pi–Pi interaction established with TYR130 also plays a key role in ligand–receptor binding. Compound 15643 forms hydrogen bonds with TYR130 and THR131. In addition, compound 14186 forms hydrogen bond interactions with GLN134, TYR130, and GLU77. Analysis showed that the rich interaction types between compound 14294 and the mPGES-1 protein resulted in the best docking results. It can be seen from the interaction analysis that the docking results are reliable, and the selected compounds can be further analyzed.

Table 1. The ten compounds with the best Glide scores and MMGBSA scores compared.

Ligand	SP Docking Score (kcal/mol)	XP Docking Score (kcal/mol)	MMGBSA_dG_Bind
4U9	−3.291	−0.374	−49.058
Compound 13295	−3.355	−5.806	−51.275
Compound 14294	−3.140	−5.558	−57.754
Compound 14186	−3.129	−5.202	−49.202
Compound 9838	−4.184	−6.185	−49.225
Compound 15643	−3.197	−4.664	−55.336
Compound 12340	−3.520	−4.007	−55.425
Compound 13836	−3.404	−4.696	−57.291
Compound 13581	−3.464	−2.227	−54.882
Compound 12553	−3.474	−3.984	−50.575
Compound 13106	−3.782	−4.692	−50.208

2.4. ADMET Analysis

Absorption, distribution, metabolism, elimination, and toxicity (ADMET) characteristics are important in determining the efficacy and safety of drug candidates [27]. It is important to predict ADMET in order to avoid drug failure in late clinical trials. After excluding three false positive structures using "False Positive Remover", drug absorption, distribution, metabolism, excretion, and toxicity were tested using "SwissADME". A total of 17 related indexes were analyzed, including lipophilicity, hydrogen bond, solubility, permeability, etc., and they were compared with positive control compounds; finally, three compounds were obtained that were more similar to drug-like molecules [28]. The results of ADMET analysis of the compounds are shown in Table 2.

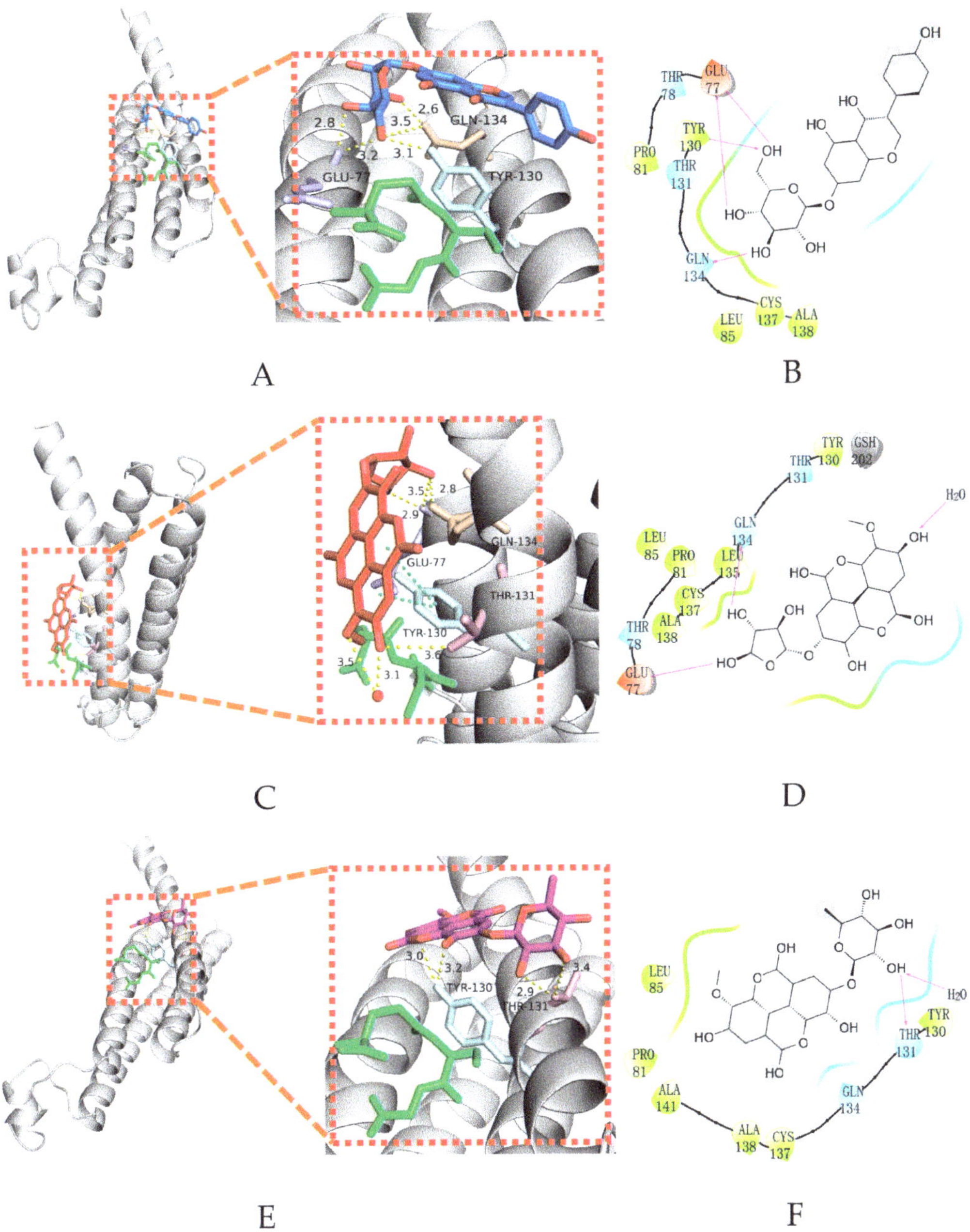

Figure 4. Interactions between protein–ligand complexes. (**A**) Three-dimensional binding pattern of protein and compound 14186; (**B**) two-dimensional binding pattern of protein and compound 14186; (**C**) three-dimensional binding pattern of protein and compound 14294; (**D**) two-dimensional binding pattern of protein and compound 14294; (**E**) three-dimensional binding pattern of protein and compound 15643; and (**F**) two-dimensional binding pattern of protein and compound 15643.

Table 2. ADMET analysis of 4U9 and three selected compounds.

	4U9	15643	14186	14294
Chemical formula	C23H11BrClF4N3	C21H18O12	C21H20O10	C19H14O12
Molecular weight	520.7	462.36	432.38	434.31
Hydrogen bond acceptors	6	12	10	12
Hydrogen bond donors	1	5	6	5
Water solubility (Log S)	−7.82	−2.95	−3.18	−2.25
Lipophilicity (Log P)	8.76	0.46	0.38	−0.04
Gastrointestinal absorption	Low	Low	Low	Low
Skin permeation (Log Kp (cm/s))	−4.5	−9.03	−8.33	−9.1
BBB permeation	NO	NO	NO	NO
P-glycoprotein substrate	YES	NO	NO	NO
Abbott bioavailability score	0.17	0.55	0.55	0.55
Toxicity of hepatotoxicity	Active	Inactive	Inactive	Inactive
Carcinogenicity	Inactive	Inactive	Inactive	Active
Immunotoxicity	Active	Active	Inactive	Active
Mutagenicity	Inactive	Active	Inactive	Active
Cytotoxicity	Inactive	Inactive	Inactive	Inactive
Predicted LD50 (mg/kg)	2000	5000	2500	1100
Predicted toxicity class	4	5	5	4

The logS value reflects the solubility of the drug. The smaller the value, the less soluble the compound is in water. When LogS < −6.0, the compound is almost insoluble in water. As shown in Table 2, the reference compound 4U9LogS < −6 is difficult to dissolve in water, while the selected compounds have better water solubility, which is related to their higher number of hydrogen bond receptors and hydrogen bond donors. The more hydrogen bond interactions, the better the hydrophilicity of the compound; the more polar interactions formed by Pi, the better the lipophilicity. The LogP value is the logarithmic value of the ratio of the partition coefficient of the compound in n-octanol and water, indicating the oil–water partition coefficient of the substance. The reference ligand 4U9 has a higher lipophilicity, while the screened compound has a lower lipophilicity. In addition, compound 4U9 and the three candidate compounds all had low gastrointestinal absorption and high skin penetration. BBB permeation is used to assess the ability of chemicals to cross the blood–brain barrier, which is a must for central drugs. The mPGES-1 inhibitors developed in this study belong to the non-central class of inhibitors, and none of the four compounds cross the blood–brain barrier. P-glycoprotein is related to the availability of compounds in the body. Compound 4U9 is P-glycoprotein substrate, so it is more likely to be pumped out of cells by P-glycoprotein in an ATP-dependent transport mode. The three candidate compounds are non-P-glycoprotein substrates and have better bioavailability. Using comprehensive analysis, compound 15643, compound 14186, and compound 14294 had higher bioavailability scores than compound 4U9.

In terms of toxicity prediction, six important toxicity assessments of Hepatotoxicity, Carcinogenicity, Immunotoxicity, Mutagenicity, Cytotoxicity, and LD50 were carried out in this study. The predicted toxicity class of compound 15643 and compound 14186 is 5, which is higher than compound 4U9.

2.5. MD Simulation

MD simulations are used to study the binding stability of small molecules and target proteins. RMSD and RMSF analysis, hydrogen bond analysis, and MM-PBSA analysis were performed on the selected small molecules. The analysis results are presented below.

2.5.1. RMSD and RMSF Analysis

The RMSD value reflects the degree to which the atoms deviate from the average position, that is, the size of the motion of each atom. The molecular RMSD is shown in Figure 5A,B. Compound 15643 showed no significant difference in RMSD values at the beginning and end of MD, but significant fluctuations were observed over a period of 7–49 ns before stabilizing. Compound 14186 showed several distinct spikes in 100 ns MD but eventually stabilized. Compound 14294 fluctuated in 100 ns MD and was ultimately not completely stable. Although the stability of RMSD is different, all three molecules have low average RMSD. The average RMSDs of 15643, 14186 and 14294 were 0.1306 nm, 0.0925 nm and 0.1476 nm, respectively.

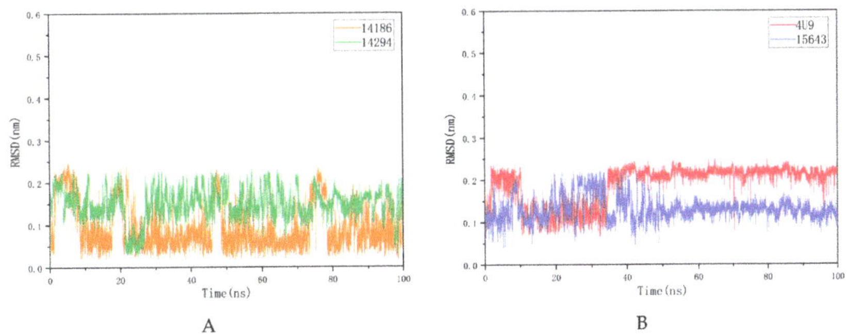

Figure 5. Compound 4U9 and three candidate compounds are plotted as root mean square deviation (RMSD) after fitting to the protein, respectively. (**A**) shows RMSDs of 4U9 and compound 15643 for 100 ns after fitting to mPGES-1, respectively. (**B**) shows RMSDs of compound 14186 and 14294 after fitting with mPGES-1 for 100 ns, respectively.

The RMSF of a protein residue represents the root mean square shift of the residue in the protein conformation, which reflects the degree of freedom of the atom. As shown in Figure 6, RMSFs of the three compounds ranged from 0.0731 nm to 0.8764 nm. In general, the overall motion trend of the three compounds was consistent. The RMSF of 14186 fluctuated at residues 118–125 but then stabilized. RMSFs of 4U9 and 15643 appeared to peak at the end of the simulation, suggesting that they may have similar binding patterns. RMSD and RMSF analysis showed that 15643 had good binding stability with proteins, while 14294 and 14186 had ordinary binding stability with proteins. Therefore, compound 15643 is considered to have better binding stability.

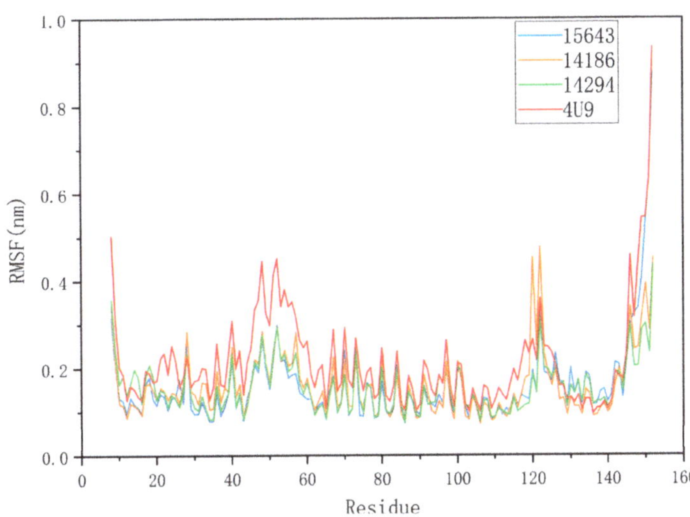

Figure 6. Root mean square fluctuation (RMSF) diagram for protein with positive compound 4U9 and three candidate compounds complexes.

2.5.2. Hydrogen Bond Analysis

The hydrogen bond between the protein and the ligand is an important factor in keeping the molecule within the active site cavity. As shown in Figure 7, compounds 14186 and mPGES-1 proprotein ligand complexes exhibited too few hydrogen bonds throughout the simulation. Compound 15643 also had fewer hydrogen bonds in the first 50 ns, while it formed more hydrogen bonds in the last 50 ns. Compound 14294 had fewer hydrogen bonds at 30 to 70 ns and more hydrogen bonds at 70 to 95 ns. As can be seen from these figures, compounds 15643 and 14294 formed more hydrogen bonds with mPGES1 during the simulation. It proves that compound 15643 has better interactions with proteins.

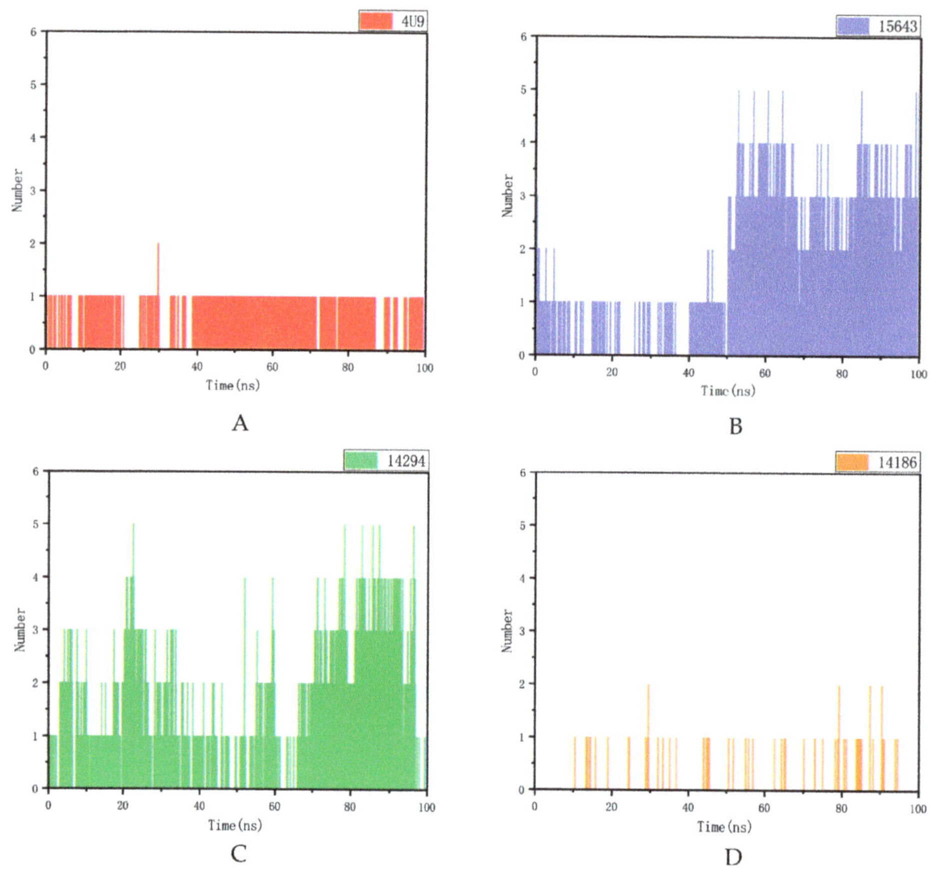

Figure 7. Hydrogen bond analysis of 4U9 and three compounds in the protein fitting process. (**A**) Positive compound 4U9 with protein. (**B**) Compound 15643 with protein. (**C**) Compound 14294 with protein. (**D**) Compound 14186 with protein.

2.5.3. MM-PBSA Analysis

Poisson–Boltzmann surface area (MM-PBSA) is an effective and reliable method in molecular mechanics research, and MM-PBSA is used to calculate the free energy of compounds bound to their protein targets. The lower the free energy produced by the binding of proteins and compounds, the better the ligand binds to proteins. As shown in Figure 8, the free energies of compound 15643 and the original ligand are −68.837 kJ/mol and −67.404 kJ/mol, respectively, which are also mainly contributed by van der Waals forces. Obviously, 15643 and 14294 have binding free energies similar to or lower than 4U9.

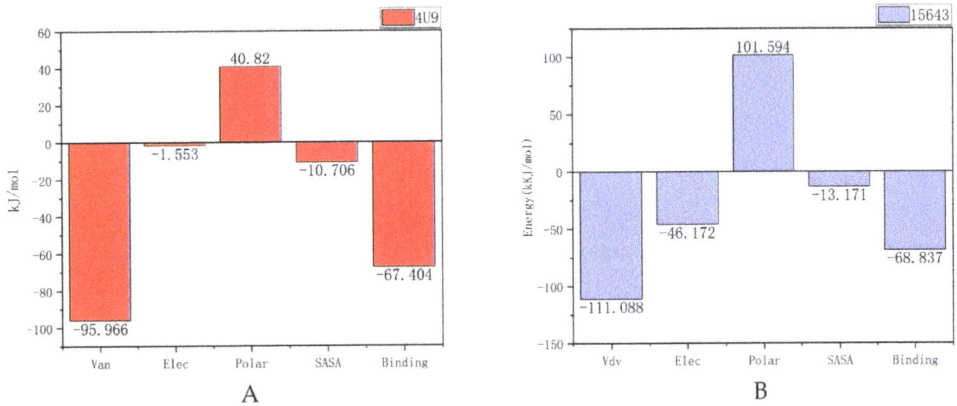

Figure 8. Binding energy analysis of 4U9 combined with protein and compound 15643. (**A**) Positive compound 4U9 with protein. (**B**) Compound 15643 with protein.

The MM/GBSA for the complex system was modeled from the corresponding 100 ns (Table 3). The analysis of the contribution of each energy component showed that the screened compound 15643 could maintain a longer interaction with the protein than the original ligand.

Table 3. Binding energy analysis of compound 4U9 and compound 15643 combined with proteins.

Criteria	Title 2	Title 3
Van der Waal energy (kJ/mol)	−95.966	−111.088
Electrostatic energy (kJ/mol)	−1.533	−46.172
Polar solvation energy (kJ/mol)	40.820	101.594
SASA energy (kJ/mol)	−10.706	−13.171
Binding energy (kJ/mol)	−67.404	−68.837

3. Discussion

Makoto Murakami first proposed prostaglandin E synthase as a novel drug target for inflammation and cancer. In particular, gene targeting studies of mPGES-1 indicate that this enzyme represents a new target for anti-inflammatory and anti-cancer drugs [29]. In the following year, AbdulHameed MD, Hamza A et al. published a study on the binding of human microsomal prostaglandin E synthase-1 (mPGES-1) to inhibitors and its quantitative structure–activity correlation [30]. Subsequently, researchers continued to explore the development of mPGES-1 inhibitors, and several mPGES-1 inhibitors were reported between 2008 and 2023. Currently, Zaloglanstat (ISC-27864), an mPGES-1 inhibitor developed by Di Micco, S et al., which can be used to study asthma, osteoarthritis, rheumatoid arthritis, acute or chronic pain, and neurodegenerative diseases, has completed a Phase II clinical trial [31]. To date, researchers have discovered a range of mPGES-1 inhibitors with different chemical structures, such as MF63 antipyretic and analgesic effects of [2-(6-chloro-1*H*-phenanthro-[9,10-d]imidazol-2-yl)isophthalonitrile] in an inflammatory animal model [32]; the effective and selective dioxane-fused mPGES-1 inhibitor tricyclic benz [d] imidazole derivatives [33], oxadiazole thione-benzimidazole derivatives [34], and PF-4693627, mPGES-1 inhibitors that can be used in studies of osteoarthritis and rheumatoid arthritis [35]; etc. The discovery of these inhibitors has enriched the selection and research of mPGES-1 inhibitors. However, despite a long history of research on mPGES-1 inhibitors, there are currently no commercially available inhibitors of mPGES-1. With the deepening of Chinese medicine research, the use of Chinese medicine is more and more closely related to the field of drug discovery and molecular biology. Considering that small molecules of traditional Chinese medicine have become one of the important sources of rich therapeutic drugs, the purpose of our study was to screen mPGES-1 inhibitors from medicines collected and integrated from the Chinese medicine database.

The rapid development of computer technology has accelerated the development of drugs, and computer-aided design (CADD) has been widely used to rapidly screen out molecules with excellent targeted binding potential from large compound databases. In this study, the structure of the molecule was based on its virtual screening. We collected a batch of compounds containing IC50 values and certain inhibitory effects on mPGES-1 by reviewing the literature as a data set. Maestro was used to construct pharmacophore models based on the common pharmacophore characteristics of multiple ligands, and the constructed pharmacophore models were verified. The validated pharmacophore model was used in the first step of virtual screening, and only compounds meeting four of the five pharmacophore characteristics were retained.

As one of the computational tools for drug design and development, the docking method can predict the affinity of direct binding modes of compounds bound within protein binding sites and generate an energy score function for evaluating ligand posture by using corresponding algorithms to guide the selection of the screening results [36]. In this study, compounds screened using pharmacophores were butt screened successively for HTVS, SP, and XP, respectively, in order to improve the accuracy of butt screening. After analyzing the docking results, compounds in the top 20% of the docking scores were selected for the preliminary binding energy calculation of MMGBSA, which avoided the inaccuracy caused by relying solely on the docking scores. Finally, 10 compounds with better butt scores and MMGBSA binding free energy scores than the reference ligand 4U9 were selected for ADMET analysis. Only three of the ten compounds, compound 14294, compound 15643, and compound 14186, were in line with Lipinski's five rules and had the lowest toxicity levels. The ADMET results of these four compounds were superior to those of the reference ligand 4U9, so they were reserved for kinetic studies.

Molecular dynamics studies are used to reflect the stability of compounds binding to proteins [37]. In this study, the crystal structure of mPGES-1 was respectively compared with three compounds of research value and the reference ligand 4U9 for 100 ns simulation locus, and RMSD, RMSF, and hydrogen bond analysis were performed. Compared with 4U9, compound 15643 had the property of more stable interaction with protein, so the other two compounds were removed. The binding energies between the protein and compounds 15643 and 4U9 were calculated and compared. The results showed that compound 15643 had better binding effect than 4U9.

Although the results of our study are scientific, there are still some shortcomings. The number of compounds in the Chinese medicine database selected for this study is 20,011, which is not enough, and the opportunity to screen better Chinese medicine compound inhibitors targeting mPGES-1 may be missed. In addition, we have not purchased the screened compounds for experimental verification, which does not make the accuracy of our research results 100% sure. At the same time, we also encountered challenges in the research process. In the screening process of the TCM database, the pharmacophore model construction had certain requirements regarding the number of compounds in the training set and the validation set, and more molecules with certain activities should be sought, to the degree possible, for the model construction and verification.

In conclusion, we identified an mPGES-1 inhibitor, compound 15643, from the traditional Chinese medicine database through multi-step virtual screening. This result provides a new reference for the development of mPGES-1 inhibitors.

4. Materials and Methods

4.1. Protein Structure

Suitable protein structures were selected from the protein database (https://www.rcsb.org/, accessed on 2 May 2022) for structure-based calculation to identify mPGES-1 inhibitors. The PDB ID of the selected protein was 4YL3. It is derived from the human protein structure with A conformational resolution of 1.41 A and complete protein conformational sequence, and the protein structure is complex with ligand (4U9), so it was selected as the object of study [38]. The complex structures of 4U9 and mPGES-1 were adjusted and optimized using the Protein Preparation module with the help of the Maestro visualization window of Schrodinger's drug screening design and molecular simulation software. The 4U9 ligand binding site within the 3D receptor structure was designated as the binding region of the screening ligand, and a grid was created to surround the region.

4.2. Data Set

In this study, a database of traditional Chinese medicine compounds was selected as the screening object, and there were 20,011 compounds in the database. In order to ensure that the selected compounds have better drug utilization properties, the Maestro Ligprep module was used to optimize the conformation and energy settings of traditional Chinese medicine compounds. The LigPrep panel is used to set up and start configuration preparation calculations [39]. Our goal with LigPrep was to take the 2D structure and produce the corresponding low-energy 3D structure that Glide and programs can use. We used Epic to generate possible ionized states at pH 7.0 ± 2.0, resulting in tautomers, forming up to 32 stereoisomers per ligand, determining chirality from 3D coordinates, and producing low energy rings. Then, the OPLS3e force field was used to minimize conformation and energy [40]. Finally, more than 400,000 databases of TCM compounds in different conformations were generated for screening studies.

4.3. Pharmacophore Analysis

Pharmacophore modeling is a widely used method for calculating the chemical characteristics of the geometry of active sites and the spatial arrangement of ligand substituents in 3D space [41]. A number of known active mPGES-1 inhibitors were collected as data sets by reviewing extensive literature [38,42–47]. These were used to construct and validate the pharmacophore model. The Maestro program was used to carry out pharmacophore models based on ligand structure. After research and analysis, the model with the highest score and five pharmacophore characteristics (AAHNR) was selected. In addition, a total of 61 compounds containing IC50 values and that were active against mPGES-1 were collected from the literature search and compound library as the training set and verification set. Eighteen of them were training sets, and the rest were verification sets. The reliability of the model was evaluated from the AUC value of the ROC, which was between 0.5 and 1. The larger the value of AUC, the more likely the current classification algorithm will rank the positive samples before the negative samples; that is, the better the classification can potentially be [48]. The AUC value can be used to determine whether the model can be used for a pharmacophore based virtual screening.

After determining the selected pharmacophore model (AAHNR), the Maestro program was used to conduct the first virtual screening of compounds in the Chinese drug bank, namely the virtual screening based on the pharmacophore model [49]. The screening conditions were set as satisfying at least four of the five pharmacophore characteristics, and finally 270,000 compounds meeting the conditions were screened out from the Chinese drug bank.

4.4. Molecular Docking

After pharmacophore-based screening, compounds with high pharmacophore evaluation scores (truncation value set as 1) were selected for docking analysis. The Maestro tool was used to interconnect them with mPGES-1 for HTVS, SP, and XP in turn [50], aiming to gradually improve the accuracy of the interconnecting process. The compounds in the top 20% of the XP interconnecting score were selected for preliminary calculation of MMGBSA binding energy [51] and compared with 4U9. We selected 10 compounds with higher Glide scores and MMGBSA scores than 4U9 for further research.

4.5. ADMET

ADMET is a subject that quantitatively studies the process (absorption, distribution, metabolism, and excretion) of drugs in the biological body and describes the dynamic law of drugs in the body. It has become an important part of drug preclinical research and clinical research [27]. The SwissADME server (http://www.swissadme.ch/, accessed on 23 September 2022) was used to evaluate elected compounds after molecular docking ligands. The SMILES format was imported into the server, and the relevant parameters such as pharmacokinetic characteristics and drug solubility were combined to further select the compounds for the next operation.

4.6. Dynamic Simulation

Molecular dynamics simulations (MDs) were performed to evaluate the binding stability to proteins of the three candidate molecules. Before MDs were run, the simulation system was built using the GROMACS 2019.1 software package (sourced by Mark Abraham et al., Uppsala University, Stockholm University, and KTH Royal Institute of Technology, Stockholm, Sweden) [37]. The topological system of proteins was constructed by AMBER99SB-ILDN force field [52]. The Bio2byte web server (https://www.bio2byte.be/, accessed on 11 February 2023) and the GAFF force field were used to generate the molecular topology file [37,53]. In the simulation system, the cube box and TIP3P water model with radius of 1.2 nm were selected to define periodic boundary conditions (PBC) [54]. In addition, 8 chloride ions were added to each system to ensure the electrical neutrality of the simulated system; this step was calculated and completed using GROMACS. At the simulated temperature of 300 K, the system energy was minimized in 50,000 steps. In order to maintain the pressure and temperature of the system, the two simulation systems were balanced using position-constrained MD simulation at 300 K, lasting 100 ps. Finally, we ran a MD simulation with a duration of 100 ns. Through MD simulation, the root mean square deviations (RMSDs) and root mean square fluctuations (RMSFs) of atomic position were analyzed by extracting trajectory coordinates. In addition, the intermolecular H bond interaction between mPGES-1 and the ligand was extracted using the gmx-H bond analysis tool.

In order to calculate the free energy of receptor–ligand binding, this study adopted the molecular mechanics-Poisson–Boltzmann surface area (MM-PBSA) method, which was implemented using the GROMACS built-in tool g_mmpbsa [55]. The basic principle is to calculate the difference between the bound and unbound free energies of two solvated molecules or to compare the free energies of different solvated conformations of the same molecule. The combined free energy of recombination is calculated according to the following formula. $G_{complex}$ Gcomplex represents the free energy of the protein–ligand complex, $G_{protein}$ represents the free energy of the protein in the solvent, and G_{ligand} represents the free energy of the ligand in the solvent.

$$G_{binding} = G_{complex} - \left(G_{protein} + G_{ligand}\right)$$

The free energy of the protein–ligand complex is represented by the $G_{complex}$, where the $G_{protein}$ represents the free energy of the protein in the solvent, and G_{ligand} represents the free energy of the ligand in the solvent.

5. Conclusions

mPGES-1 plays an important role in inflammatory response and anti-tumor drug development. In this study, pharmacophore models were established based on the common pharmacodynamic characteristics of multiple ligands, and virtual screening was conducted based on the crystal structure of mPGES-1 and combined small-molecule inhibitor 4U9. A novel mPGES-1 inhibitor compound 15643 was identified by ADMET, dynamic simulation, and multi-step virtual screening. Compared with compound 4U9, compound 15643, which was finally screened in this study, had similar binding ability to the protein mPGES-1, and ADMET analysis showed that it had better pharmaceutical potential. In addition, compound 15643 has a more stable binding energy. Therefore, we can assume that compound 15643 has the potential to be an inhibitor of mPGES-1 in this study. This potential small-molecule inhibitor provides a new direction for the development of anti-cancer and anti-inflammatory drugs. The small molecule can be further evaluated via different laboratory-based experimental techniques to help determine the activity of the compound and provide reference values for the study of novel mPGES-1 inhibitors.

Supplementary Materials: The following supporting information can be downloaded at: https://www.mdpi.com/article/10.3390/molecules28166059/s1, Figure S1: Relationship between PG and inflammation and cancer; Figure S2: The virtual screening workflow (VSW) used in this study was used to identify targeted mPGES-1 inhibitors; Figure S3: Compound structure and IC50 values for pharmacophore modeling.

Author Contributions: Conceptualization, L.L.; Data curation, T.L.; Formal analysis, Q.W.; Funding acquisition, L.L.; Investigation, T.L. and Q.H.; Methodology, Q.W.; Project administration, L.L.; Resources, Q.H.; Software, Q.H. and T.L.; Supervision, L.L.; Validation, Q.H.; Writing—original draft, Q.H. and L.L.; Writing—review and editing, L.L. All authors have read and agreed to the published version of the manuscript.

Funding: This research was funded by the Science and Technology Special Project of Zhanjiang (2022A01034); The Guangdong Provincial Department of Education Research Project (2022KTSCX); and The Science and Technology Program of Guangdong Province (2023A1515010850).

Institutional Review Board Statement: Not applicable.

Informed Consent Statement: Not applicable.

Data Availability Statement: The data used to support the findings of this study are included within the article.

Acknowledgments: Thank you to everyone who contributed to this article.

Conflicts of Interest: The authors declare no conflict of interest.

Sample Availability: Not applicable.

References

1. Ricciotti, E.; FitzGerald, G.A. Prostaglandins and inflammation. *Arter. Thromb. Vasc. Biol.* **2011**, *31*, 986–1000. [CrossRef]
2. Takeuchi, K. [Development and repair of NSAIDs-induced small intestinal lesions: Relation to COX isozymes and EP receptor subtypes]. *Nihon. Rinsho.* **2011**, *69*, 1123–1128.
3. Bogdan, D.; Falcone, J.; Kanjiya, M.P.; Park, S.H.; Carbonetti, G.; Studholme, K.; Gomez, M.; Lu, Y.; Elmes, M.W.; Smietalo, N.; et al. Fatty acid-binding protein 5 controls microsomal prostaglandin E synthase 1 (mPGES-1) induction during inflammation. *J. Biol. Chem.* **2018**, *293*, 5295–5306. [CrossRef]
4. Takahashi, R.; Amano, H.; Ito, Y.; Eshima, K.; Satoh, T.; Iwamura, M.; Nakamura, M.; Kitasato, H.; Uematsu, S.; Raouf, J.; et al. Microsomal prostaglandin E synthase-1 promotes lung metastasis via SDF-1/CXCR4-mediated recruitment of CD11b(+)Gr1(+)MDSCs from bone marrow. *Biomed. Pharmacother.* **2020**, *121*, 109581. [CrossRef]
5. van Rees, B.P.; Sivula, A.; Thorén, S.; Yokozaki, H.; Jakobsson, P.J.; Offerhaus, G.J.; Ristimäki, A. Expression of microsomal prostaglandin E synthase-1 in intestinal type gastric adenocarcinoma and in gastric cancer cell lines. *Int. J. Cancer* **2003**, *107*, 551–556. [CrossRef]
6. Yoshimatsu, K.; Golijanin, D.; Paty, P.B.; Soslow, R.A.; Jakobsson, P.J.; DeLellis, R.A.; Subbaramaiah, K.; Dannenberg, A.J. Inducible microsomal prostaglandin E synthase is overexpressed in colorectal adenomas and cancer. *Clin. Cancer Res.* **2001**, *7*, 3971–3976.
7. Xu, L.W.; Qian, M.; Jia, R.P.; Xu, Z.; Wu, J.P.; Li, W.C.; Huang, W.B.; Chen, X.G. Expression and significance of microsomal prostaglandin synthase-1 (mPGES-1) and Beclin-1 in the development of prostate cancer. *Asian Pac. J. Cancer Prev.* **2012**, *13*, 1639–1644. [CrossRef]
8. Mehrotra, S.; Morimiya, A.; Agarwal, B.; Konger, R.; Badve, S. Microsomal prostaglandin E2 synthase-1 in breast cancer: A potential target for therapy. *J. Pathol.* **2006**, *208*, 356–363. [CrossRef]
9. Herfs, M.; Herman, L.; Hubert, P.; Minner, F.; Arafa, M.; Roncarati, P.; Henrotin, Y.; Boniver, J.; Delvenne, P. High expression of PGE2 enzymatic pathways in cervical (pre)neoplastic lesions and functional consequences for antigen-presenting cells. *Cancer Immunol. Immunother.* **2009**, *58*, 603–614. [CrossRef]
10. Hasan, S.; Satake, M.; Dawson, D.W.; Funahashi, H.; Angst, E.; Go, V.L.; Reber, H.A.; Hines, O.J.; Eibl, G. Expression analysis of the prostaglandin E2 production pathway in human pancreatic cancers. *Pancreas* **2008**, *37*, 121–127. [CrossRef]

11. Kim, S.H.; Roszik, J.; Cho, S.N.; Ogata, D.; Milton, D.R.; Peng, W.; Menter, D.G.; Ekmekcioglu, S.; Grimm, E.A. The COX2 Effector Microsomal PGE2 Synthase 1 is a Regulator of Immunosuppression in Cutaneous Melanoma. *Clin. Cancer Res.* **2019**, *25*, 1650–1663. [CrossRef]
12. Leoncini, E.; Ricciardi, W.; Cadoni, G.; Arzani, D.; Petrelli, L.; Paludetti, G.; Brennan, P.; Luce, D.; Stucker, I.; Matsuo, K.; et al. Adult height and head and neck cancer: A pooled analysis within the INHANCE Consortium. *Eur. J. Epidemiol.* **2014**, *29*, 35–48. [CrossRef]
13. Omi, Y.; Shibata, N.; Okamoto, T.; Obara, T.; Kobayashi, M. Immunohistochemical demonstration of membrane-bound prostaglandin E2 synthase-1 in papillary thyroid carcinoma. *Acta Histochem. Cytochem.* **2009**, *42*, 105–109. [CrossRef]
14. Mattila, S.; Tuominen, H.; Koivukangas, J.; Stenbäck, F. The terminal prostaglandin synthases mPGES-1, mPGES-2, and cPGES are all overexpressed in human gliomas. *Neuropathology* **2009**, *29*, 156–165. [CrossRef]
15. Chini, M.G.; Giordano, A.; Potenza, M.; Terracciano, S.; Fischer, K.; Vaccaro, M.C.; Colarusso, E.; Bruno, I.; Riccio, R.; Koeberle, A.; et al. Targeting mPGES-1 by a Combinatorial Approach: Identification of the Aminobenzothiazole Scaffold to Suppress PGE(2) Levels. *ACS Med. Chem Lett.* **2020**, *11*, 783–789. [CrossRef]
16. Korotkova, M.; Jakobsson, P.J. Characterization of microsomal prostaglandin E synthase 1 inhibitors. *Basic Clin Pharmacol. Toxicol.* **2014**, *114*, 64–69. [CrossRef]
17. Sun, Y. The role of Chinese medicine in clinical oncology. *Chin. J. Integr. Med.* **2014**, *20*, 3–10. [CrossRef]
18. Huang, Y.; Cai, T.; Xia, X.; Cai, Y.; Wu, X.Y. Research Advances in the Intervention of Inflammation and Cancer by Active Ingredients of Traditional Chinese Medicine. *J. Pharm. Pharm. Sci.* **2016**, *19*, 114–126. [CrossRef]
19. Kumar Bhardwaj, V.; Purohit, R.; Kumar, S. Himalayan bioactive molecules as potential entry inhibitors for the human immunodeficiency virus. *Food Chem.* **2021**, *347*, 128932. [CrossRef]
20. Tanwar, G.; Mazumder, A.G.; Bhardwaj, V.; Kumari, S.; Bharti, R.; Yamini; Singh, D.; Das, P.; Purohit, R. Target identification, screening and in vivo evaluation of pyrrolone-fused benzosuberene compounds against human epilepsy using Zebrafish model of pentylenetetrazol-induced seizures. *Sci. Rep.* **2019**, *9*, 7904. [CrossRef]
21. Yu, W.; MacKerell, A.D., Jr. Computer-Aided Drug Design Methods. *Methods Mol. Biol.* **2017**, *1520*, 85–106. [CrossRef]
22. Sun, H. Pharmacophore-based virtual screening. *Curr. Med. Chem.* **2008**, *15*, 1018–1024. [CrossRef]
23. Luo, L.; Zhong, A.; Wang, Q.; Zheng, T. Structure-Based Pharmacophore Modeling, Virtual Screening, Molecular Docking, ADMET, and Molecular Dynamics (MD) Simulation of Potential Inhibitors of PD-L1 from the Library of Marine Natural Products. *Mar. Drugs* **2021**, *20*, 29. [CrossRef]
24. Tai, W.; Lu, T.; Yuan, H.; Wang, F.; Liu, H.; Lu, S.; Leng, Y.; Zhang, W.; Jiang, Y.; Chen, Y. Pharmacophore modeling and virtual screening studies to identify new c-Met inhibitors. *J. Mol. Model.* **2012**, *18*, 3087–3100. [CrossRef]
25. Martínez Pérez, J.A.; Pérez Martin, P.S. [ROC curve]. *Semergen* **2023**, *49*, 101821. [CrossRef]
26. Friesner, R.A.; Banks, J.L.; Murphy, R.B.; Halgren, T.A.; Klicic, J.J.; Mainz, D.T.; Repasky, M.P.; Knoll, E.H.; Shelley, M.; Perry, J.K.; et al. Glide: A new approach for rapid, accurate docking and scoring. 1. Method and assessment of docking accuracy. *J. Med. Chem.* **2004**, *47*, 1739–1749. [CrossRef]
27. Ferreira, L.L.G.; Andricopulo, A.D. ADMET modeling approaches in drug discovery. *Drug Discov. Today* **2019**, *24*, 1157–1165. [CrossRef]
28. Daina, A.; Michielin, O.; Zoete, V. SwissADME: A free web tool to evaluate pharmacokinetics, drug-likeness and medicinal chemistry friendliness of small molecules. *Sci. Rep.* **2017**, *7*, 42717. [CrossRef]
29. Murakami, M.; Kudo, I. Prostaglandin E synthase: A novel drug target for inflammation and cancer. *Curr. Pharm. Des.* **2006**, *12*, 943–954. [CrossRef]
30. AbdulHameed, M.D.; Hamza, A.; Liu, J.; Huang, X.; Zhan, C.G. Human microsomal prostaglandin E synthase-1 (mPGES-1) binding with inhibitors and the quantitative structure-activity correlation. *J. Chem. Inf. Model.* **2008**, *48*, 179–185. [CrossRef]
31. Di Micco, S.; Terracciano, S.; Ruggiero, D.; Potenza, M.; Vaccaro, M.C.; Fischer, K.; Werz, O.; Bruno, I.; Bifulco, G. Identification of 2-(thiophen-2-yl)acetic Acid-Based Lead Compound for mPGES-1 Inhibition. *Front. Chem.* **2021**, *9*, 676631. [CrossRef]
32. Xu, D.; Rowland, S.E.; Clark, P.; Giroux, A.; Côté, B.; Guiral, S.; Salem, M.; Ducharme, Y.; Friesen, R.W.; Méthot, N.; et al. MF63 [2-(6-chloro-1H-phenanthro[9,10-d]imidazol-2-yl)-isophthalonitrile], a selective microsomal prostaglandin E synthase-1 inhibitor, relieves pyresis and pain in preclinical models of inflammation. *J. Pharmacol. Exp. Ther.* **2008**, *326*, 754–763. [CrossRef]
33. Muthukaman, N.; Deshmukh, S.; Sarode, N.; Tondlekar, S.; Tambe, M.; Pisal, D.; Shaikh, M.; Kattige, V.G.; Honnegowda, S.; Karande, V.; et al. Discovery of 2-((2-chloro-6-fluorophenyl)amino)-N-(3-fluoro-5-(trifluoromethyl)phenyl)-1-methyl-7,8-dihydro-1H-[1,4]dioxino[2′,3′:3,4]benzo[1,2-d]imidazole-5-carboxamide as potent, selective and efficacious microsomal prostaglandin E(2) synthase-1 (mPGES-1) inhibitor. *Bioorg. Med. Chem. Lett.* **2016**, *26*, 5977–5984. [CrossRef]
34. Ergül, A.G.; Maz, T.G.; Kretzer, C.; Olğaç, A.; Jordan, P.M.; Çalışkan, B.; Werz, O.; Banoglu, E. Novel potent benzimidazole-based microsomal prostaglandin E(2) synthase-1 (mPGES-1) inhibitors derived from BRP-201 that also inhibit leukotriene C(4) synthase. *Eur. J. Med. Chem.* **2022**, *231*, 114167. [CrossRef]
35. Arhancet, G.B.; Walker, D.P.; Metz, S.; Fobian, Y.M.; Heasley, S.E.; Carter, J.S.; Springer, J.R.; Jones, D.E.; Hayes, M.J.; Shaffer, A.F.; et al. Discovery and SAR of PF-4693627, a potent, selective and orally bioavailable mPGES-1 inhibitor for the potential treatment of inflammation. *Bioorg. Med. Chem. Lett.* **2013**, *23*, 1114–1119. [CrossRef]
36. Guedes, I.A.; de Magalhães, C.S.; Dardenne, L.E. Receptor-ligand molecular docking. *Biophys. Rev.* **2014**, *6*, 75–87. [CrossRef]

37. Van Der Spoel, D.; Lindahl, E.; Hess, B.; Groenhof, G.; Mark, A.E.; Berendsen, H.J. GROMACS: Fast, flexible, and free. *J. Comput. Chem.* **2005**, *26*, 1701–1718. [CrossRef]
38. Luz, J.G.; Antonysamy, S.; Kuklish, S.L.; Condon, B.; Lee, M.R.; Allison, D.; Yu, X.P.; Chandrasekhar, S.; Backer, R.; Zhang, A.; et al. Crystal Structures of mPGES-1 Inhibitor Complexes Form a Basis for the Rational Design of Potent Analgesic and Anti-Inflammatory Therapeutics. *J. Med. Chem.* **2015**, *58*, 4727–4737. [CrossRef]
39. Sheikh, I.A.; Jiffri, E.H.; Ashraf, G.M.; Kamal, M.A.; Beg, M.A. Structural studies on inhibitory mechanisms of antibiotic, corticosteroid and catecholamine molecules on lactoperoxidase. *Life Sci.* **2018**, *207*, 412–419. [CrossRef]
40. Babaoglu, Z.Y.; Kilic, D. Virtual screening, molecular simulations and bioassays: Discovering novel microsomal prostaglandin E Synthase-1 (mPGES-1) inhibitors. *Comput. Biol. Med.* **2023**, *155*, 106616. [CrossRef]
41. Yadav, D.K.; Kumar, S.; Teli, M.K.; Kim, M.H. Ligand-based pharmacophore modeling and docking studies on vitamin D receptor inhibitors. *J. Cell Biochem.* **2020**, *121*, 3570–3583. [CrossRef]
42. Schiffler, M.A.; Antonysamy, S.; Bhattachar, S.N.; Campanale, K.M.; Chandrasekhar, S.; Condon, B.; Desai, P.V.; Fisher, M.J.; Groshong, C.; Harvey, A.; et al. Discovery and Characterization of 2-Acylaminoimidazole Microsomal Prostaglandin E Synthase-1 Inhibitors. *J. Med. Chem.* **2016**, *59*, 194–205. [CrossRef]
43. Shekfeh, S.; Caliskan, B.; Fischer, K.; Yalcin, T.; Garscha, U.; Werz, O.; Banoglu, E. A Multi-step Virtual Screening Protocol for the Identification of Novel Non-acidic Microsomal Prostaglandin E(2) Synthase-1 (mPGES-1) Inhibitors. *ChemMedChem* **2019**, *14*, 273–281. [CrossRef]
44. Misra, S.; Saini, M.; Ojha, H.; Sharma, D.; Sharma, K. Pharmacophore modelling, atom-based 3D-QSAR generation and virtual screening of molecules projected for mPGES-1 inhibitory activity. *SAR QSAR Environ. Res.* **2017**, *28*, 17–39. [CrossRef]
45. Lauro, G.; Cantone, V.; Potenza, M.; Fischer, K.; Koeberle, A.; Werz, O.; Riccio, R.; Bifulco, G. Discovery of 3-hydroxy-3-pyrrolin-2-one-based mPGES-1 inhibitors using a multi-step virtual screening protocol. *Medchemcomm* **2018**, *9*, 2028–2036. [CrossRef]
46. Psarra, A.; Nikolaou, A.; Kokotou, M.G.; Limnios, D.; Kokotos, G. Microsomal prostaglandin E(2) synthase-1 inhibitors: A patent review. *Expert Opin. Ther. Pat.* **2017**, *27*, 1047–1059. [CrossRef]
47. Waltenberger, B.; Wiechmann, K.; Bauer, J.; Markt, P.; Noha, S.M.; Wolber, G.; Rollinger, J.M.; Werz, O.; Schuster, D.; Stuppner, H. Pharmacophore modeling and virtual screening for novel acidic inhibitors of microsomal prostaglandin E(2) synthase-1 (mPGES-1). *J. Med. Chem.* **2011**, *54*, 3163–3174. [CrossRef]
48. Muschelli, J. ROC and AUC with a Binary Predictor: A Potentially Misleading Metric. *J. Classif.* **2020**, *37*, 696–708. [CrossRef]
49. Löwer, M.; Proschak, E. Structure-Based Pharmacophores for Virtual Screening. *Mol. Inform.* **2011**, *30*, 398–404. [CrossRef]
50. Alogheli, H.; Olanders, G.; Schaal, W.; Brandt, P.; Karlén, A. Docking of Macrocycles: Comparing Rigid and Flexible Docking in Glide. *J. Chem. Inf. Model.* **2017**, *57*, 190–202. [CrossRef]
51. Bathula, R.; Lanka, G.; Muddagoni, N.; Dasari, M.; Nakkala, S.; Bhargavi, M.; Somadi, G.; Sivan, S.K.; Rajender Potlapally, S. Identification of potential Aurora kinase-C protein inhibitors: An amalgamation of energy minimization, virtual screening, prime MMGBSA and AutoDock. *J. Biomol. Struct. Dyn.* **2020**, *38*, 2314–2325. [CrossRef] [PubMed]
52. Wang, J.; Wolf, R.M.; Caldwell, J.W.; Kollman, P.A.; Case, D.A. Development and testing of a general amber force field. *J. Comput. Chem.* **2004**, *25*, 1157–1174. [CrossRef] [PubMed]
53. Hornak, V.; Abel, R.; Okur, A.; Strockbine, B.; Roitberg, A.; Simmerling, C. Comparison of multiple Amber force fields and development of improved protein backbone parameters. *Proteins* **2006**, *65*, 712–725. [CrossRef]
54. Harrach, M.F.; Drossel, B. Structure and dynamics of TIP3P, TIP4P, and TIP5P water near smooth and atomistic walls of different hydroaffinity. *J. Chem. Phys.* **2014**, *140*, 174501. [CrossRef] [PubMed]
55. Kumari, R.; Kumar, R.; Lynn, A. g_mmpbsa--a GROMACS tool for high-throughput MM-PBSA calculations. *J. Chem. Inf. Model* **2014**, *54*, 1951–1962. [CrossRef] [PubMed]

molecules

MDPI

Article

Collection of Partition Coefficients in Hexadecyltrimethylammonium Bromide, Sodium Cholate, and Lithium Perfluorooctanesulfonate Micellar Solutions: Experimental Determination and Computational Predictions

Leila Saranjam [1], Miroslava Nedyalkova [2,*], Elisabet Fuguet [3,4], Vasil Simeonov [2], Francesc Mas [1] and Sergio Madurga [1,*]

1 Department of Material Science and Physical Chemistry, Research Institute of Theoretical and Computational Chemistry (IQTCUB), University of Barcelona, C/Martí i Franquès 1, 08028 Barcelona, Spain; lsaransa19@alumnes.ub.edu (L.S.); fmas@ub.edu (F.M.)

2 Faculty of Chemistry and Pharmacy, University of Sofia "St. Kl. Ohridski", 1 James Bourchier Blvd., 1164 Sofia, Bulgaria; ahvs@chem.uni-sofia.bg

3 Department of Chemical Engineering and Analytical Chemistry, Institute of Biomedicine (IBUB), University of Barcelona, C/Martí i Franquès 1, 08028 Barcelona, Spain; elifuguetj@ub.edu

4 Serra Húnter Programme, Generalitat de Catalunya, 08017 Barcelona, Spain

* Correspondence: miroslava.nedyalkova@unifr.ch (M.N.); s.madurga@ub.edu (S.M.)

Abstract: This study focuses on determining the partition coefficients (logP) of a diverse set of 63 molecules in three distinct micellar systems: hexadecyltrimethylammonium bromide (HTAB), sodium cholate (SC), and lithium perfluorooctanesulfonate (LPFOS). The experimental $\log p$ values were obtained through micellar electrokinetic chromatography (MEKC) experiments, conducted under controlled pH conditions. Then, Quantum Mechanics (QM) and machine learning approaches are proposed for the prediction of the partition coefficients in these three micellar systems. In the applied QM approach, the experimentally obtained partition coefficients were correlated with the calculated values for the case of the 15 solvent mixtures. Using Density Function Theory (DFT) with the B3LYP functional, we calculated the solvation free energies of 63 molecules in these 16 solvents. The combined data from the experimental partition coefficients in the three micellar formulations showed that the 1-propanol/water combination demonstrated the best agreement with the experimental partition coefficients for the SC and HTAB micelles. Moreover, we employed the SVM approach and k-means clustering based on the generation of the chemical descriptor space. The analysis revealed distinct partitioning patterns associated with specific characteristic features within each identified class. These results indicate the utility of the combined techniques when we want an efficient and quicker model for predicting partition coefficients in diverse micelles.

Keywords: partition coefficient; micelle; hexadecyltrimethylammonium bromide (HTAB); sodium cholate (SC); lithium perfluorooctanesulfonate (LPFOS); SVM; DFT; k-means clustering

1. Introduction

The partition coefficient (logP) is a significant physicochemical parameter used in various fields such as drug and pharmaceutical product design, substance toxicology, and environmental fate modeling of organic compounds [1]. It measures the solute's solubility in two immiscible solvents, providing valuable insights into solute distribution. In drug delivery systems, the partition coefficient is crucial in determining the system's ability to distribute molecules between the aqueous phase and micelles [2]. Regular micelles consist of a polar head and a nonpolar tail, enabling the dissolution of both polar and nonpolar molecules. Hydrophilic solutes preferentially interact with the polar, hydrophilic surface of the micelle, while hydrophobic solutes tend to accumulate in the nonpolar, hydrophobic core of the micelle [3,4]. Understanding the partitioning behavior within micelles can contribute to enhancing the efficacy and safety of drug delivery systems. Moreover, this knowledge can be leveraged to optimize the design and performance of such systems, leading to improved therapeutic outcomes [5,6]. When hydrophobic compounds are introduced into micellar solutions, they have a higher tendency to associate with the micelles because the hydrophobic regions of the micelles (the inner hydrophobic tails) can provide a more favorable environment for the hydrophobic molecules. As a result, more hydrophobic molecules have higher values of partition coefficients in micellar systems. In this context, the use of pluronic micelles for delivering hydrophobic drugs presents an interesting and alternative approach [7,8].

Various experimental methods can be used to estimate the micelle–water partition coefficient, such as solubility analysis, micellar-enhanced ultrafiltration, micellar liquid chromatography [9,10], and cloud-point extraction [11]. In this study, the micelle–water partition coefficients were estimated from the retention times of micellar electrokinetic chromatography (MEKC) experiments [12,13].

The MEKC technique is widely employed for the separation and identification of components within a mixture. This technique utilizes as a pseudo-stationary phase a surfactant above its critical micellar concentration (CMC) to facilitate the formation of micelles in an aqueous solution. By applying an electric field, the components within the mixture are partitioned between the aqueous and the micellar phases, leading to their separation. MEKC separations present high resolution and efficiency in the analysis of both neutral and charged compounds. Moreover, the separations can be easily optimized just by changing the nature of the surfactant [14,15]. This technique has been used for the determination of partition coefficients in micelles in many different fields [16–20].

From a computational point of view, molecular dynamics (MD) simulations could provide valuable insights into the transfer of solutes between different phases, such as from the aqueous phase to the micellar phase [4]. These simulations allow free energy profiles to be obtained, which quantitatively describe the energetic changes associated with solute transfer. In the context of drug delivery systems, these profiles help in understanding the distribution of drugs within micelles and optimize their design. However, it is important to note that molecular dynamics simulations can be computationally costly because of the need for long converged trajectories.

MD simulations combined with the COSMOmic method have been shown to be a promising alternative [5,21,22] to experimental methods for predicting the partition coefficient in micellar systems. Previous studies have demonstrated good correlation between predicted and experimental data for a variety of micelles [23], including sodium dodecyl sulfate (SDS), hexadecyltrimethylammonium bromide (HTAB, also known as Cetyltrimethylammonium Bromide, CTAB), sodium cholate (SC), lithium perfluorooctanesulfonate (LPFOS), $C_{12}E_{10}$, Brij35, Triton X-114, and Triton X-100.

Recently, a study on mixed micelles formed by sodium laureth sulfate (SLES) and fatty acids, using molecular dynamics simulations, shows that the micelle–water partition coefficients of neutral and charged fatty acids could be calculated using the COSMOmic and the MD approach [24]. Based on the potential of mean force (PMF) calculations performed using umbrella sampling (US), the study shows that the partition coefficients for neutral solutes can be accurately calculated using both the COSMOmic and additive CGenFF US/PMF approaches, while the Drude polarizable force field is needed to accurately calculate the experimental partition coefficient of the charged solute. There are other examples of MD simulations with US and COSMOmic [25], demonstrating the utility of these methods for predicting partition equilibria in micellar systems.

Moreover, the fragmental constant method (FCA) has also been applied to determine partition coefficients. The FCA model defines a micelle–water partition coefficient as the sum of the partition coefficients of the component's atomic/molecular fragments, determined by fragmental constant values [26]. Fragmental techniques are ineffective at estimating the parameters of other solvents and are only appropriate for a narrow range of solvents (typically octanol/water).

Alternatively, other ways of categorizing logP predictors use parametric models, which employ methods such as least squares estimation or multiple linear regression to fit the parameters governing the relative contributions of different input features. Machine-learning-based methods, including Support Vector Machines (SVM) [27,28], Neural Networks (NNs) [29], and Graph Convolutional Networks (GCN) [30], have also been utilized for logP prediction. In a recent study by Dickson et al. [31], various methods for predicting logP values in a dataset of small molecules were examined. The study focused on transforming atomic properties, such as radius and partial charge, which are commonly employed as force field parameters in classical molecular dynamics simulations. These attributes were converted into index-invariant molecular features using a recently developed technique known as geometric scattering for graphs (GSG) [31]. The results obtained from this investigation demonstrate that the most accurate predictions were achieved using atomic attributes generated with the CHARMM generalized force field and 2D molecular structures. This highlights the significance of employing appropriate molecular representations and force field parameters for accurate logP prediction.

Here, the focus is on the use of density functional theory (DFT) calculations and SVM calculations for predicting the partition coefficients of compounds in micellar systems. DFT calculations are a Quantum Mechanics (QM) computational method that can be used to predict the partition coefficients of compounds in micellar systems. They are faster and less computationally demanding than molecular dynamics simulations, making them an attractive alternative for predicting the properties of drugs in micellar systems. By calculating the energy changes associated with transferring a compound from the aqueous phase to a solvent phase that resembles the behavior of the micellar phase, DFT calculations can provide an estimate of the compound's partition coefficient in a micellar system. This makes DFT calculations a valuable tool for drug delivery design and optimization. The first step is to use DFT calculations to identify the combination of solvents that can best predict the experimental partition coefficients of compounds in a specific micellar system. This study aims to apply the DFT calculation approach to predict the partition coefficients of 63 compounds in HTAB, SC, and LPFOS micellar solutions (Table 1). The study will compare the predicted partition coefficients with experimental data to assess the accuracy of the DFT approach. The prediction of 15 solvent–water partition coefficients is achieved by applying DFT with the B3LYP method [32] with a 6-31++G** basis set. The solvation model based on the density (SMD) is applied to evaluate the free energy of solvation [33,34]. This model divides the solvation free energy into two main contributions—the bulk electrostatic contribution and the cavity dispersion contribution—and it can be applied to any charged or uncharged solute in any type of solvent as a universal solvation model. Using this approach, correlations with micellar partition coefficients in SC, HTAB, and LPFOS micellar systems are performed.

Finally, SVM calculations are performed using the experimental values obtained from the three micellar systems. SVM calculations involve the application of a supervised machine learning algorithm widely utilized in pattern recognition and regression tasks. SVM-based models can capture complex relationships between molecular descriptors and partition coefficients, thereby enabling the prediction of partition coefficients for a diverse range of compounds. The utilization of SVM calculations in partition coefficient prediction offers several advantages. Firstly, it enables the rapid and cost-effective screening of large compound libraries, facilitating the identification of promising candidates for drug development or the assessment of environmental impact. Additionally, SVM models can accommodate a wide range of chemical structures and properties, making them applicable to various classes of compounds. Moreover, SVM-based models can incorporate both structural and physicochemical descriptors, providing a comprehensive representation of the molecular characteristics that influence partitioning behavior. This facilitates the exploration of structure–activity relationships and the identification of key features contributing to partition coefficients, thereby assisting in the design and optimization of compounds with desired properties. Furthermore, the predictive accuracy of SVM models can be continuously improved by incorporating more diverse and high-quality data, as well as by optimizing the selection and combination of molecular descriptors. As a result of this iterative process, the models are refined and their reliability and robustness are enhanced.

Table 1. Chemical structure of the molecules that form the micelles of this study.

Micelle Name	Symbol	Structure	Schematic Representation of Formed Micelles
Hexadecyltrimethyl-ammonium bromide	HTAB	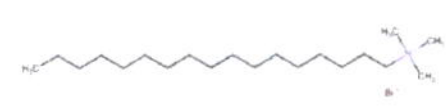	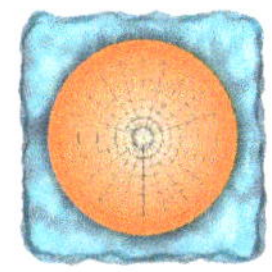
Lithium perfluorooc-tanesulfonate	LPFOS	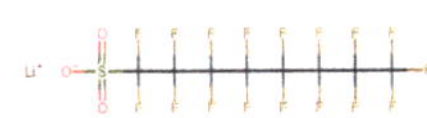	
Sodium cholate	SC	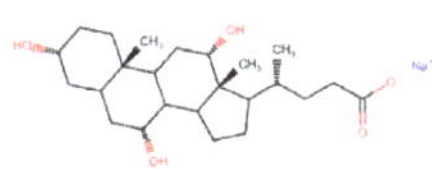	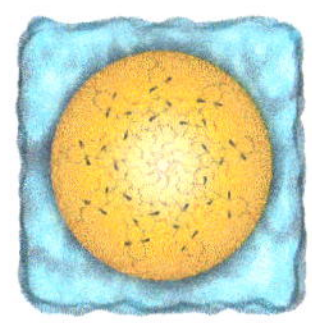

2. Results and Discussion

The experimental values of logP obtained from the SC, LPFOS, and HTAB micelles were analyzed and used to parametrize the computational methodology applied for each type of micelle. Initially, the logP values were estimated based on simple DFT calculations of the molecules in different solvents. Subsequently, SVM predictions were made after conducting a study on the most relevant descriptors using k-means clustering and PCA.

2.1. Experimental logP Values of SC, HTAB, and LPFOS Micelles

The experimental partition coefficients (logP values) in three different types of micelles, namely SC, LPFOS, and HTAB, are presented in Table 2. These logP values were determined by measuring the retention factors of the compounds in 80 mM SC micelles in 20 mM phosphate buffer, 40 mM LPFOS micelles in 20 mM phosphate buffer, and 20 mM HTAB micelles in 20 mM phosphate buffer at pH 7 and 25 °C. The logP values of 63 compounds, representing a diverse set of compounds including benzene derivatives, nitrogen-containing heterocycles, pesticides, hormones, and pharmaceutical compounds, are displayed in Table 2. The selection of compounds was performed according to a previous study [13]. Basically, to obtain a representative set of compounds that cover a wide chemical space, the Abraham descriptor values (excess molar refraction, dipolarity/polarizability, hydrogen bond acidity and basicity, and McGowan volume) of the compounds were considered [35]. To this end, a total of 2975 compounds of different natures were analyzed according to their descriptor values through a principal component analysis. Then, the 2975 compounds were plotted according to the two main principal component values (which represent the highest variance in the system). This plot provided a map of compounds distributed according to their physicochemical properties. The final selection of 63 compounds was performed, trying to cover all the regions of the plot. Additional requirements were that the selected compounds must have a chromophore group to be compatible with the detection system, and must be neutral at the pH of the determination.

Table 2. List of experimental partition coefficients of compounds in SC, LPFOS, and HTAB micelles ($LogP_{SC}$, $LogP_{LPFOS}$, $LogP_{HTAB}$) determined from retention factors obtained from MECK experiments with 80 mM of SC, 40 mM of LPFOS, and 20 mM of HTAB, all in 20 mM phosphate buffer at pH 7 at 25 °C.

Compound	$logP_{SC}$	$logP_{HTAB}$	$logP_{LPFOS}$
Ethylbenzene	2.50	3.00	2.06
Propylbenzene	2.94	3.42	2.39
Butylbenzene	3.26	3.71	2.71
1-Phenylethanone	1.33	2.03	2.19
1-Phenylpropan-1-one	1.65	2.42	2.44
1-Phenylbutan-1-one	2.01	2.80	2.72
1-Phenylpentan-1-one	2.41	3.24	3.01
1-Phenylheptan-1-one	3.15	-	3.68
Furan	0.77	1.48	1.19
2-Nitroaniline	1.59	2.67	1.80
2,3-Benzofuran	2.12	2.82	1.82
Diphenylmethanone	2.48	3.28	3.01
Benzamide	1.06	1.72	1.50
4-Chloroaniline	1.69	2.69	1.44
2,3-Dimethylphenol	1.90	3.15	1.66
Naphtalen-2-ol	2.31	-	1.73
4-Aminobenzamide	0.98	1.11	1.76
3-Methylphenol	1.53	2.78	1.43
2,4-Dimethylphenol	1.93	3.17	1.02
Naphthalene	2.67	3.47	2.09
Pyrimidine	0.56	-	1.27
Benzaldehyde	1.20	1.91	1.91
3-Chloroaniline	1.63	2.72	1.41
Pyrrole	0.68	1.65	0.72
3-Nitroaniline	1.38	2.42	1.53
4-Chlorophenol	2.00	3.24	1.30
Phenol	1.21	2.35	1.08
Methylbenzoate	1.71	2.39	2.36
Bromobenzene	2.37	2.95	1.80
1,4-Xylene	2.51	3.04	2.10
Benzene-1,3-diol	1.21	2.48	0.75
2-Methylaniline	1.17	2.15	1.59
Aniline	0.92	1.83	1.34
Nitrobenzene	1.47	2.21	1.94
Chlorobenzene	2.21	2.77	1.77
N-4-chlorophenylacetamide	2.03	2.80	1.84
N-Phenylacetamide	1.25	1.98	1.58
4-Nitroaniline	1.52	2.50	1.45
Anisole	1.66	2.31	1.83

Table 2. *Cont.*

Compound	$logP_{SC}$	$logP_{HTAB}$	$logP_{LPFOS}$
Benzonitrile	1.21	1.96	1.95
1-Ethyl-4-nitrobenzene	2.19	3.02	2.68
Benzyl benzoate	2.99	-	3.18
Caffeine	1.11	1.32	1.85
Corticosterone	1.94	3.69	3.64
Cortisone	1.72	3.16	3.37
β-Estradiol	2.77	-	2.84
Estriol	2.32	3.52	2.01
Cortisol	1.83	3.39	2.89
Hydroquinone	1.09	1.94	0.19
Quinoline	1.65	2.36	2.68
Atrazine	1.86	1.90	2.71
Diuron	2.46	2.19	2.34
Isoproturon	2.19	1.95	2.61
Linuron	2.59	2.24	2.50
Metobromuron	2.16	2.03	2.22
Monuron	1.81	1.73	2.03
Metoxuron	1.69	1.46	2.34
Phenylurea	1.20	1.20	1.38
Propazine	2.02	2.08	3.03
Fluometuron	2.01	1.92	2.57
N,N-Diethyl-4-nitroaniline	2.44	3.56	3.36
1-Methoxy-4-nitrobenzene	1.69	2.58	2.20
1-Methoxy-2-nitrobenzene	1.55	2.37	2.26

The substances displaying the highest logP values in this table are butylbenzene for both $logP_{SC}$ and $logP_{HTAB}$, and 1-phenylheptan-1-one for $logP_{LPFOS}$. On the other hand, pyrimidine shows the lowest logP value for $logP_{SC}$, 4-aminobenzamide for $logP_{HTAB}$, and hydroquinone for $logP_{LPFOS}$. The $logP_{SC}$ and $logP_{LPFOS}$ exhibit more similar values compared to $logP_{HTAB}$, indicating a possible correlation between $logP_{SC}$ and $logP_{LPFOS}$. Furthermore, a general trend is noticed: compounds with higher hydrophobicity tend to have higher logP values, while those with lower values are more hydrophilic. Therefore, the logP values for the three types of micelles serve as measures of the lipophilicity or hydrophobicity of the respective compounds.

2.2. Correlation of logP Values in Micelles Using DFT Calculations

Figure 1 shows the correlation coefficient values among the experimental and calculated logP in 15 different solvent–water combinations, where a darker color represents a higher correlation coefficient. Molecular representation of all solvents used in DFT calculations are shown in Table S1. It can be seen in Figure 1a that the experimental logPs in SC and HTAB show high correlation between them and with some calculated $logP_{solv/water}$ values. However, the logP in HTAB is not correlated with any combination of computed logP values. With respect to the experimental logPs in SC and LPFOS, the highest correlation of computed logP is obtained with propan-1-ol or propan-2-ol solvents. It seems that a curious pattern can be observed for the calculated $logP_{octanol/water}$. While it is highly correlated with the experimental $\log P_{SC}$ values, it also exhibits a high correlation with all other calculated logP values for different solvent combinations.

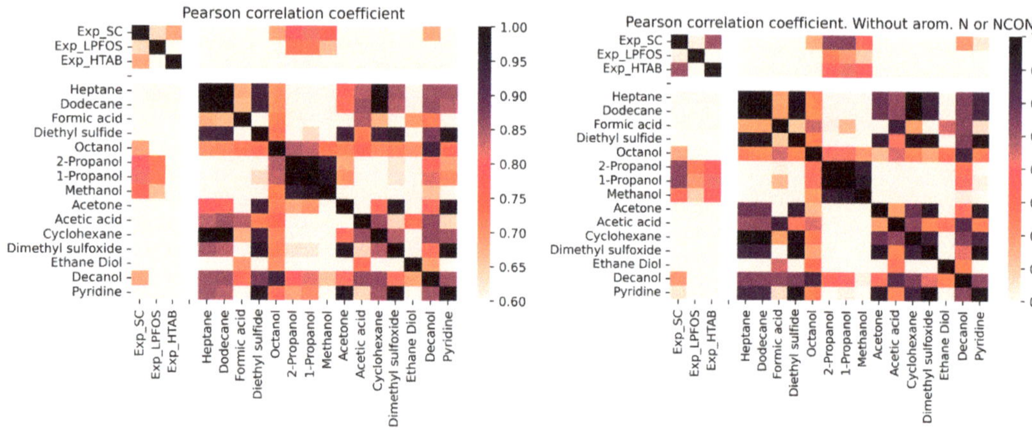

(a) (b)

Figure 1. Heatmap of pairwise correlation of logP values for experimental and B3LYP calculated predictions. Three first logP values are the experimental values in SC, LPFOS and HTAB micelles, respectively. The heat map is colored by the significance of Pearson coefficient, where a darker red indicates a higher degree of correlation. In (**a**) all compounds are used, in (**b**) molecules with Nitrogen in an aromatic ring or with the urea (carbamide) group are excluded.

In Figure 1b, a new heatmap is presented that shows the pairwise correlation between experimental and calculated logP values, but with the exclusion of compounds containing nitrogen in an aromatic ring or the urea group. It is observed that all experimental logP values, including $\log P_{HTAB}$, show a high correlation with propan-2-ol and propan-1-ol. Additionally, for $\log P_{SC}$ and $\log P_{HTAB}$, a high correlation with methanol is also observed. This suggests that the excluded compounds may have a different mechanism for describing the partition coefficient of the HTAB micelle.

An analysis is performed comparing calculated and experimental partition coefficients for the HTAB, SC, and LPFOS micelles. The results of the linear regression analysis for the partition coefficients of propan-1-ol/water, propan-2-ol/water, and methanol/water are presented in Table 3. The best correlation is observed for SC micelles. The partition coefficient calculated for propan-1-ol compared to the experimental partition coefficient of SC micelles provided the best correlation (R^2 of 0.67). It can be seen that the SC and LPFOS micelles behave similarly to aqueous mixtures with alcoholic solvents with dielectric constants ranging from 20 to 33. It needs to be mentioned that because these solvents are miscible with water, the partition coefficient of these solvents cannot be evaluated using the traditional shake flask technique. Alternatively, these coefficients can be determined through the application of appropriate thermodynamic cycles and using immiscible solvents. With respect to HTAB micelles, the prediction is improved for compounds that do not contain nitrogen in an aromatic ring or the urea group.

Table 3 presents a predictive tool that facilitates the identification of the most suitable micellar system for carrying a specific drug. By employing the equations provided in this table, it becomes possible to make a comparison of the LogP values among the three types of micelles (HTAB, SC, and LPFOS). This comparison enables the determination of which micellar system would yield a higher LogP value for the particular drug being considered. Finally, this predictive approach helps in selecting the most appropriate micelle for drug delivery and optimizing drug formulation and efficacy.

Table 3. Best linear regressions obtained to predict the logP in SC, LPFOS, and HTAB micelles using DFT calculations. Results from B3LYP functional with 6-31++G** basis set using SC, LPFOS, and HTAB for propan-1-ol, propan-2-ol, and methanol are indicated. x refers to predicted logP alcohol/water, and y refers to the predicted logP in micelles. *N set: compounds containing nitrogen in an aromatic ring or the urea group are excluded.

Micelle	Solvent	B3LYP
LPFOS	Propan-1-ol	y = 0.46x + 0.77 R^2 = 0.52 MAE = 0.87
	Propan-2-ol	y = 0.49x + 0.61 R^2 = 0.53 MAE = 0.92
	Methanol	y = 0.41x + 0.90 R^2 = 0.43 MAE = 0.86
SC	Propan-1-ol	y = 0.47x + 0.55 R^2 = 0.67 MAE = 0.92
	Propan-2-ol	y = 0.46x + 0.51 R^2 = 0.64 MAE = 1.08
	Methanol	y = 0.41x + 0.68 R^2 = 0.58 MAE = 0.89
HTAB	Propan-1-ol	y = 0.23x + 1.80 R^2 = 0.13 MAE = 0.74
	Propan-2-ol	y = 0.22x + 1.83 R^2 = 0.1 MAE = 0.72
	Methanol	y = 0.24x + 1.78 R^2 = 0.13 MAE = 0.72
HTAB without N set *	Propan-1-ol	y = 0.56x + 1.24 R^2 = 0.66 MAE = 0.45
	Propan-2-ol	y = 0.54x + 1.24 R^2 = 0.62 MAE = 0.43
	Methanol	y = 0.56x + 1.26 R^2 = 0.63 MAE = 0.46

* Molecules with Nitrogen in an aromatic ring or with the urea (carbamide) group are excluded.

2.3. Estimation of logP Values in Micelles Using SVM Calculations

A k-means clustering was performed on the set of compounds using a collection of 85 chemical descriptors to analyze the data (Table S3). In the present study, the determined number of partitioning patterns (clusters) was three. It can be concluded that the partitioning into three categories is related to specific features characteristic of each obtained class.

Cluster 1 contains 45 out of all 63 compounds (approximately 70% of the cases). The members do not differ substantially with respect to their structural and molecular descriptors, whose values are on a medium level (see Figure 2) without an expressed minima or maxima of their absolute (standardized) values. It may be assumed that this pattern of objects is a specific "medium" with respect to the descriptor values, is characterized by a good consistency of indication, and could be called, conditionally, a "mixed compound" pattern.

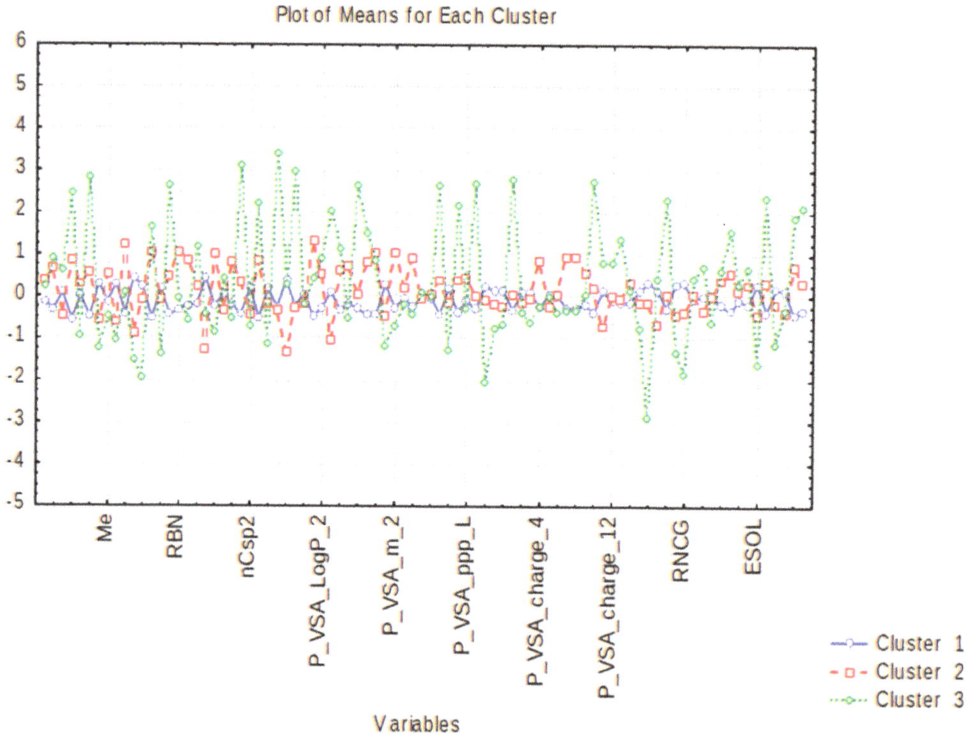

Figure 2. Plot of means for each variable for each identified cluster. Due to lack of space, only 10 of the descriptor names are plotted but the order is the same as in the input matrix (of variables); the distance between the plotted variables is 8 spaces.

Cluster 2 consists of 13 members out of a total of 63 cases, accounting for approximately 20% of all cases. It is important to emphasize that this cluster predominantly comprises representative pesticide compounds and can be conditionally referred to as the "pesticide compounds" pattern. The pattern exhibits specific feature characteristics (Figure 2) that are responsible for its partitioning. These characteristics include maximal levels for the descriptors GD, RBN, H%, PVS_A_m2, P_VSA_e2, P_VSA_i2, P_VSA_charge3, P_VSA_charge8, and P_VSA_charge9, as well as minimal levels for the descriptors N%, MCD, and P_VSA_LogP_3.

Cluster 3 consists of only 5 members out of a total of 63 cases, accounting for approximately 7%. All the members belong to hormone compounds, and the conditional name for this cluster should be the "hormonal compounds" pattern. It is characterized by significantly different values of the descriptors compared to those of clusters 1 and 2. Twenty-six of the descriptors indicate maximal values, while the other twenty-six indicate minimal values. This represents a typical case of an "outlying cluster", further supporting the conclusion that this group of objects is markedly different from the rest.

Figure 2 displays the averages for each descriptor of the three identified clusters, effectively demonstrating the differences between them and the descriptors responsible for this partitioning.

Furthermore, a PCA was conducted to explore the partitioning among the 85 descriptors. PCA is a widely used chemometric technique that involves projecting the original variables onto new, orthogonal directions known as latent factors. These factors are linear combinations of the original variables, and their associated factor loadings determine their impact on the analysis. The resulting factor scores represent the new coordinates of the objects in the reduced-dimensional space. The PCA analysis revealed that three latent factors accounted for more than 70% of the total variance, as shown in Table S3 (factor loadings table). Figure 3 shows that the most significant set of cases is explained by the highest loadings in factor 1 (all three experimental parameters are included in one factor). The second factor consists of the highest loadings for the second big set of objects, and the outlying set of objects is related to the highest loadings in factor 3.

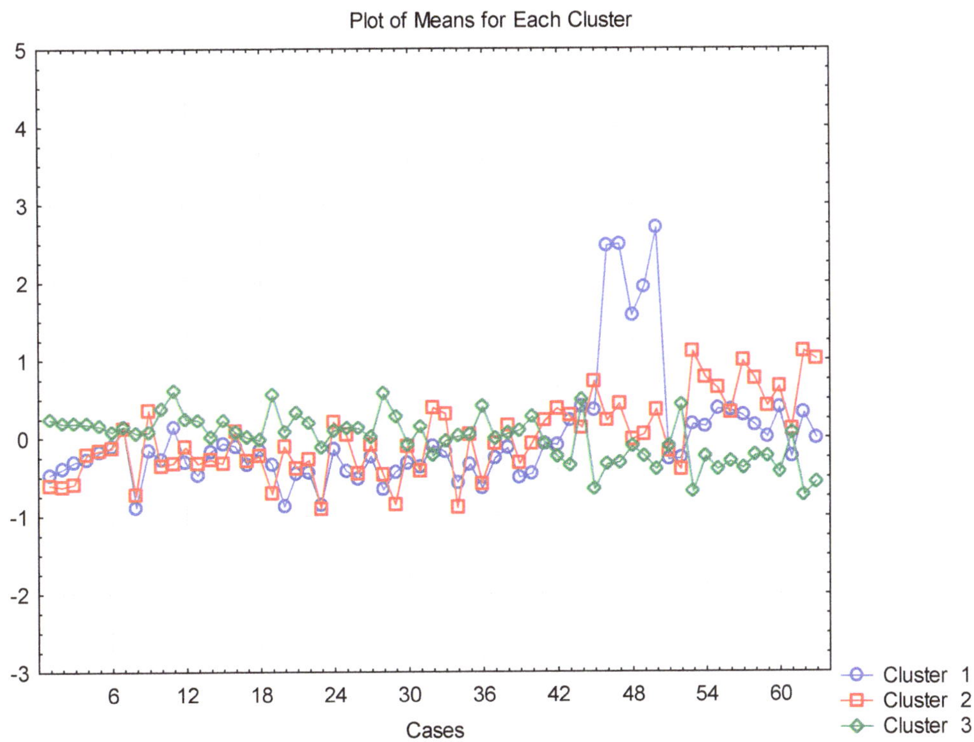

Figure 3. Plot of means (after standardization of the input matrix) of each of the cases for each of the identified patterns of variables; latent factor 1 is marked with blue color, latent factor 2—with red color, latent factor 3—with green color. The plot gives an idea of the relationships between cases and variables.

This work uses supervised and unsupervised machine learning methods to predict the logP values for different micelle formations. SVM calculations were applied as a regression method. The obtained results are presented in Table 4. The regression model was developed based on the list of descriptors presented in Table 4. Grid searches were performed using 10-fold cross-validation. The main descriptors were obtained from the class of molecular descriptors known as P_VSA descriptors, which quantify the van der Waals surface area (VSA) with a specific property P within a certain range [36]. For model development, 85% of the data was used for training, while the remaining 15% was reserved for testing. It is evident that the selected SVM model and the list of desired features yield a significantly high prediction rate for logP.

Table 4. SVM regression results for LogP prediction in SC, HTAB, and LPFOS micellar systems.

	SC	HTAB	LPFOS
Variables (descriptors)	Mv, RBN, RBF, H%, N%, O%, NRS, nR09, nR10, X4Av, P_VSA_LogP_2, P_VSA_s_4, P_VSA_ppp_P, P_VSA_charge_1, P_VSA_charge_3, P_VSA_charge_4, P_VSA_charge_5, P_VSA_charge_13, P_VSA_charge_14	nSK, nH, N%, Xu, S1K, DELS, BAC, X0, X0sol, P_VSA_LogP_1, P_VSA_LogP_4, P_VSA_LogP_6, P_VSA_LogP_8, P_VSA_MR_5, P_VSA_m_5, P_VSA_s_3, P_VSA_ppp_D, P_VSA_charge_2, P_VSA_charge_4, P_VSA_charge_5, P_VSA_charge_6, P_VSA_charge_9, P_VSA_charge_12, P_VSA_charge_14, qpmax, qnmax, Qpos, Qneg, Qtot, Qmean, Q2, RPCG, RNCG, TPSA(NO), TPSA(Tot)	RBF, nTB, MaxTD, P_VSA_LogP_2, P_VSA_LogP_3, P_VSA_LogP_4, P_VSA_MR_2, P_VSA_s_3, P_VSA_charge_2, P_VSA_charge_6, P_VSA_charge_7, P_VSA_charge_9, P_VSA_charge_13, P_VSA_charge_14, Qmean
R^2	0.693	0.565	0.783
RMSE	0.369	0.248	0.202
MAE	0.318	0.241	0.304

Our study demonstrates that SVM is a powerful machine learning model capable of predicting logP values from both high-dimensional and low-dimensional data spaces based on the selective nature of the descriptors.

3. Materials and Methods

3.1. Regents and Materials

Phosphoric acid (85% in water), lithium hydroxide (98%), sodium dihydrogen phosphate monohydrate (G.R.), disodium hydrogen phosphate (G.R.), sodium hydroxide (G.R.), phenyl-undecyl ketone (98%), and methanol (for chromatography) were obtained from Merck. SC (>98%), HTAB (>99%), and LPFOS (25% in water) were obtained from Fluka. Water was Milli-Q plus (Millipore) with resistivity of 18.2 MΩ cm. The test solutes were reagent-grade or better and obtained from several makers.

3.2. Determination of Partition Coefficients in Systems of SC, LPFOS, and HTAB Micelles

MEKC analyses were conducted using a UV diode array detector in a Beckman P/ACE System 5500 capillary electrophoresis instrument, with a fused silica capillary of 47 cm total length (40 cm effective length) and 50 μm internal diameter. The measurements were carried out at 25 °C and +15 kV for the anionic surfactants (SC and LPFOS) and −15 kV for the cationic one (HTAB). Detection was set at 214 nm. To inject the test compounds into the capillary, a pressure of 0.5 p.s.i. was applied for 1 s.

The capillary was prepared through a conditioning process, which involved flushing with water for 5 min, treating with 1 M sodium hydroxide solution for 20 min, followed by a rinse with water for 10 min, treatment with 0.1 M sodium hydroxide solution for 10 min, and, finally, treatment with separation buffer for 20 min. Before each injection, the capillary was rinsed with a separation buffer for 5 min.

Three different micelle solutions were prepared at pH 7: 80 mM of SC, 40 mM of LPFOS, and 20 mM of HTAB, all three in 20 mM phosphate buffer. Test compounds were dissolved in a methanol solution (used as an electro-osmotic flow marker), which already contained 2 mg mL^{-1} of phenyl-undecyl ketone (used as a micellar marker). The concentration of the test compounds was 2 mg mL^{-1}. All solutions were filtered through 0.45 μm nylon syringe filters (Albet). All measurements were performed in triplicate.

In MEKC, the separation of neutral molecules occurs based on their partitioning between the micellar phase and the aqueous phase. The retention factor (k) of a compound can be determined using the following formula:

$$k = \frac{t_R - t_0}{t_0(1 - t_R/t_m)} \tag{1}$$

The retention time (t_R) for the specific compound being analyzed is measured, while the retention times of the electro-osmotic flow and micellar markers (methanol and phenyl-undecyl ketone, respectively) are denoted as t_0 and t_m.

In this particular study, partition coefficients between water and SC, LPFOS, and HTAB micelles were determined by utilizing previously obtained retention times [13] and applying the following formula:

$$k = \frac{v(C_T - CMC)}{1 - v(C_T - CMC)} \tag{2}$$

where P is the partition coefficient, and C_T is the total surfactant concentration (80 mM for SC, 40 mM for LPFOS, and 20 mM for HTAB). The *CMC* values in 20 mM phosphate buffer (pH 7) were experimentally determined in previous works and are 12.4 mM for SC, 3.27 mM for LFPOS, and 0.34 mM for HTAB [37]. U is the partial molar volume of the surfactants and has a value of 0.317 L mol^{-1} for SC [38], 0.285 L mol^{-1} for LPFOS [39], and 0.324 L mol^{-1} for HTAB [40].

3.3. QM Computational Determination of Partition Coefficients

The computations in this study were conducted using the Gaussian 16 (Revision C.01) [41] quantum chemistry software package to calculate solvation free energies and various molecular properties. The Avogadro cross-platform molecule editor was utilized to generate all molecular structures, and only the more extended conformation was used for each compound. Specifically, the focus of this work was to determine the solvation free energy of 63 compounds in 16 different solvents.

In this study, the B3LYP calculations were performed with the 6.311++G** basis set to optimize the geometries of all compounds. The solvation model based on SMD was employed to predict the partition coefficient of the molecules in different solvents. This model divides the solvation free energy into two main contributions, bulk electrostatic and cavity dispersion contributions, making it a widely applicable and universal solvation model that can be used for any solute (neutral and charged) in a variety of parametrized solvents.

In order to determine the solvent–water partition coefficient, the compounds were optimized to obtain their minimum energy at a pressure of 1 atm and temperature of 298.15 K, while ensuring that all vibrational frequencies were positive. The calculation of the Gibbs free energy associated with the transfer of solutes between the solvent and water phases is fundamental in determining the partition coefficient. The logarithm of the partition coefficient (logP) is directly proportional to the difference in solvation free energies ($\Delta G^{\circ}{}_{solv/wat}$):

$$\Delta G^{\circ}{}_{\frac{solv}{wat}} = \Delta G^{\circ}{}_{solv} - \Delta G^{\circ}{}_{wat} \tag{3}$$

$$\text{logP} = \frac{-\Delta G^{\circ}{}_{\frac{solv}{wat}}}{RTln(10)} \tag{4}$$

where *R* is the molar gas constant and *T* is the temperature (298 K). We employed the same procedure for SC, LPFOS, and HTAB that we applied for SDS micelles [42].

3.4. Correlation Analysis

Heatmaps of Pearson correlation coefficients were produced to identify correlation between variables. Each variable is represented by a colored square, with the color indicating the strength and direction of the correlation between that variable and every other variable in the dataset. The Pearson correlation coefficient ranges from −1 to +1, with −1 indicating a perfect negative correlation, 0 indicating no correlation, and +1 indicating a perfect positive correlation. The heatmap allows us to quickly identify patterns and relationships between variables, as highly correlated variables will appear as blocks of similar color in the heatmap.

3.5. Supervised and Unsupervised Methods

Linear regression analysis was conducted using python tools to calculate coefficients, confidence intervals, standard errors, F statistics, significant data of partition coefficient and F, as well as Pearson's correlation coefficient. Additionally, the accuracy of the regression model's prediction of experimental octanol/water logP values was evaluated through the computation of statistical measures, including the mean absolute error (MAE), mean square error (MSE), and root mean square error (RMSE).

3.6. K-Means Clustering

K-means clustering is a well-documented supervised machine learning pattern recognition procedure [43,44]. It requires an a priori determined number of clusters to which the objects of interest should be partitioned. The hypothesis for the predetermined number of clusters follows expert opinion or specific reasons of the researcher (preliminary information, preliminary testing, etc.). The major goal of this statistical method is to partition the objects of interest into patterns of similarity (clusters) whose number is in line with the preliminary hypothesis. The algorithm used relies on minimization of the within-group distances (usually squared Euclidean distances). Cluster centers (centroids) are used to find groups of comparable special distribution.

The input matrix is of the dimension 63 cases × 85 variables. The raw data were subject to a standardization procedure (z-transform) to avoid differences in variable dimensions. The goal of the partitioning procedure was to reveal patterns of similarity within the 63 compounds of interest and, further, to determine the descriptors contributing mostly to the partitioning. The structure of the dataset was based on the generated descriptors retrieved using the AlvaDesc v.2 software (Milano, Italy) [45].

3.7. Principal Component Analysis (PCA)

PCA is a typical projection method based on the reduction of the dimensionality of the system under consideration. This makes it possible to present on a plot the relationship between the variables (using the values of the factor loadings) or between the objects (using the calculated factor scores as the new coordinates of the objects). The reduction of the dimensionality of the initial large dataset enhances the interpretability of the original system while preserving a high amount of explained variance from the original set. The algorithm achieves this through the decomposition of the starting large data matrix into a smaller number of principal components (latent variables), which are linear combinations of the original variables representing directions in space.

4. Conclusions

This study aimed to determine a methodology for predicting the experimental partition coefficients (Log P) for a diverse set of compounds for three different micelle formulations: SC, LPFOS, and HTAB. The obtained LogP values were used to parametrize computational methodologies for each type of micelle. Correlations between experimental and calculated logP values were examined using simple DFT calculations, and the SVM regression model was built based on relevant descriptor space. This predictive approach could have significant implications for drug delivery and formulation optimization. It can be used to identify the micellar system that offers a higher Log P value, thus enhancing the potential for successful therapeutic applications.

When considering the entire set of compounds, the results revealed an increased correlation between experimental logP values and the DFT predictions for the SC and LPFOS micelle systems obtained for the propan-1-ol/water or propan-2-ol/water solvent mixtures. However, the logP values in HTAB were not correlated with any of the calculated logP values. It has been found that compounds containing nitrogen in an aromatic ring, or the urea group, exhibited a different mechanism in describing the partition coefficient of the HTAB micelle. Excluding these compounds from the set, the best correlation was observed for all micelles in the propan-1-ol/water or propan-2-ol/water solvent mixtures.

Furthermore, SVM calculations and k-means clustering were conducted using a set of 85 descriptors. The findings imply that the partitioning into three classes is coupled to specific features that are characteristic of each obtained class. These results provide valuable insights into the behavior of different types of micelles and can contribute to the development of more accurate computational methods for predicting partition coefficients in micelles.

Supplementary Materials: The following supporting information can be downloaded at https://www.mdpi.com/article/10.3390/molecules28155729/s1, Table S1: Molecular representation of non-aqueous solvents used in DFT; Table S2: Linear regression parameters obtained for the correlation of the calculated LogP in 15 different solvents with respect to the experimental partition coefficients in SC, LPFOS, and HTAB micelles; Table S3: Factor loadings (Varimax-normalized) of the chemical descriptors obtained from PCA.

Author Contributions: Conceptualization, S.M. and M.N.; methodology, S.M., M.N. and E.F.; software, S.M. and M.N.; formal analysis, S.M. and M.N.; investigation, L.S. and E.F.; resources, F.M. and S.M.; data curation, L.S. and S.M.; writing—original draft preparation, L.S., S.M. and M.N.; writing—review and editing, E.F., V.S. and F.M.; visualization, S.M. and M.N.; project administration, F.M. and S.M.; funding acquisition, F.M. and S.M. All authors have read and agreed to the published version of the manuscript.

Funding: S.M. and F.M. acknowledge the financial support from Generalitat de Catalunya (Grant 2021SGR00350). S.M. and F.M. acknowledge the Spanish Structures of Excellence María de Maeztu program through Grant CEX2021-001202-M. M.N. and V.S. gratefully acknowledge the financial support from Bulgarian Science Found with grant number KP-06-H59/10-19.11.2021. M.N. gratefully acknowledges the support provided by the project UNITe—BG05M2OP001-1.001-0004 funded by the OP SESG and co-funded by the EU Structural and Investment Funds. M.N. was supported by the European Union—NextGenerationEU, through the National Recovery and Resilience Plan of the Republic of Bulgaria, project No. BG-RRP-2.004-0008-C01.

Institutional Review Board Statement: Not applicable.

Informed Consent Statement: Not applicable.

Data Availability Statement: The data that support the findings of this study are available on request from the corresponding author.

Conflicts of Interest: The authors declare no conflict of interest.

Sample Availability: Not applicable.

References

1. Paprikar, A.; Soni, A.; Kaushal, N.; Lin, S. Polymeric Micelles for Drug Delivery. In *Smart Nanomaterials in Biomedical Applications. Nanotechnology in the Life Sciences*; Kim, J.C., Alle, M., Husen, A., Eds.; Springer: Cham, Switzerland, 2021; pp. 345–372. [CrossRef]
2. Kedar, U.; Phutane, P.; Shidhaye, S.; Kadam, V. Advances in polymeric micelles for drug delivery and tumor targeting. *Nanomed. Nanotechnol. Biol. Med.* **2010**, *6*, 714–729. [CrossRef] [PubMed]
3. Rangel-Yagui, C.O.; Pessoa, A.; Tavares, L.C.; Becke, A.D. Micellar solubilization of drugs. *J. Pharm. Pharm. Sci.* **2005**, *8*, 147–163.

4. Torchilin, V.P. Structure and design of polymeric surfactant-based drug delivery systems. *J. Control. Release* **2001**, *73*, 137–172. [CrossRef]
5. Ingram, T.; Storm, S.; Kloss, L.; Mehling, T.; Jakobtorweihen, S.; Smirnova, I. Prediction of micelle/water and liposome/water partition coefficients based on molecular dynamics simulations, COSMO-RS, and COSMOmic. *Langmuir* **2013**, *29*, 3527–3537. [CrossRef] [PubMed]
6. Mehling, T.; Ingram, T.; Smirnova, I. Experimental methods and prediction with COSMO-RS to determine partition coefficients in complex surfactant systems. *Langmuir* **2012**, *28*, 118–124. [CrossRef] [PubMed]
7. Singla, P.; Singh, O.; Sharma, S.; Betlem, K.; Aswal, V.K.; Peeters, M.; Mahajan, R.K. Temperature-Dependent Solubilization of the Hydrophobic Antiepileptic Drug Lamotrigine in Different Pluronic Micelles—A Spectroscopic, Heat Transfer Method, Small Angle Neutron Scattering, Dynamic Light Scattering, and in Vitro Release Study. *ACS Omega* **2019**, *4*, 11251–11262. [CrossRef]
8. Singla, P.; Singh, O.; Chabba, S.; Aswal, V.K.; Mahajan, R.K. Sodium deoxycholate mediated enhanced solubilization and stability of hydrophobic drug Clozapine in pluronic micelles. *Spectrochim. Acta Part A Mol. Biomol. Spectrosc.* **2018**, *191*, 143–154. [CrossRef]
9. Mehling, T.; Kloss, L.; Ingram, T.; Smirnova, I. Partition coefficients of ionizable solutes in mixed nonionic/ionic micellar systems. *Langmuir* **2013**, *29*, 1035–1044. [CrossRef]
10. Khan, A.M.; Bashir, S.; Shah, A.; Nazar, M.F.; Rahman, H.M.A.; Shah, S.S.; Khan, A.Y.; Khan, A.R.; Shah, F. Spectroscopically probing the effects of Holmium(III) based complex counterion on the dye-cationic surfactant interactions. *Colloids Surf. A Physicochem. Eng. Asp.* **2018**, *539*, 407–415. [CrossRef]
11. Materna, K.; Goralska, E.; Sobczynska, A.; Szymanowski, J. Recovery of various phenols and phenylamines by micellar enhanced ultrafiltration and cloud point separation. *Green Chem.* **2004**, *6*, 176–182. [CrossRef]
12. Fuguet, E.; Ràfols, C.; Torres-Lapasió, J.R.; García-Álvarez-Coque, M.C.; Bosch, E.; Rosés, M. Solute-solvent interactions in micellar electrokinetic chromatography. 6. Optimization of the selectivity of lithium dodecyl sulfate-lithium perfluorooctanesulfonate mixed micellar buffers. *Anal. Chem.* **2002**, *74*, 4447–4455. [CrossRef] [PubMed]
13. Fuguet, E.; Rafols, C.; Bosch, E.; Abraham, M.H.; Roses, M. Solute-solvent interactions in micellar electrokinetic chromatography III. Characterization of the selectivity of micellar electrokinetic chromatography systems. *J. Chromatogr. A* **2002**, *942*, 237–248. [CrossRef] [PubMed]
14. Terabe, S.; Otsuka, K.; Ichikawa, K.; Tsuchiya, A.; Teiichi, A. Electrokinetic separations with micellar solutions and open-tubular capillaries. *Anal. Chem.* **1984**, *56*, 111–113. [CrossRef]
15. Baker, D.R. *Capillary Electrophoresis*; Wiley–Interscience: New York, NY, USA, 1995.
16. Azeem, W.; John, P.; Nazar, M.F.; Khan, I.U.; Riaz, A.; Sharif, S. Spectral and chromatographic characterization of fixed dose combination norfloxacin and metronidazole interacting with cetyltrimethylammonium bromide. *J. Mol. Liquids* **2017**, *244*, 135–140. [CrossRef]
17. Maeder, C.; Beaudoin III, G.M.J.; Hsu, E.; Escobar, V.A.; Chambers, S.M.; Kurtin, W.E.; Bushey, M.M. Measurement of bilirubin partition coefficients in bile salt micelle/aqueous buffer solutions by micellar electrokinetic chromatography. *Electrophoresis* **2000**, *21*, 706–714. [CrossRef]
18. Herbert, B.J.; Dorsey, J.G. n-Octanol-water partition coefficient estimation by micellar electrokinetic capillary chromatography. *Anal. Chem.* **1995**, *67*, 744–749. [CrossRef]
19. Mrestani, Y.; Marestani, Z.; Neubert, R.H.H. Characterization of micellar solubilization of antibiotics using micellar electrokinetic chromatography. *J. Pharm. Biomed. Anal.* **2001**, *26*, 883–889. [CrossRef]
20. Godyn, J.; Hebda, M.; Wieckowska, A.; Wieckowski, K.; Malawska, B.; Bajda, M. Lipophilic properties of anti-Alzheimer's agents determined by micellar electrokinetic chromatography and reversed-phase thin-layer chromatography. *Electrophoresis* **2017**, *38*, 1268–1275. [CrossRef]
21. Storm, S.; Jakobtorweihen, S.; Smirnova, I. Solubilization in mixed micelles studied by molecular dynamics simulations and COSMOmic. *J. Phys. Chem. B* **2014**, *118*, 3593–3604. [CrossRef]
22. Storm, S.; Jakobtorweihen, S.; Smirnova, I.; Panagiotopoulos, A.Z. Molecular dynamics simulation of SDS and CTAB micellization and prediction of partition equilibria with COSMOmic. *Langmuir* **2013**, *29*, 11582–11592. [CrossRef]
23. Yordanova, D.; Smirnova, I.; Jakobtorweihen, S. Molecular modeling of Triton X micelles: Force field parameters, self-assembly, and partition equilibria. *J. Chem. Theory Comput.* **2015**, *11*, 2329–2340. [CrossRef] [PubMed]
24. Turchi, M.; Kognole, A.A.; Kumar, A.; Cai, Q.; Lian, G.; Mackerell, A.D. Predicting Partition Coefficients of Neutral and Charged Solutes in the Mixed SLES-Fatty Acid Micellar System. *J. Phys. Chem. B* **2020**, *124*, 1653–1664. [CrossRef]
25. Yordanova, D.; Ritter, E.; Gerlach, T.; Jensen, J.H.; Smirnova, I.; Jakobtorweihen, S. Solute Partitioning in Micelles: Combining Molecular Dynamics Simulations, COSMOmic, and Experiments. *J. Phys. Chem. B* **2017**, *121*, 5794–5809. [CrossRef]
26. Burns, S.T.; Khaledi, M.G. Predictions of Micelle-Water Partition Coefficients and Retention in Micellar Electrokinetic Chromatography from Solute Structure. 2. Fragmental Constant Approach. *Anal. Chem.* **2004**, *76*, 5451–5458. [CrossRef] [PubMed]
27. Chen, H.F. In Silico Log P Prediction for a Large Data Set with Support Vector Machines, Radial Basis Neural Networks and Multiple Linear Regression. *Chem. Biol. Drug Des.* **2009**, *74*, 142–147. [CrossRef]
28. Liao, Q.; Yao, J.; Yuan, S. SVM approach for predicting LogP. *Mol. Divers.* **2006**, *10*, 301–309. [CrossRef]
29. Wu, K.; Zhao, Z.; Wang, R.; Wei, G.-W. TopP–S: Persistent Homology-Based Multi-Task DeepNeural Networks for Simultaneous Predictions of PartitionCoefficient and Aqueous Solubility. *J. Comput. Chem.* **2018**, *39*, 1444–1454. [CrossRef]
30. Popova, M.; Isayev, O.; Tropsha, A. Deep reinforcement learning for de novo drug design. *Sci. Adv.* **2018**, *4*, eaap7885. [CrossRef]

31. Donyapour, N.; Hirn, M.; Dickson, A. ClassicalGSG: Prediction of log P using classical molecular force fields and geometric scattering for graphs. *J. Comput. Chem.* **2021**, *42*, 1006–1017. [CrossRef] [PubMed]
32. Becke, A.D. Density-functional exchange-energy approximation with correct asympotic behaviour. *Phys. Rev. A* **1998**, *38*, 3098–3100. [CrossRef]
33. Marenich, A.V.; Cramer, C.J.; Truhlar, D.G. Perspective on foundations of solvation modeling: The electrostatic contribution to the free energy. *J. Chem. Theory Comput.* **2008**, *4*, 877–887. [CrossRef] [PubMed]
34. Marenich, A.V.; Cramer, C.J.; Truhlar, D.G. Universal Solvation Model Based on Solute Electron Density and on a Continuum Model of the Solvent Defined by the Bulk Dielectric Constant and Atomic Surface Tensions. *J. Phys. Chem. B* **2009**, *113*, 6378–6396. [CrossRef]
35. Abraham, M.H. Scales of hydrogen bonding: Their construction and application to physicochemical and biochemical processes. *Chem. Soc. Rev.* **1993**, *22*, 73–83. [CrossRef]
36. Labute, P. A widely applicable set of descriptors. *J. Mol. Graph. Model.* **2000**, *18*, 464–477. [CrossRef] [PubMed]
37. Fuguet, E.; Ràfols, C.; Rosés, M.; Bosch, E. Critical micelle concentration of surfactants in aqueous buffered and unbuffered systems. *Anal. Chim. Acta.* **2005**, *548*, 95–100. [CrossRef]
38. Zana, R. The role of hydrogen bonding in the formation of bile salt micelles. *J. Phys. Chem.* **1978**, *82*, 2440–2443. [CrossRef]
39. Fuguet, E.; Ràfols, C.; Rosés, M. Characterization of the solvation properties of surfactants by solvatochromic indicators. *Langmuir* **2003**, *19*, 6685–6692. [CrossRef]
40. Velegol, S.B.; Fleming, B.D.; Biggs, S.; Wanless, E.J.; Tilton, R.D. Counterion effects on hexadecyltrimethylammonium surfactant adsorption and self-assembly on silica. *Langmuir* **2000**, *16*, 2548–2556. [CrossRef]
41. Frisch, M.J.; Trucks, G.W.; Schlegel, H.B.; Scuseria, G.E.; Robb, M.A.; Cheeseman, J.R.; Scalmani, G.; Barone, V.; Petersson, G.A.; Nakatsuji, H.; et al. *Gaussian 16, Revision, C.01*; Gaussian, Inc.: Wallingford, CT, USA, 2016.
42. Saranjam, L.; Fuguet, E.; Nedyalkova, M.; Simeonov, V.; Mas, F.; Madurga, S. Prediction of Partition Coefficients in SDS Micelles by DFT Calculations. *Symmetry* **2021**, *13*, 1750. [CrossRef]
43. Ikotun, A.M.; Ezugwu, A.E.; Abualigah, L.; Abuhaija, B.; Heming, J. K-means clustering algorithms: A comprehensive review, variants analysis, and advances in the era of big data. *Inf. Sci.* **2023**, *622*, 178–210. [CrossRef]
44. Nedyalkova, M.; Madurga, S.; Simeonov, V. Combinatorial K-Means Clustering as a Machine Learning Tool Applied to Diabetes Mellitus Type 2. *Int. J. Environ. Res. Public Health* **2021**, *18*, 1919. [CrossRef] [PubMed]
45. Mauri, A. alvaDesc: A Tool to Calculate and Analyze Molecular Descriptors and Fingerprints. In *Ecotoxicological QSARs*; Springer: Berlin/Heidelberg, Germany, 2020; pp. 801–820. [CrossRef]

MDPI AG
Grosspeteranlage 5
4052 Basel
Switzerland
Tel.: +41 61 683 77 34

Molecules Editorial Office
E-mail: molecules@mdpi.com
www.mdpi.com/journal/molecules

www.ingramcontent.com/pod-product-compliance
Lightning Source LLC
Chambersburg PA
CBHW061626210926
51850CB00009B/2449
9783725868988